UNDERSTANDING FOOD SYSTEMS

science &
technology books

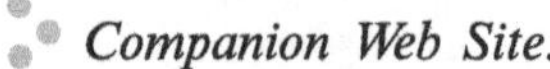

Companion Web Site:

https://textbooks.elsevier.com/web/Manuals.aspx?isbn=9780128044452

Understanding Food Systems: Agriculture, Food Science, and Nutrition in the United States
Ruth MacDonald and Cheryll Reitmeier

Resources available:

- Assignments
- Learning Activities and Discussion Topics
- Resources for instructors
- Slides Deck

UNDERSTANDING FOOD SYSTEMS

Agriculture, Food Science, and Nutrition in the United States

RUTH MACDONALD
Professor and Chair, Food Science and Human Nutrition,
Iowa State University, Aimes, IA, United States

CHERYLL REITMEIER
Professor Emeritus, Food Science and Human Nutrition,
Iowa State University, Aimes, IA, United States

Academic Press is an imprint of Elsevier
125 London Wall, London EC2Y 5AS, United Kingdom
525 B Street, Suite 1800, San Diego, CA 92101-4495, United States
50 Hampshire Street, 5th Floor, Cambridge, MA 02139, United States
The Boulevard, Langford Lane, Kidlington, Oxford OX5 1GB, United Kingdom

British Library Cataloguing-in-Publication Data
A catalogue record for this book is available from the British Library

Library of Congress Cataloging-in-Publication Data
A catalog record for this book is available from the Library of Congress

ISBN: 978-0-12-804445-2

For Information on all Academic Press publications
visit our website at https://www.elsevier.com/books-and-journals

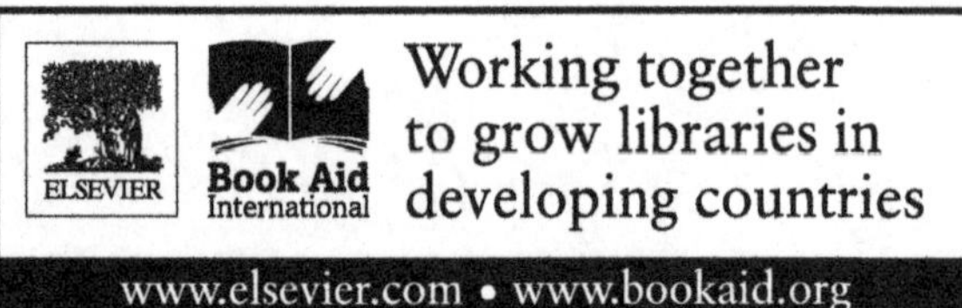

Publisher: Andre G. Wolff
Acquisition Editor: Nancy Maragioglio
Editorial Project Manager: Billie Jean Fernandez
Production Project Manager: Nicky Carter
Cover Designer: Mark Rogers

Typeset by MPS Limited, Chennai, India

Contents

6. Food Processing

7. Nutrition and Food Access

8. Sustainability of the Food System

Acknowledgments by Ruth MacDonald

This book was a result of the many interactions and discussions I have had with my coauthor Cheryll Reitmeier over the past decade. I deeply respect her perspectives about food science and agriculture and her high standards for teaching quality. She continually inspires me to be a better teacher and to encourage students to think creatively and independently. The material in this book also reflects the input of the many students that have provided their candid feedback and suggestions.

Much thanks to Jeni Maiers, graphic designer for the Iowa State University (ISU) Center for Crops Utilization Research, for her valuable assistance with all of the graphic design and photos used in the text. She was always available to provide assistance with questions about images and provided her technical talents throughout the development of the book. The illustrations in the text were created by Reannon Overbey, an ISU student majoring in graphic design. I greatly appreciate her talent and hard work to meet the publication deadlines and her willingness to do many revisions and changes.

Many colleagues were called upon to review text and provide information for this book, and for their willing assistance I am very grateful. Dr. Paul Lasley, Dr. Dana Dinnes, Dr. Cornelia Flora, Mr. Doug Svendsen, Dr. Stephanie Clark, Dr. Pamela Riney-Kehrberg, and Dr. Gary Munkvold each provided critical assessment of sections or made themselves available to me to discuss aspects of the text. In addition, all of my colleagues in the department of Food Science and Human Nutrition and other departments at ISU have provided me both directly and indirectly with information about food systems that has contributed to my understanding and perspectives while preparing this book.

I wish to express special gratitude to my husband, Dr. Ted MacDonald, for his patience and support throughout the writing of this book. He spent many hours reviewing text and assisting me with research on a wide range of topics. My two sons, Neal and Scott, also deserve thanks for their encouragement and support as I undertook this significant project.

Acknowledgments by Cheryll Reitmeier

I am thankful to the students in the food systems course at Iowa State University who, over the past 6 years, have asked intelligent questions, listened respectfully, struggled with ethical dilemmas, and developed their scientific thinking skills. Their concerns and interests provided the basis for this book. I am hopeful for the future because of these students' thoughtful and creative solutions to problems of the food system.

I am personally grateful to Linda Svendsen for listening to me every week and providing valuable feedback about topics in this book. Reviews of Chapter 6, Food Processing, by Dr. Julie Goldman and Dr. Pat Murphy were much appreciated. Farmers Don Andringa, Paul Reitmeier, and Doug Svendsen helped interpret technical information about agricultural methods, government programs, and economics; any mistakes in translation of their explanations are mine alone. Their honesty, integrity, and investments in sustainable practices, and that of many farmers like them, confirm my confidence in the US food system.

I also thank helpful librarians at the Detroit Lakes Public Library (Lake Agassiz Regional Library), Iowa State University, North Dakota State University, and especially, the National Agricultural Library, US Department of Agriculture.

Introduction to Understanding Food Systems: Agriculture, Food Science, and Nutrition in the United States

Throughout the day we engage in activities associated with food. We start the day with breakfast, maybe pack a lunch, meet friends for dinner, and grab a midnight snack. Trips to the local pizzeria, grocery store, or farmer's market are regular social outings for many people. We know that food is necessary for life, but we also recognize that what and how much we eat affects our health. The choices we make about food are determined by our social and cultural backgrounds, preferences, and experiences. Where and how food is produced may not be the first thing on people's minds, but increasingly the term "food system" has become integrated into everyday conversations. Understanding of the complex interconnections that comprise the US food system requires a broad perspective of agriculture, food science, human nutrition, and the environment, which are the topics of this textbook.

We have spent many years teaching food science and human nutrition, at a land-grant institution (Iowa State University), in a state (Iowa) where agriculture is a dominant economic and social component. This has given us a broad awareness of the interactions among these disciplines. The curricula offered to students in agriculture, food science, and human nutrition at most academic institutions have developed over time to become self-contained. Rarely do students engage in discussions or expand their coursework to learn about the interconnectedness of these areas. Within our students we saw a lack of knowledge about how food is grown and produced, misunderstandings about food processing technology, and a mistrust of nutrition and diet recommendations. Furthermore, the public discussions about food and health or the role of agriculture are politically and socially polarized. This has led to a unique social environment in which decisions about food and the food system are made with diminishing regard for facts or evidence.

To provide a structure for students to learn about food systems, we created an undergraduate course that brought together scientifically based information about agriculture, food science, human nutrition, and the environment. This textbook was built from the content we developed for the course along with input from many students and colleagues. Our goal is to facilitate understanding of how the US food system evolved and to raise awareness among students of the imperative to make sound, scientifically based decisions about food.

We begin our course and this textbook with a discussion about ethics. Ethical perspectives are an essential component of the food system at all levels. Any topic raised within a food system discussion will have an ethical component. Our application of ethical principles is by necessity superficial, and we strongly encourage further study of these principles. Only with a deep appreciation for ethics can we hope to address the complexities of a sustainable and equitable food system for the United States and the world.

We also reintroduce the concepts of scientific thinking. While most students are taught scientific thinking in grade school, we find that a reminder of the process of the scientific method and providing a framework to apply these methods are helpful to learn about the food system. The ready access to information on the Internet and expanding number of self-identified experts have blurred the distinctions about factual information. We encourage thoughtful assessment of sources of information and provide standards to gauge the reliability of those sources.

The main chapters of the textbook address the historical perspectives and current status of agriculture in the United States, animal food production, human labor in agriculture, food processing and technology, human nutrition, and the connection between food and the environment. By providing the background of historical events, we aim to develop an understanding of how political, economic, and social situations have influenced the food system. We created a table, provided in the Appendix, that outlines important historical events that have impacted the US food system. Students are encouraged to refer to this table frequently to gain a solid perspective of the ways social, political, economic, and cultural events have impacted the food system. It is essential that students, who will be future leaders and decision makers, have this comprehensive perspective. Throughout these chapters, we attempt to present factual information and refrain from personal perspective or judgment about topics or issues. In the supplemental information provided for instructors, we recommend discussion topics and assignments that allow students to engage in personal reflections and debate controversial issues. We strongly encourage that this textbook be used as a framework from which further discussion takes place.

There is clear evidence that US food production has been highly effective and efficient and that Americans enjoy safe, abundant, and low-cost food. We have applied technology to our food production and processing systems that have enhanced the quality, safety, and nutritional value of foods. Yet chronic illnesses associated with diet and lifestyle choices including obesity, cardiovascular disease, and type 2 diabetes are significant public health concerns. Concurrently, a significant number of Americans, including children, are food insecure. The balance of creating a food system that ensures the right nutritional needs for everyone is a challenge.

On a global scale, concerns about rising world population, climate change, environmental damage, and water scarcity have a direct impact on the US food system. While our focus in this textbook is on the US food system, we fully recognize that food is global. The world situation has and always will influence the US food system. The last chapter covers some of these topics and provides context from which the US food system will need to function as part of the global food system. We have provided the background from which the concept of a sustainable food system may be defined. Sustainability is complex and multifaceted and will require integration of many agricultural practices, food production systems, and consumer behaviors. It will be critical that scientific thinking and ethics are well applied as we attempt to address the future of food production.

Throughout the textbook, we encourage students to reflect and consider their personal food choices and perspectives about food. Consumers have a significant voice in the food system, which needs to be used wisely. Applying scientific thinking and considering ethical principles are as important at the individual level as at the global level. Making demands on the food system that are not based on scientific evidence may do more harm than good in the long run. Food provides enjoyment as well as nourishment and is an integral part of our daily lives. By working together across the disciplines of agriculture, food science, nutrition, and environmental science we can ensure a sustainable, safe, abundant, nutritious, and enjoyable food supply today and into the future.

CHAPTER

1

Ethics and Scientific Thinking

1.1 INTRODUCTION TO THE FOOD SYSTEM

Human survival depends on an adequate intake of foods that provide essential nutrients and energy. That simple statement belies the much wider and diverse influence of food on human lives. Food defines the history, culture, religion, and identity of populations. We readily recognize the Chinese culture of food as being distinct from the Italian culture of food. Special foods are linked to celebrations and events. What would Thanksgiving in the United States be without turkey or Christmas in the United Kingdom without mince pies? If you love gumbo, you are probably from Louisiana and if you think lutefisk is great, you may be from Minnesota.

Food brings comfort, enhances celebrations, and connects us to each other. Sharing food and meals are integral components of social networks. Potluck suppers are the foundations of many church events and neighborhood gatherings. We celebrate birthdays with cake and ice cream and the 4th of July with hot dogs and hamburgers. Research has shown that families that eat dinner together most nights of the week have stronger connections and children perform better in school.

Wars have been waged over food and access to land to grow food. Population migrations have resulted from famines due to crop failures, such as the Irish potato blight in the 1800s, when nearly one million people fled Ireland and came to the United States. Adequately feeding military personnel led to either success or failure of conquests throughout history. Napoleon Bonaparte is credited with saying, "An army marches on its stomach." French scientists found a way to preserve food by canning, which allowed Bonaparte's army to be successful. Also, English sailors survived long sea voyages to reach distant lands because they drank a citrus concoction containing vitamin C that prevented them from succumbing to scurvy, earning them the nickname "limeys."

The history of human civilization is defined by food. Early humans were hunter-gatherers and their survival depended on securing food every day. Over time, humans learned how to cultivate crops and raise animals to produce a more consistent and reliable source of food. A higher-quality food supply may have contributed to advances in human brain development and certainly allowed more time for other activities and intellectual pursuits. Food production technology has advanced significantly leading to better quality and higher yields. But commercial food production has environmental impacts and consumes natural resources. The capacity to feed a growing population, expected to reach nine billion people by 2050, will depend on

Understanding Food Systems.
DOI: http://dx.doi.org/10.1016/B978-0-12-804445-2.00001-6

making the right choices in how we produce food and use our natural resources.

1.1.1 Definition of the Food System

A food system encompasses all the components involved in providing food to a population (Fig. 1.1). "From field to fork" is a popular phrase that describes the food system. The production, processing, distribution, marketing, and consumption of foods constitute the food system and within each aspect are multiple layers of complexity. Production encompasses the interaction between food crops and animals

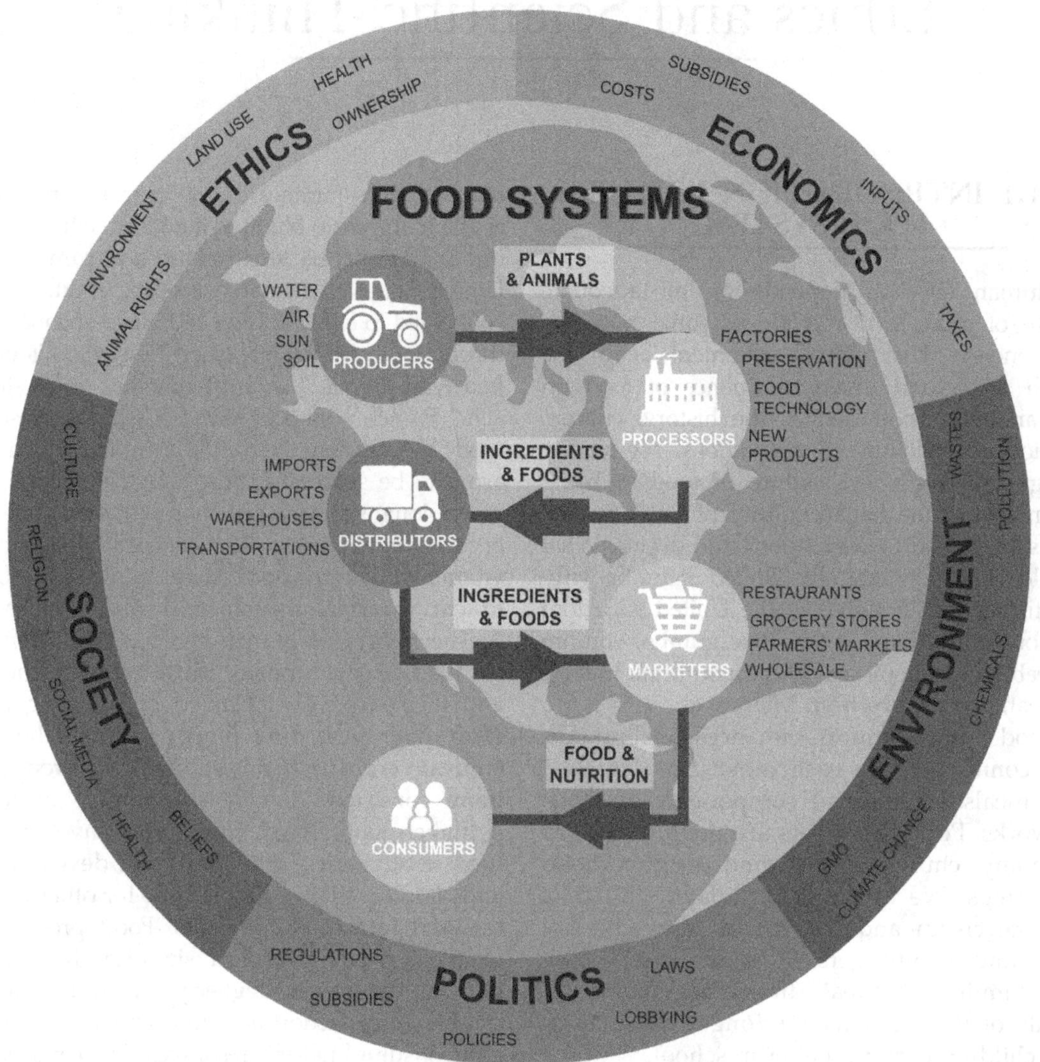

FIGURE 1.1 A food system is a complex network that starts with the producers of food, processors that preserve and modify the food, transportation and distribution systems, and markets that make food available to consumers. Influences from the realm of economics, ethics, politics, environment, and society are part of the food system and must be considered. *Source: Illustration by Reannon Overbey.*

with natural resources and climate, types and locations of food production, and human engagement in agriculture. Processing includes harvesting, storing, manufacturing of foods and food ingredients, and controlling food waste and loss. Distribution and marketing involve getting food from producers and processors to consumers. And food consumption spans personal choices to human health. Integrated within the food system are political, sociological, ethical, economic, environmental, and cultural pressures and influences.

A stakeholder is a person or group who has a share, an investment, or an interest in an issue. In the case of the food system, there are many stakeholders. In fact, because everyone eats, everyone is a stakeholder in the food system. Groups that are directly involved with the food system include producers (farmers, growers, ranchers, and workers), processors (who convert raw materials into foods and food ingredients), distributors (transportation, warehouses, and storage), marketers (farmers markets, grocery stores, restaurants), and finally consumers (people who eat the food). In addition, government, corporations, private foundations, universities, healthcare organizations, local organizations, schools, and a wide range of advocacy groups are engaged both directly and indirectly in the US food system.

As an illustration of the food system, consider what has to happen for you to get a sausage pizza delivered from your local pizzeria. Wheat must be planted, tended, harvested, and transported to a mill, where it is ground and processed into flour. The flour is then packaged and delivered to the dough processor, who prepares it with yeast and other ingredients. Tomatoes must be planted, tended, harvested, and transported to a processor to make the sauce (consider also the added spices that are grown, packaged, and transported). Hogs must be bred, fed, and cared for, and then harvested and processed into sausage. Dairy cattle must also be bred, fed, and cared for; milk must be collected and processed into mozzarella cheese. The pizza ingredients must be delivered to the shop (which would have to be built, equipped, and managed), the pizza prepared and cooked (by trained workers), packaged, and delivered to your door (most likely via a person driving a car on roads). Each ingredient has a production system that requires multiple economic and environmental decision points and people, as well as infrastructure and transportation. Some steps may involve following government regulations, health codes, and industry standards. Various, possibly controversial, inputs may be required, such as irrigation of the wheat, migrant workers to pick the tomatoes, or confinement housing for the hogs, which have social, ethical, political, environmental, and cultural implications. These decision points have broad and lingering implications. So while you are enjoying your next pizza, stop to think about everything that had to happen before it arrived on your table.

1.1.2 Influences on the Food System

The concept of food systems has entered the public discussion as people realize the implications of their own food choices and want to understand where their food comes from and how it is produced. While farmers and food producers, food scientists, and food manufacturers are obviously engaged in food systems, the implications of understanding food reach far beyond these professions. Government leaders and policy makers, economists, environmentalists, and healthcare professionals including nutritionists and dietitians, nurses, physicians, and research scientists make decisions and set policies that influence the broad food system. Journalists, chefs, and even celebrities influence how food is perceived and create trends that may have a wide impact on the food system. And individual consumers, who make food choices each and every day, directly impact the food system through their economic and political influence. Discussions about food, how it is produced, and how it affects health likely arise daily in your life. Controversies in the food system are abundant, complicated, and often draw strong emotional responses. Many of these will be highlighted in the following chapters. Providing a solid reference, based on the available scientific evidence, for framing decisions about food is the intent of this book.

Our primary goal is to provide an overview of the US food system within the context of societal influences. From an historical perspective to present day, we explore how the US food system developed. There are increasing complexities within our food production systems that influence how much food and what types of food we produce; how food is grown, processed, marketed, and consumed; and how food affects human health. We begin in Chapter 2, History of US Agriculture and Food Production, with defining agricultural history and the evolution to our modern farming systems. In Chapter 3, Innovations in US Agriculture, innovations in farming practices are discussed including advances in technology and biotechnology in agriculture. The production of food from animal sources is covered in Chapter 4, Animals in the Food System including discussion of animal rights. The role of human agricultural workers and the political, ethical, and economic factors surrounding labor are covered in Chapter 5, Human Resources in the Food System, In Chapter 6, Food Processing, the role of food science in the current food system will include information about techniques for food preservation, additives, nutrients, safety, labeling, waste, and health and economic benefits, as well as concerns regarding the food supply. The integration of food with human health and disease and the complexity of food security are discussed in Chapter 7, Nutrition and Food Access. Finally, the sustainability of food production, processing, and accessibility and current environmental and climate change challenges will be addressed in Chapter 8, Sustainability of the Food System. Throughout, we encourage the use of scientific thinking and the consideration of ethical principles in each area of discussion. We will describe policies that affect food, outline the connection of these complex issues, and suggest the impacts on the overall food system. Readers are encouraged to challenge their beliefs and understanding of food within their own lives and to consider how their political views, cultural, ethical, and religious beliefs and personal choices impact the overall food system.

Because all aspects of the food system are interconnected, a change in one aspect may have broad, even global, repercussions. For example, the grain quinoa (pronounced *keen-wah*) became a trendy food in the United States because of its high nutrient value and perceived health benefits. Quinoa is a staple food of native South Americans living in the high plains of the Andes Mountains, primarily in Bolivia and Peru, and these countries account for over 90% of the world's production. Traditionally, quinoa has been produced by small-scale farmers for their own consumption. Increasing the global demand for quinoa raises concerns as to whether commercial production for export will squeeze out these farmers and alter the economic and environmental balance of the region. There is evidence that, as the local price of quinoa increased, indigenous populations were unable to afford quinoa and forced to consume less nutritionally balanced foods. Debates have arisen between governments as to who owns the right to the quinoa germplasm. Many countries want to develop quinoa production but Bolivia controls much of the quinoa seed and has set treaty limits on what it will share, concerned that the country will suffer economically if quinoa production becomes widespread elsewhere. The social-cultural and political issues around this one food item are very complex and most consumers may be unaware of the implications of their decision to purchase a box of quinoa at their local grocery store.

Much attention has been focused on creating a sustainable food system. In addition to the aspects of food production, processing, distribution, and access, the importance of natural resources (genetics, soil, land, water, air, and energy) is emphasized in a sustainable food system model. Sustainability must include the interactions between society (human resources, values, and trends), economics (capital resources), and the environment (natural resources). A system that fails in any one of these areas cannot be sustained. A policy that enhances an economic return but damages the environment is not sustainable; similarly one that protects the environment but cannot be achieved economically is not sustainable. And policies that are socially unacceptable, regardless of their economic or environmental value, will not be sustained. Interactions between people and their resources are delineated in policy, education, politics and government, and research and technology. These interactions influence each of the spheres and determine the priorities for the uses of natural, human, and capital resources for food. Sustainability can occur only when the competing interactions between society, economics, and the environment are in balance (Fig. 1.2).

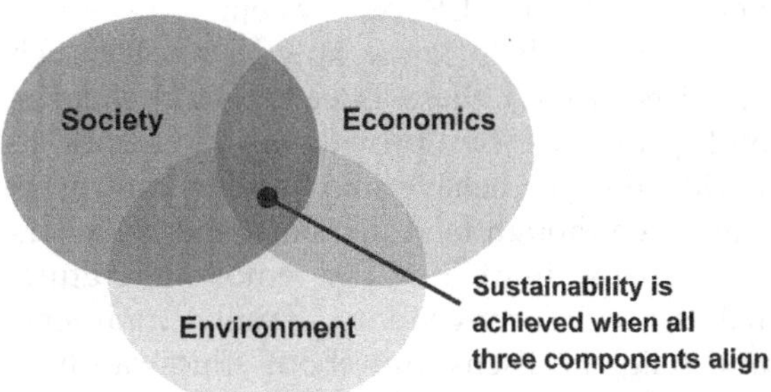

FIGURE 1.2 A sustainable food system must satisfy environmental, economic, and societal demands. *Source: Illustration by Reannon Overbey.*

1.1.3 Consumer Engagement With the Food System

Everyone is a consumer of the food system and as such yields great influence. An individual's buying power, needs, and desires regarding food will change over a lifetime, but people will always need healthy and safe food to survive. It is important for all consumers to have knowledge about food, not only what is on the plate in front of you but also a view of the larger food system. Your food decisions have an impact well beyond your kitchen.

Consumers make decisions based on inputs from a wide range of sources so it is very important to assess the credibility of those sources. One of the primary goals of this book is to provide a framework from which recommendations about food and diet choices offered by popular books, television, movies, magazines, and social networks can be evaluated. There are countless documentaries, books, movies, blogs, and websites about agriculture, food, and health. Celebrities have taken up these topics and used their access to the public to promote various diets and philosophies and promote agendas. While she was First Lady, Michelle Obama used her influence to encourage exercise and eating vegetables from a home garden. Many researchers have investigated the reasons for consumer behaviors relative to their food choices. The major influences for most consumers when making decisions about food are messages from the media and advice from friends and family, according to a 2016 International Food Information Council (IFIC) survey. Healthcare professionals; dietitians; agricultural, nutrition, and food scientists; and governmental agencies have a lesser influence on consumer choices (Fig. 1.3).

Consumers also exert influence on the behavior of other aspects of the food system. They demand, via their purchasing choices, convenient, low cost, healthful foods, and recently, organic and local products. Food

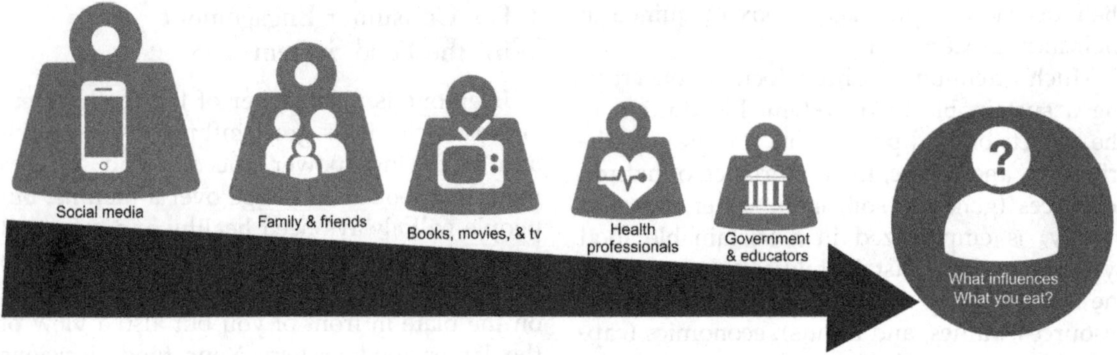

FIGURE 1.3 Decisions consumers make about food are influenced more by their family and friends, social networks, and the media than by healthcare professionals, government resources, and educators. *Source: Illustration by Reannon Overbey.*

companies, as businesses seeking profits, respond to these demands by marketing foods that consumers will buy. When low-fat foods were demanded, thousands of new products appeared on the market seemingly overnight. Similar trends were seen in the marketplace with food products labeled organic, gluten-free, non-GMO, natural, sustainably raised, local, or other appealing descriptors. In many cases, consumer trends outpaced government regulations. For example, consumers were demanding these products long before regulators had formulated a means of appropriately labeling them. Without standards for labeling products as natural, sustainably raised, or local, consumers were left on their own to determine the accuracy of these messages.

The US food system is highly integrated with the global food system and influenced by social, economic, and political factors in other countries as well as trade agreements, environmental treaties, and business decisions between US and foreign entities. Energy policy, gas and oil production, transportation, weather, and other global factors have significant effects on US food policy and trade. Although the US food system is the focus of this book, the US food system does not function in isolation but both influences and interacts with global forces. The fact that all parts of the world participate in the global food system should be continually kept in mind while considering the contents of this book.

1.2 ETHICAL THEORIES AND PRINCIPLES IN FOOD SYSTEMS

Meatless Mondays! Locally grown food! Organic food! No GMOs! Confinement crates! Pesticides! Chemical additives! Migrant workers! SNAP benefits! No CAFO in my neighborhood! School lunch! Processed foods! These terms highlight some of the many complex, emotional, and often contentious, controversies within the US food system.

It is difficult to know how to resolve such conflicts around these issues. Ethical theories and principles can provide a framework to decide how to behave and resolve controversies. Even though food decisions and behaviors are greatly influenced by emotion, culture, habits, social conventions, family traditions, and religion, discussions about ethics can help determine common values and lead to resolution of problems by encouraging the exchange of opinions, sharing ideas and experiences, and recognizing the ultimate purpose and goal.

1.2.1 Definition of Ethics

Ethics is a moral operating system to determine right and wrong. It is a branch of philosophy that involves the study of arguments and theories about what actions are right or wrong, moral intuitions, and rules of moral reasoning, but transcends differences in religion, law, and customs. Laws may be based on ethics but do not define an ethical position. In some people's view, some things are legal but may not be ethical (abortion); others may be ethical but not legal (assisted suicide). It is common to assume ethics is the same as religious belief or doctrine, but these tenets do not comprise an ethical argument. Decisions about right and wrong require serious thinking and hard work and are not based on opinion, religious conviction, or "gut feelings."

Ethics should inform, seek facts, and propose norms for behavior. The intent behind actions and the surrounding environment that influences behavior must also be considered: "Ethics is a discipline for asking better questions about societal dilemmas and certainly should be applied to food" (Thompson, 2015, p. 7).

There are many factors that influence personal decisions about food (Fig. 1.4). The majority of people in the world are constrained in the availability of food by their life situation (subsistence farmers or poverty). Obtaining enough food is the primary goal rather than being selective about how the food was grown or processed. Economically secure people living in stable environments are free to engage in broader issues about their food choices. Consideration of lifestyles, preferences, and ability above economics and access arise in affluent societies. This generates debate about how food choices made by privileged societies impact food access of underprivileged societies: "Food choices become ethical when they intersect with complex economic supply chains in ways that cause better or worse outcomes for other people, for nonhuman animals, or for the environment" (Thompson, 2015, p. 5). Local and personal decisions about how and what food is produced, processed, distributed, and consumed have global impacts and must be made with full understanding of these outcomes.

A dilemma is an argument or situation necessitating a choice between equally unfavorable alternatives. Ethical dilemmas are conflicting issues or situations that are not easily

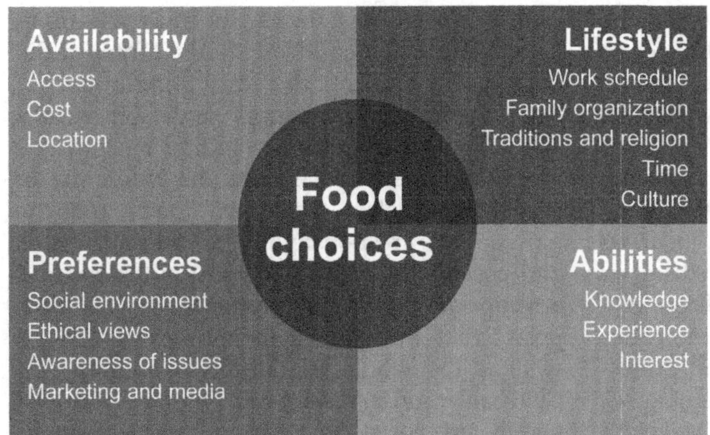

FIGURE 1.4 Many factors influence the decisions people make about food. Family and cultural factors tend to be the earliest influencers on food choices, and people may change their preferences and habits as they move through life. *Source: Illustration by Reannon Overbey.*

resolved. Many food system decisions are ethical dilemmas. The decision about whether or not to eat meat poses an ethical dilemma. Meat provides essential nutrients that are important for health and meat is an enjoyable food. Large-scale meat production systems require substantial amounts of grain and water and can have negative effects on the environment. Should the nutritional benefits outweigh the environmental damage? This discussion has many other layers that further add to the ethical dilemma, such as animal rights, antibiotic resistance, and greenhouse gases in contrast to right to livelihoods for cattlemen, production of manure for fertilizer, and efficient use of pasture land. Applying ethical principles can help address such ethical dilemmas.

1.2.2 Ethical Principles

A principle is a standard for behavior or general rule of conduct. Principles explain approved practices for everyday situations as well as for difficult problems. For example, children are taught to tell the truth in spite of the consequences. This ethical principle, truthfulness, is generally seen as a positive behavior and rule of conduct. Beneficence, nonmaleficence, autonomy, justice, and paternalism are the primary ethical principles involved in discussions about food system issues. There are other ethical principles, no less important, such as charity, mercy, peace, fidelity, compassion, integrity, honesty, courage, honor, respect, and responsibility that may also be considered.

Beneficence is doing good, nonmaleficence is doing no harm or avoiding harm, autonomy is allowing people self-determination, justice is treating people fairly, and paternalism is deciding for others when necessary. These ethical principles figure importantly in the codes of ethics for food science and nutrition professionals (Institute of Food Technologists, Academy of Nutrition and Dietetics, American Society for Nutritional Sciences, and others) as well as in codes of behavior for veterinary, medical, and healthcare professionals.

Beneficence is the principle involved when providing food to the needy and hungry or when teaching nutrition to school children. Food companies who assure that preserved food is safe to consume are practicing the principle of nonmaleficence. Allowing consumers to choose the types of food they prefer is an example of autonomy and providing farm workers an equitable wage is an example of the ethical principle of justice. Parents who select wholesome and nutritious foods for their children are practicing paternalism.

The interrelationship among components of the food system and the occurrence of obesity is one example of an ethical dilemma that requires the application of ethical principles. From the position of autonomy, some will argue that individuals should be able to make their own decisions about what, when, and how they consume food and there should be no oversight to direct their food choices. From the position of nonmaleficence, others argue that the harm to individuals' health and costs of obesity to society are significant, and oversight of the types, portion sizes, or times/locations for availability of certain foods is needed to reduce calorie consumption and benefit the majority. The position of paternalism may be more acceptable when children are the target of the obesity issue but not the adult population.

Whether or not snacks and soft drinks should be sold in school vending machines is an ethical dilemma between the autonomy of students to select their own foods and the paternalistic responsibility of administrators and parents to provide the most nutritious and appropriate foods for students. It was proposed that individuals in New York City as well as the whole community would benefit (with an improved quality of life and reduced healthcare costs) if residents reduced their consumption of sugar-sweetened beverages, but adoption of laws (which did not pass) to reduce the size of soft drinks may violate the autonomy of citizens to consume the types of drinks they want. Overlain on these ethical dilemmas is evidence of efficacy of the proposed policies. It would be important to know if limiting the access of students to sodas and snacks or NYC residents to large sodas would have a positive effect on body weight or overall health. Scientific thinking must also be part of the ethical decision process.

Other aspects of food choice dilemmas include the autonomy of food producers (those that make cookies and doughnuts) or restaurants (fast food outlets, bakeries, candy stores) to make a living and earn a profit by selling what some might consider "junk food." Is it the role of the government, parents, school superintendents, or employers to determine the acceptability of foods? Proposed bans on junk food arise frequently in discussions about the food system. Defining when a food falls into the junk category is complicated. Clearly, foods that are low in nutritive value and high in calories, fat, sugar, and salt should be consumed in moderation. Moderate intake of cookies, candy, salty snacks, and fast food is unlikely to cause obesity or chronic disease in most people, but the risks increase with overconsumption, especially for children. Because some people overconsume these foods, should those who do not overconsume or those who do not suffer from obesity or chronic disease related to their consumption be prevented from consuming them? The decisions regarding food choice involve complex ethical questions and are playing out in society today. Well-meaning laws designed to protect consumers from themselves, and/or the food system, imply that paternalism and beneficence should be implemented. But are the principles of autonomy and justice being equally served? As food system changes are made in an attempt to provide optimal foods, it is important to include all members of society, especially those who are usually unheard, in the conversation about needed changes.

1.2.3 Ethical Theories

Ethical principles are the tenets or basic truths upon which ethical theories are built. Ethical theories can provide guidance for making personal decisions about food and oversight of the entire food system. A theory is a doctrine based on observations and reasoning. Mathematics, physics, medicine, language, psychology, and other fields of arts and sciences use theories to understand, explain, and predict ideas and concepts about people and the natural world. Theories can be constructed from hypotheses, evidence, and scientific experimentation, leading to a fact-based framework for the description of a phenomenon such as gravity and evolution. Philosophical theories, formed through discussion, questioning, argument, and logic, are developed into systematic methods to solve problems. Ethics and ethical theories are one branch of philosophy. Ethical theories, similar to theories in other areas of scientific inquiry, have been developed over centuries by many practitioners and have resulted in different schools of thought. Consider ethics a guide for discussion of questions, dilemmas, and problems related to human behavior. A review of ethical theories can provide a basis for discussion of the many-faceted issues related to our obligations, as food and nutrition professionals, citizens, parents, and consumers, to provide adequate and nutritious food for our families and communities.

Access to food is considered a basic human right, so changes in the food system do have ethical dimensions.

> Everyone has the right to a standard of living adequate for the health and well-being of self and of family, including food, clothing, housing and medical care and necessary social services, and the right

to security in the event of unemployment, sickness, disability, widowhood, old age or other lack of livelihood in circumstances beyond personal control. Motherhood and childhood are entitled to special care and assistance. All children, whether born in or out of wedlock, shall enjoy the same social protection.

United Nations Universal Declaration of Human Rights (http://www.un.org/en/universal-declaration-human-rights/)

A right to food does not mean free food for all but rather it is an "opportunity right" or a claim that a just society owes its citizens the opportunity to nourish themselves. A just society would also support providing food to the poor and needy to prevent starvation and human suffering.

Issues of world hunger, obesity, malnutrition, food safety, and environmental well-being require serious consideration by people, communities, and nations. In order to have rational and reasoned discussions about complex issues, an understanding of the morals and values that are important to the people involved is necessary. A civil society, by definition, solves problems peacefully, using persuasion rather than coercion, as well as sound reasoning and factual information developed through the scientific process.

Ethical theories provide a framework for making ethically based decisions. An understanding of each ethical theory with its strengths and weaknesses will lead to informed decisions about ethical dilemmas. Some believe that no ethical theory fits all situations and that each situation depends on the context, culture, or time. Certainly ethical decisions about food made during war or famine will differ from those made during peace or abundance. This is referred to as "relativism." Often, in discussions of food system issues, many variables will influence the outcome. However, even if one theory does not provide succinct answers, knowledge of ethical theories and the arguments used to support them will provide guidance for analysis of the food system, a basis for discussion of complex issues and valid reasons for decisions that need to be made.

A brief overview of four ethical theories is presented here for the beginning student of food systems. For more detailed discussion, readers are encouraged to read Comstock (2002) and Thompson (2015). Ethical theories for human behavior have been debated and analyzed for centuries and have formed the basis of many civilizations. Rights, utilitarianism, virtue and environmental theories, or parts of these ethical theories, will likely be the most useful in discussions of the food system (Table 1.1). Several other ethical theories exist (deontology, egalitarianism, and

TABLE 1.1 Ethical Theories Useful in Discussions of the Food System

Ethical theory	Philosopher	Premise
Rights	Locke	Society determines inherent rights for each citizen.
Utilitarianism	Mill, Singer, Bentham	Choose the greatest good for the greatest number, increase happiness and reduce pain, weigh the benefits and harm.
Virtue	Plato, Aristotle, MacIntyre	Moral behavior is dependent on the goodness of the individual.
Environmental ethics	Callicott, Norton	Nature, species, and the ecosystem have rights and are intrinsically valuable.

care ethics) but are beyond the scope of this discussion.

Rights theory holds that certain protected privileges are universally applied to everyone and are inherent for human beings, such as the right to life, to self-determination, to not be injured, and to not be exploited. Society determines which rights, duties, and responsibilities are ethically correct and valid. These rights must be based on the society's goals and ethical priorities, yet within a society each human is a free and autonomous agent. For example, rights theory would uphold the right of farmers to determine which agricultural methods they use and how they use the land they own. Some philosophers may accord rights to animals, plants, nature, and the environment. Critics contend that rights theory offers little guidance for complex situations when there are conflicting obligations and problems may arise if society determines rights for certain individuals or groups to the exclusion of others. In the example of farmers' rights, it is not clear if one farmer's right to apply synthetic chemicals to a field using an airplane would override the rights of a neighboring farmer who uses organic practices and whose crops might be contaminated by the spraying.

Utilitarianism is a type of consequentialism as it is based on the ability to predict the consequences of an action. The choice that yields the greatest benefit to the most people is seen as the choice that is ethically correct. Utilitarianism finds the greatest good for the greatest number and optimizes happiness in society while minimizing pain. Similar predicted solutions can be evaluated by a point system, which provides a logical and rational argument for each decision. This theory promotes evaluation of expected consequences and chooses the best consequence for the majority.

John Stuart Mill, a 19th century English philosopher, revised utilitarianism to consider pleasure as the good and pain as the evil. Mill stressed the importance of individual autonomy and said that intellectual and moral pleasures were superior to physical pleasures, indicating that happiness was quantifiable. An "act-utilitarian" performs acts that benefit the most people, regardless of laws or personal feelings. A "rule-utilitarian" seeks to benefit the most people but through the fairest and most just means available, meaning that justice, autonomy, and beneficence are part of the benefit/harm calculation.

Utilitarianism strives for the most good or least possible evil in the universe to produce the most pleasure and least pain for the greatest number of people. The amount of happiness and suffering created by a person's actions is considered most important. Utilitarianism is useful for making balance sheets or checklists about the benefits and risks of food system problems but does not always consider rights, justice, relationships, and feelings that are more highly valued in other theories. It is difficult to use utilitarianism for some decisions that require future predictions (such as the benefits and impacts of certain pesticides) or when various consequences cannot be compared on a similar scale (short-term livelihoods of farmers versus long-term risks to the environment). Some acts determined by utilitarian theory may violate individual rights (one or few people) in order to benefit society (many people). For example, use of synthetic pesticides and fertilizers by conventional farmers may increase the amount of food produced and provide good for many people but could violate the rights of organic farmers, supplying foods to a small group of local consumers, whose crops might be contaminated. Utilitarian theory supports optimization of food production to feed billions of people as the most important goal, but, critics say, if implemented for money or glory associated with scientific prestige, it is in opposition to virtue and other ethical theories.

Virtue theory asserts that ethical decisions will be made if a person is kind, fair, wise, courageous, and of good moral character. Character matters above all else and good character will lead to ethical behavior. Living an ethical life, or acting rightly, requires developing and demonstrating the virtues of courage, compassion, wisdom, and temperance. It also requires the avoidance of vices like greed, jealousy, and selfishness. A person's morals, reputation, and motivation are evaluated when rating ethical behavior. The Greeks considered gluttony (eating for pleasure rather than nourishment) a mortal sin and certainly not virtuous.

The Greek philosopher Plato is credited with saying that if you "know good, you will do good" and that there were objective and universal truths. Another Greek philosopher, Aristotle, thought that good character was developed through practice and that excellence and personal perfection were desirable while pleasures and emotions should be tempered. Aristotle believed ethics could not be calculated like math theorems (as did Plato) but depended on judgments and the context of the situation. Self and the relationship with others are of primary importance rather than following rules. Be a good person, have good character, be generous, affectionate, and courageous; this behavior leads to happiness. In this view, some people are more moral than others. Thomas Jefferson's view of agrarian democracy, based on farmers and rural communities, is compatible with virtue ethics. Farmers were virtuous in their goals and intentions because they provided food and nourishment to others, albeit tarnished by the unjust and immoral practice of slavery. The presentation of the family farm as a virtuous enterprise continues to be prominent in advertisements for seed and herbicides and promotional messages about agriculture. This theory may be difficult to apply to food system issues because moral character is hard to uncover and separate from other motivations. But increasingly, consumers are demanding truthfulness and integrity within the food system and stakeholders are taking heed. Being recognized as virtuous is essential to gaining consumer trust.

Environmental ethics extends ethical consideration to the nonhuman world and nature. It is a relatively new area of philosophy that grew out of the environmental movement in the 1970s. There are different ethical approaches to environmental ethics. Aldo Leopold described "The Land Ethic" in *A Sand County Almanac* (1949) and considered the natural world, including farms, as an ethical entity. Frances Moore Lappé advocated vegetarianism in *Diet for a Small Planet* (1975) in a utilitarian argument that stressed efficient use of natural resources and offered the greatest benefits to the most people. Lappé's reasoning now includes the impact of livestock production on climate change and illustrates a shift in thinking about the environment itself as worthy of ethical consideration.

Examples of ethical dilemmas related to food and nutrition with the ethical theories that support each are summarized in Table 1.2. There is little argument against providing food for the hungry. How food should be provided to those in need, and who should pay for it, are different but related ethical arguments. When it comes to the ethics of consuming meat and genetically modified foods, well-established ethical theories provide support for each contrasting position. Some theories provide support for both sides of the issue, depending on reasoning, the methods used to tabulate the costs or harms and benefits, and the constituents to be considered. This examination of ethical positions illustrates the complexity of food system issues and the necessity of discussing values before resolution of food dilemmas can occur.

TABLE 1.2 Ethical Justification for Food System Dilemmas

Dilemma	Ethical theories and principles to support	Ethical theories to not support
Should society provide food/ money to feed poor children?	We have a duty to help the less fortunate (rights, virtue).	
	There should be equitable distribution of food (utilitarianism).	
	All citizens have a right to food (rights).	
	Our relationships in the community require us to help (virtue).	
	Children should be cared for (paternalism, beneficence).	
Should people eat meat?	The benefits to humans outweigh the costs to animals (utilitarianism).	Animals have a right to life (rights).
		We have a duty to protect animals and the environment (virtue, rights).
Should we eat genetically modified foods?	The benefits to farmers, environment, and hungry outweigh the risks (utilitarianism).	Costs to the environment outweigh the benefits (utilitarianism).
		Consumers should have the right to choose (rights).
		Control of seeds and plants is unjust (rights, justice).

Unfortunately, ethical philosophy is not the basis for typical conversation when problems or dilemmas arise around the food system. Too often, public discussions, even about food, tend to become polarized into "us versus them." Identifying values in political discourse may aid resolution of disagreements and, in turn, lead to thoughtful and effective governmental policies. Democratic Party tenets promote fairness and equality with statements such as "everyone gets a fair shot, everyone does their fair share, and everyone plays by the same rules," which is the application of utilitarianism and beneficence. The Republican Party defends independence and personal rights in their party platform with statements such as "free from government intrusion" and "healthcare decisions should be made by us and our doctors," which is the application of autonomy and rights. Recognition of the ethical principles that form the foundations of both parties might reduce the divisiveness of many political debates. Civil society must continually strive for compromise between conflicting values and opposing interests. The premise that decisions about the food system must be rooted in sound scientific evidence is also essential. Ethical dimensions must be always in mind as we make personal decisions about the food we eat and as we define public policy, laws, and regulations. Becoming educated about food and all of the components that comprise the food system is the responsibility of everyone and is essential for finding solutions to food system problems. Each rational and moral being has a duty and obligation to do something!

EXPANSION BOX 1.1

CAN MY EATING HABITS BE JUDGED ETHICALLY?

Kevin, a vegetarian and Eagle Scout from New York City, was paired with Nancy, a former County Pork Queen and 4-H participant from Denison, Iowa, for a project in their nutrition class. The objective of the project was to evaluate the nutritional adequacy of each of their diets. In their discussion about their eating habits, Kevin and Nancy began to argue about whether or not it was ethical to eat meat. Kevin said eating meat was barbaric while Nancy claimed eating meat was natural. Their teacher intervened and suggested they first determine the "goodness" of their diets. They reviewed the ethical principles they had previously learned and tried to determine if eating had an ethical dimension. From their research and discussion they made several observations.

Four ethical principles are beneficence, nonmaleficence, justice, and autonomy. Autonomy means that I can make decisions for myself, as long as I don't interfere with the autonomy of others. Kevin said killing animals for food interferes with the autonomy of animals, which are sentient beings and have a right to life. Nancy replied that animals used for food are treated humanely during their lives and respectfully at slaughter. They had a lively discussion about methods of animal production but Nancy contended that animals have a good life on her parents' farm and people who choose to eat meat should be free to do so.

Beneficence is the principle of doing good for others. Kevin and Nancy both agree that their eating habits are healthful and, by staying fit, they can do the most good for others. As students preparing to be registered dietitians, they are very aware of the types of food they eat and want to help others eat well too. Not consuming more than one's share and not being wasteful is beneficial for the community. Nancy says that raising animals for food utilizes land that is not suitable for crop production and is a beneficial use of resources. Kevin provided statistics that support the idea that a vegetarian diet uses fewer resources than a meat-based diet, and that a meat-based diet is not sustainable. They agree that they should investigate the validity of these statistics about utilization of resources (discussed in Chapter 8: Sustainability of the Food System).

Kevin and Nancy wonder which type of diet would work for everyone on Earth, that is, what kind of diet would treat everyone fairly (justice). They could send aid for those who have less food than they do, but does it matter what they themselves eat every day? Nancy knows that Heifer International helps farmers in less developed countries obtain animals for food production. Kevin mentions that every culture has different types of food and there are many economic and political factors that affect the availability of food. The students agree that they are fortunate to be able to afford and choose the food that they eat.

Utilitarianism and rights are philosophical theories that can be used to sort out ethical decisions. Utilitarians would contend that Kevin and Nancy should weigh the risks and benefits of specific dietary habits. Consuming meat provides important nutrients and supports the livelihood of producers and processors, yet does present some environmental concerns. How serious is the environmental impact and do some people benefit more than others?

Those who support animal rights believe that animals feel pain, can suffer harms, and "have a life," which make them morally

EXPANSION BOX 1.1 (*cont'd*)

significant. The degree to which all types of animals have moral status may vary among philosophers. Animal rights proponents oppose animal agriculture and the use of animals for scientific experiments and favor vegetarian diets. Animal rights theories are generally in conflict with animal welfare ideas, which are based on utilitarian views.

Kevin and Nancy agree that choosing "the right foods" to eat for themselves is confusing and determining diets for an entire society would be a complicated matter. They also agree that they should learn more about ethical philosophy; find more facts about agriculture, food, and nutrition; and be open to new ideas and ask more questions.

Suggested reading: Life Science Ethics *(Comstock, 2002).*

1.3 THE SCIENTIFIC METHOD

> At the heart of science is an essential tension between two seemingly contradictory attitudes—an openness to new ideas, no matter how bizarre or counterintuitive they may be, and the most ruthless skeptical scrutiny of all ideas, old and new. This is how deep truths are winnowed from deep nonsense.
>
> *Carl Sagan (1997, p. 304).*

As children, we learn that science is a process that involves asking questions or stating a hypothesis, carrying out an experiment or observation that leads to collecting data, analysis of the data typically with a statistical assessment, and interpretation of the data to either support or refute the stated hypothesis. At the core of science is the testing of ideas, which begins with gathering data. Science creates, builds, and organizes knowledge in testable explanations and predictions. Scientific method is a disciplined way to study the natural world.

The basic components of the scientific method are (1) establishing a hypothesis usually based on an observation or question; (2) predicting the expected results (what should happen, based on the hypothesis) and designing ways to collect actual measurements or observations (what actually happens); (3) interpreting the information or data (which usually involves application of statistical analysis); (4) comparing findings to previous knowledge and observations and to the stated hypothesis; and (5) accepting or rejecting the hypothesis and asking new questions (Fig. 1.5). Science is an iterative process of continually refining hypotheses, defining new and better experiments and technologies, and building on previous work. Scientists ask questions and seek information, then interpret and debate their findings.

When enough data and scientists agree, a theory or theorem may be proposed, but even these are subject to continued refinement, or even dismissal, as new ideas and information are obtained. For these reasons, some may consider science to be confusing and unreliable. Skepticism about science often arises in food systems discussions. For example, since the 1960s scientists recommended people eat fewer eggs because they were high in cholesterol. With new information about heart disease and recognition that eggs are a good source of nutrients, this recommendation was removed from the 2015–2020 Dietary Guidelines. Because of the nature of the scientific process, new ideas and concepts will continually arise. This doesn't mean that

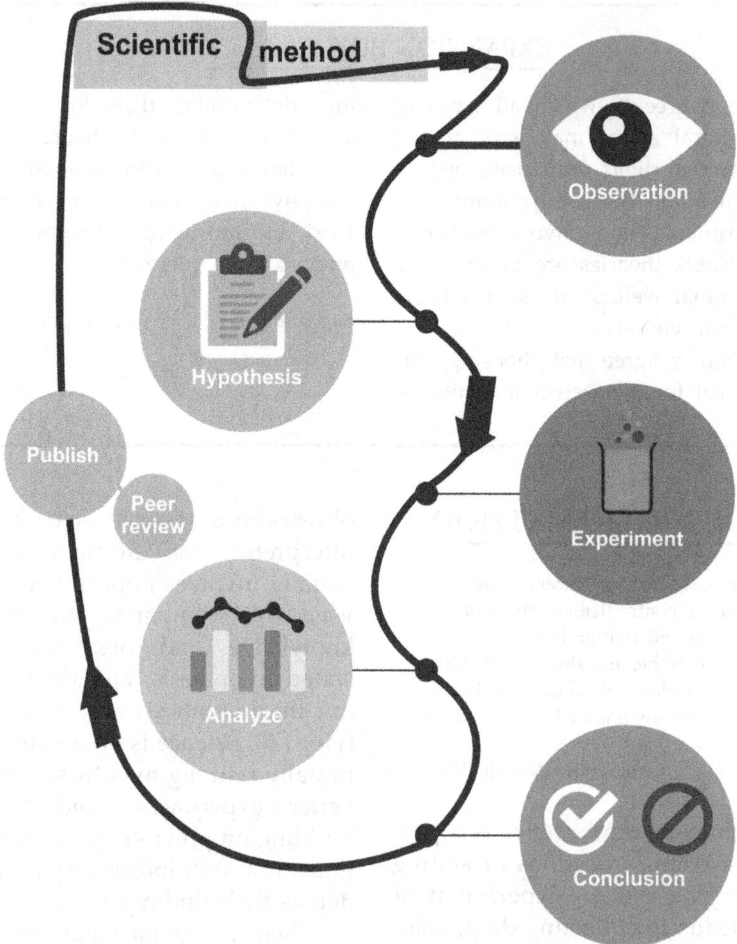

FIGURE 1.5 The scientific method is a defined series of steps that provide a means for testing a hypothesis. Scientific inquiry is iterative and never completed because new questions or ways to measure outcomes are always developing. *Source: Illustration by Reannon Overbey.*

scientists "got it wrong," but rather that the scientific process worked by allowing new ideas and information to refine the conclusions.

1.3.1 Peer Review

Of great importance is the understanding that science occurs in a community. Reporting and communicating experimental outcomes is essential for science to progress so that data can be used and interpreted by others. Within the scientific community, the process of peer review is a core component of the reporting process. Scientists submit their findings, typically to a scientific journal or to a professional meeting, and other scientists who have expertise in a similar area evaluate the quality of the work and the appropriateness of the interpretation. The peer reviewers' identity is not shared with the authors of the work to allow them to be unbiased in their assessment. These peer reviewers will look for a clearly defined hypothesis, suitable and well-controlled experimental design, accuracy in the data, and the quality of statistical interpretation. They will also determine if the authors have appropriately interpreted their data within the larger context of previously published work. The peer reviewers may suggest the authors of the study redo some of their experiments, change their description of the outcomes, or modify their interpretation. In some cases, the peer reviewers may find the work to be flawed and not acceptable for publication or presentation.

This approach of peer evaluation within the scientific community is intended to ensure the scientific method is maintained and the quality of published research is high. Within any discipline of science, there is a hierarchy of journals that are considered by scientists in that discipline as being the most well-respected places to have their work published. Peer-reviewed journals are managed by an editorial board and a core of scientists who provide manuscript reviews, typically on a volunteer, unpaid basis. These individuals are usually listed on the journal's website or cover. Some journals, such as *Science*, *Nature*, or the *New England Journal of Medicine* are very highly regarded and publish some of the most influential research. And each scientific discipline has a list of top tier publications, such as the *Journal of Food Science*, *American Journal of Nutrition*, *Journal of Animal Science*, and *Journal of Horticulture*. Peer review is an important and essential part of the scientific method, but does not completely prevent inaccurate or flawed research from being published. Newer information or ways of studying some aspect of the research may completely change how the original work is understood. However, the open peer assessment process allows for ongoing dialog and interpretation to ensure advancement of knowledge.

With the ready availability of the Internet, online publications have become common. These journals may not have the same rigor of peer evaluation as those published by professional scientific organizations. Articles that are published by non-science-based groups are called "lay publications." These include magazines, newspapers, or newsletters. Being able to determine the quality of information in a publication is an important component of scientific thinking.

Maybe you are a runner and wonder if eating breakfast would improve your performance at racing events. A scientific study to define the effect of eating breakfast on performance would include a large number of subjects, a well-defined and standard meal, test and control groups, verified measures of performance, quantification of the data, and statistical analysis. From such a study, it could be concluded, within the parameters of the study, that people who ate breakfast were more productive than those who did not. Notice that the previous sentence included the qualifying phrase "within the parameters of the study." This phrase is seldom seen in newspaper headlines. Instead, we are more likely to read "Run faster by eating bacon for breakfast!" The interpretation of scientific publications by the media may be superficial and not provide enough information to adequately understand the work. It is therefore important to seek out the original publication and read the study for yourself rather than rely on summaries provided by others. PubMed, which is freely available online, provides searchable access to the scientific literature related to human health (US National Library of Medicine).

1.3.2 Credible Sources

In modern society, there are many avenues to publish ideas and information without the peer review process. Websites, blogs, and various other social media postings allow people to share their thoughts and ideas freely. These avenues are clearly important for society and allow a wide range of information to be quickly and broadly disseminated. However, without the process of peer assessment, the integrity of the information may be questionable. Finding ways to use and interpret information is critically important when considering these sources of information. Some factors to consider when assessing information on the web include:

1. Who has generated the website—an individual, a group/organization, a government agency?
2. How is the website funded?
3. Does the website provide contact information?
4. When was the website updated?
5. Does the website provide links to references or sources of information? Are these linked to the original website authors?

It is essential that websites list a reference and include authors and date of publication if they are to be considered credible (Table 1.3). A credible reference is defined by their education and work experience from which they base their expert interpretation and assessment. Food and nutrition websites by the Mayo Clinic, USDA, food and nutrition departments at universities, the Academy of Nutrition and Dietetics, and the Institute of Food Technologists have websites with reliable information. Dietary information does change so using current data is recommended.

TABLE 1.3 Characteristics of a Credible Reference

Person gained expertise by education or relevant experience
Person is recognized as an expert by others
Person or organization has gained reputation as a knowledgeable source
Person has current or recent publications in peer-reviewed journals
Person or organization relies on tested information, not opinion only
Person or organization has no conflict of interest
Person or organization cites original sources and uses valid evidence

These same criteria can be used to evaluate popular magazines, movies, television programs, and books. Many books are written by journalists whose intent is to make the public aware of a particular issue, and to sell books. Some explain valid criticisms of the US food system but may rely on emotion or anecdotal evidence, or only tell one side of the story. We recommend when reading books such as *The Omnivore's Dilemma* (2007) by Michael Pollan, *Fast Food Nation* (2001) by Eric Schlosser, *Salt Sugar Fat* (2013) by Michael Moss, *Pandora's Lunchbox* (2013) by Melanie Warner, or *A Bone to Pick* (2015) by Mark Bittman, that critical evaluation and scientific thinking are used to keep the presented information in context with the complex relationships that make up the food system.

1.3.3 Conflicting Studies

Scientific findings from research studies are being generated at an increasing rate and recommendations about supplements, diets, and components of food and their role in health are constantly evolving. Consumers are often frustrated with conflicting food and nutrition information in the news and confused about lifestyle changes they should make. Research has shown that confused consumers tend to ignore all health recommendations because they lose confidence in the scientific method.

It is within the nature of scientific inquiry for studies to generate conflicting results. When an experiment is designed, the researcher must make important decisions about which variables will, or can be, controlled, and which variables cannot be controlled. Decisions about the study population demographics (e.g., if humans will be studied, the age, gender, ethnicity, state of health, lifestyle, geographic location, economic status, and education level) are made. Furthermore, it is necessary for studies involving humans to determine if they will be allowed to remain "free-living" during the study or if they would need to be more carefully monitored using a clinical testing site. The type of data to be collected and how it will be collected must be defined. Questions such as *Will the subjects be fasted or fed prior to drawing blood?* or *Will the subjects be allowed to exercise during the study?* must be addressed. If there will be laboratory analyses conducted, for example, quantifying a metabolite in the blood, the type of assay used and its reliability must be determined. An important consideration is also how many subjects will be studied. This is referred to as the "power" of the study design. If there are too few subjects from whom data is obtained, the study would be "underpowered" and even if differences are observed they may not be "statistically significant" based on the degree of variation in the population or experimental measurement. The outcome of the study will be dependent upon those decisions. Differences in study design and variables can, in part, lead to conflicting results in research.

Interpretation of research must be done with all of these potential variables in mind, and with an understanding that no one experiment will be sufficient to define a theory or prove or disprove a hypothesis. Scientists must integrate the results from many experiments, taking into consideration the strengths and limitations of each, before drawing a conclusion. Research findings must be continually reassessed based on new studies, better techniques to measure variables, and clearer understanding of molecular, cellular, or organismal behavior. This is the iterative process of the scientific method.

1.3.4 Antiscience and Pseudoscience

Antiscience arguments reject the scientific method as an objective way to learn about the world. Throughout history, there have been various movements and conspiracy theories that criticize scientific methods as materialistic, heretical, antireligion, and antispiritual. Galileo Galilei (1564–1642), often called the "father of science," was only one of many scientists criticized or even put to death for professing theories contrary to common knowledge of the time. Galileo was condemned by the Catholic Church for supporting the theory, originally put forth by Copernicus, that Earth moved around the sun because the theory conflicted with biblical references. Most people accept that science and spiritual or religious beliefs are necessary for understanding human existence and can be compatible aspects of thought, but clear distinction must be maintained between belief and evidence.

A current antiscience trend is that of pseudoscience. One of the earliest people to discuss the philosophical aspects of scientific inquiry was Karl Popper, who wrote extensively in the 1930–60s about the process of distinguishing scientific from nonscientific theories. In his view, those people who develop scientific theories create testable ideas and understand that the theory may eventually be found to be false when new data or observations are made (the scientific method). In contrast, those people who develop nonscientific theories attempt to make observations fit the predictions of the theory. Current pseudoscience ideas include discounting the theory of evolution or denying that climate change has been influenced by human activity. Opposition to considering the evidence that supports these theories and refusal to examine new evidence are examples of nonscientific thinking. Scientific inquiry requires rigorous and careful verification of information and facts followed by thoughtful interpretation.

1.3.5 Applying Scientific Thinking

Scientific method is a very organized and structured activity that is essential for conducting research. While most people do not apply the scientific method in their daily lives, most of us do use scientific thinking without fully recognizing it. For example, you may have learned, through experience, that eating breakfast in the morning gives you more energy for running and an ability to focus during the day, and that certain types of foods make you feel better than others. Using scientific thinking to address issues we face in our lives is helpful because it separates evidence and facts from belief and emotion.

There are many scientific thinking (or critical thinking or rational thinking) techniques to help sort accurate from erroneous information and make sense of the physical world. Educators use analysis, reflection, and critical thinking strategies to help students evaluate information. Skills and behaviors such as observing, questioning, and hypothesizing can be taught. These skills can be learned in the context of disciplines as diverse as chemistry and communication, by integrating thinking skills with scientific concepts, and by doing experiments or research projects.

Costa and Kallick (2000) proposed the concept of scientific habits of mind, which encourages the use of scientific thinking (rationality or intelligence) to learn and solve everyday problems. Asking questions, managing emotions, striving for accuracy, being flexible and persistent, and applying past knowledge are key components of scientific thinking. When teamwork is required, scientific thinkers will listen actively, communicate clearly and concisely, and respect and maintain the dignity of others.

Scientific habits of mind include using all the senses to gather data, being creative, responding with wonderment and awe, using humor, taking risks, thinking interdependently, being open to learning, exerting effort and allowing thinking to continually grow. Scientific thinkers make decisions based on evidence, consider applications and implications of decisions, and incorporate multiple perspectives. Practicing these suggestions to address facets of complex problems will help you understand the problem more fully and be able to make good decisions. Application of scientific thinking skills allows more effective responses to problems and questions for which the solutions and answers are not immediately known.

Consider how you might use scientific thinking in this scenario. Your friend is consuming coconut oil because "it's supposed to be healthy" and s/he thinks you ought to try it too. You want to lose a few pounds and the websites you have viewed say it increases the body's metabolic rate by reduction of stress on the pancreas therefore causing you to burn more calories. Plus people in tropical countries who consume coconut oil aren't overweight. On YouTube, Dr. Oz described a research study that showed women who consumed coconut oil had reduced abdominal fat. You read that coconut oil has more saturated fatty acids than lard and a high content of medium chain triglycerides.

Here are some steps to apply scientific thinking to make a decision about coconut oil.

- *Learn* the facts about the composition of coconut oil and compare it to other types of fat (use the USDA food composition website: www.ndb.nal.usda.gov).
- *Seek* reliable information from experts such as a registered dietitian nutritionist (www.eatright.org), healthcare professional, or nutrition scientist.
- *Read* the scientific literature about coconut oil and health (PubMed (www.ncbi.nlm.nih.gov/pubmed) or Google Scholar).
- *Evaluate* and assess your current diet and learn about weight loss strategies using the USDA SuperTracker (www.supertracker.usda.gov).

It is important to not rely on the opinion of others as your health can be at risk. Practice scientific thinking skills and learn to make independent decisions about your health and the foods you eat.

1.3.6 Ethics and Scientific Thinking

Ethics uses reason and logic and is based on facts, not on opinion. Philosophers use debate and argument to convey their thinking about ethical behavior. These arguments rely on factual information and are supported by valid scientific processes. Determining the accuracy of facts requires scientific method as well as scientific thinking. Aristotle used the scientific method to deduce universal rules from his observations (data). Ethicists may use thought experiments to test and explain ethical theories, similar to hypotheses used by scientists.

Ethical and scientific thinking concepts rely on questioning to reach conclusions. Many food system issues, such as use of biotechnology, ways to reduce food waste, and rules for advertising nutrition messages to children present problems for which even experts cannot agree on the best solution. Skills required for addressing such problems include seeking and evaluating new information, integrating new information into what is known, and organizing information for presentation and discussion with others, which are all key components of scientific thinking. The goal of this book is provide the framework to construct and defend reasonable solutions to issues regarding food production, processing, access, health, and sustainability through application of ethical theories and principles and the use of scientific thinking. A well-educated consumer who is able to apply these tools is best suited to make decisions about food that will create a sustainable food system for themselves and future generations.

References

Bittman, M. (2015). *A bone to pick* (p. 247). New York, NY: Penguin Random House LLC.

Comstock, G. L. (Ed.), (2002). *Life science ethics* Ames, IA: Iowa State University Press, Blackwell Publishing Co.

Costa, A. L., & Kallick, B. (2000). *Habits of mind* (p. 108). Alexandria, VA: ASCD.

Moss, M. (2013). *Salt, sugar, fat: How the food giants hooked us* (p. 423). New York, NY: Random House Trade Paperbacks.

Pollan, M. (2007). *The omnivore's dilemma: A natural history of four meals* (p. 455). London: Bloomsbury Publishing Plc.

Sagan, C. (1997). *The demon-haunted world: Science as a candle in the dark* (p. 457). New York, NY: Ballantine Books.

Schlosser, E. (2001). *Fast food nation: The dark side of the all-American meal* (p. 345). New York, NY: Houghton Mifflin Company.

Thompson, P. B. (2015). *From field to fork-food ethics for everyone* (p. 329). New York, NY: Oxford University Press.

Warner, M. (2013). *Pandora's lunchbox: How processed food took over the American meal* (p. 251). New York, NY: Simon & Schuster, Inc.

Further Reading

ADA (2007). Position of the American Dietetic Association: Food and nutrition professionals can implement practices to conserve natural resources and support ecological sustainability. *Journal of the American Dietetic Association, 107*(6), 1033–1043.

Anderson, M. D. (2014). Beyond food security to realizing food rights in the U.S. *Journal of Rural Studies, 29*, 113–222.

Barnhill, A., King, K. F., Kass, N., & Faden, R. (2014). The value of unhealthy eating and the ethics of healthy eating policies. *Kennedy Institute of Ethics Journal*, 24(3), 187–217.

Brand, S. (2010). *Whole earth discipline: Why dense cities, nuclear power, transgenic crops, restored wildlands and geoengineering are necessary* (p. 337). London: Penguin Books.

Democratic Party. (2015). *Democratic party platform 2012*. Available from <http://www.democrats.org/party-platform>.

FAO (2004). *The ethics of sustainable agricultural intensification* (p. 37). Rome: Food and Agriculture Organization of the United Nations.

FAO (2013). *Food outlook: A biannual report on global food markets* (p. 140). Rome: Food and Agriculture Organization of the United Nations.

Gottwald, F., Ingensiep, H. W., & Meinhardt, M. (Eds.), (2010). *Food ethics* New York, NY: Springer Science + Business Media.

Grynbaum, M. (September 13, 2012). Health panel approves restriction on sale of large sugary drinks. New York Times.

Horowitz, D. (2011). *We need a "moral operating system"*. 6:11 min, Stanford, CA. Available from <http://www.ted.com/talks/damon_horowitz?language=en>.

Huba, M. E., & Freed, J. E. (2000). *Learner-centered assessment on college campuses* (p. 286). Needham Heights, MA: Allyn & Bacon.

International Food Information Council. (2016). *2016 Food and health survey, food decision 2016: The impact of a growing national food dialogue.* Washington, DC. Available from <http://www.foodinsight.org/articles/2016-food-and-health-survey-food-decision-2016-impact-growing-national-food-dialogue>.

Korthals, M. (2008). Ethics and politics of food: Toward a deliberative perspective. *Journal of Social Philosophy, 39* (3), 445–463.

Mepha, B. (Ed.), (1996). *Food ethics* London, UK: Routledge.

Ogien, R. (2015). *Human kindness and the smell of warm croissants: An introduction to ethics* (p. 230). New York, NY: Columbia University Press.

Rainbow, C. (2002). *Description of ethical theories and principles.* Available from <http://www.bio.davidson.edu/people/kabernd/indep/carainbow/Theories.htm>.

Republican Party. (2012). *Republican platform.* Available from <https://www.gop.com/platform/>.

United Nations. (1948). *The universal declaration of human rights.* Available from <http://www.un.org/en/universal-declaration-human-rights/index.html>.

Wiersema, J. A., Licklider, B., Thompson, J. R., Hendrich, S., Haynes, C., & Thompson, K. (2015). Mindset about intelligence and meaningful and mindful effort: It's not my hardest class any more!. *Learning Communities Research and Practice, 3*(2). Article 3. Available from <http://washingtoncenter.evergreen.edu/lcrpjournal/vol3/iss2/3>.

CHAPTER

2

History of US Agriculture and Food Production

2.1 THE BEGINNINGS OF AGRICULTURE

About 4 million years ago, early humans survived by searching for food in the environment and migrating to follow food sources. When food was abundant, life was good, but when food was scarce, survival was threatened. Humans eventually learned to manage and use plants and animals for their benefit. This domestication of plants and animals allowed for a more consistent and reliable food supply and reduced the need for migrations. It also reduced the amount of time and attention needed for gathering food, allowing other skills to be developed. Some historians mark the start of domestication in agriculture at around 12,000 years ago. *Domestication* refers to changes from native forms that make production of crops or livestock dependent on human intervention. It has been estimated that humans were altering characteristics of plants to improve yield as early as 11,000 years ago, and breeding animals (sheep, goats, cattle) for specific traits as early as 10,500 years ago.

The process of domesticating crops and food animals can be geographically mapped from DNA comparison analysis of current plants and animals. The spread of agriculture likely began in the Fertile Crescent (which is now Iraq and Syria) around 9500 BC, where wheat, barley, flax, lentils, figs, dates, grapes, olives, lettuce, onions, cucumbers, melons, fruits, and nuts were cultivated. Later, around 8500 BC, crops such as rice, soybeans, citrus, coconut, taro, yams, bananas, and sugarcane were being grown in the Far East (China and Indonesia). Around 7800 BC, millet and soybeans were crops raised in China, and in 7200 BC, squash, beans, peppers, amaranth, and maize were grown in Mesoamerica (Central America).

The first approach to domesticating crops involved the process of hybrid selection. Of course, early adopters of these techniques were unaware of the scientific principles of trait transfer or the genetic code, but did recognize that selective breeding led to stronger plants with desirable characteristics. A classic example is the breeding of teosinte, which Native Americans transformed over thousands of years from a thin, grass-like plant to modern maize plants (Fig. 2.1).

The many varieties of wheat are thought to have evolved from crossing of two or three different plants, with careful selection over time by humans. Understanding of the scientific basis for hybrid transfer began in 1866 with

Understanding Food Systems.
DOI: http://dx.doi.org/10.1016/B978-0-12-804445-2.00002-8

FIGURE 2.1 Early civilizations used selective breeding to transform the native grass called teosinte into maize. The corn we recognize today is very different from the native plant, illustrated here, which had many stalks and very small seeds. *Source: Photo from Dr. Sherry Flint-Garcia, University of Missouri-Columbia.*

Gregor Mendel, who carefully cataloged how traits were passed from parent to offspring in his famous sweet pea experiments, but understanding the genetic code did not occur until the mid-1950s with the pioneering work of Watson, Crick, and Franklin, who described the structure of DNA.

2.2 COLONIAL ERA

Europeans arriving in North America in the 1600s settled along the East Coast and established colonies that were largely self-sufficient, although food shortages were common in the early years. Sir Thomas Dale established the Bermuda Hundred settlement ("hundred" is a colonial term for a tract of land that could support 100 households) in Virginia in 1613 as the first system of free farming and private land ownership in the New World. This idea of private, rather than communal, land ownership became the basis for agricultural development and expansion throughout US history.

Most of the early settlers did not come to America seeking to become farmers. Rather they hoped to get rich, and spent more time looking for gold and other precious metals than growing food. But colonists in the northern regions of New England eventually learned to fish, hunt, and grow crops with the help of Native Americans. Because of the poor soil and harsh growing conditions, farming emerged as a subsistence lifestyle in the New England colonies. Crops included maize, sweet potatoes, tomatoes, pumpkins, gourds, squash, watermelons, beans, grapes, berries, pecans, black walnuts, peanuts, and maple sugar. As more European settlers, many of whom were farmers, became drawn to the colonies, they brought domesticated animals including pigs, chickens, cattle, and sheep. In the late 1700s, Merino sheep were imported to New England and as demand for wool sharply increased, this led to one of the first specialized agricultural markets in the United States.

In the southern colonies, the production of tobacco, which was being grown by Native Americans, and a process to cure it were established soon after settlers arrived. When the first African slaves were brought to Virginia to work the fields in 1619, tobacco production rapidly increased. Tobacco was the major cash crop of the Chesapeake Bay colonies and the Carolinas for most of the 1700s, along with rice and indigo (a blue dye extracted from *Indigofera* plant). The techniques for growing rice, indigo, and cotton were established by African slaves who were experienced in these crops. As wealth increased, cotton and sugar plantations developed. Typical plantation planters, or owners, were noted by the amount of land and number of slaves they owned. Many of these planters became very wealthy and influential. Perhaps one of the most well-known planters was Thomas Jefferson (1743–1826), who owned the Monticello plantation in Virginia and became the third president of the United States. The expansive plantations of the antebellum South (in the pre-Civil War era) were noted for their architecture and aristocratic lifestyles. But the majority of southern farmers did not own slaves or live on large plantations. Rather they were struggling to survive on small acres. These frontier farmers were known for their independence and self-reliance.

2.3 REVOLUTIONARY WAR ERA

Between the time the first settlers arrived in the United States and the start of the Revolutionary War (1775–83), agriculture in the 13 colonies flourished. Soon production exceeded demand, due to the small population. Exporting routes to Europe, the West Indies, Africa, and the Mediterranean became profitable. In 1772 exports from the Colonies included large amounts of tobacco, corn, wheat, rice, beef and pork, and indigo (Fig. 2.2).

But Great Britain imposed strict regulations on the colonists. For example, all tobacco and indigo had to be shipped directly to Britain, while fish, grain, and other food crops were forbidden to enter Britain to prevent competition. American merchants had to access markets for their agricultural products across the ocean, which required avoiding the British navy and bands of pirates. The fight for independence from Great Britain was, in part, initiated because of growing frustrations with taxes and trade restrictions imposed by Great Britain. During the war years, some farmers benefited by selling their products to both British and American armies, but others suffered from the destruction of crops and loss of thousands of slaves from southern plantations. After the war ended, the nascent US economy was in shambles, mainly from debt incurred during the war and the absence of a stable currency. Many of the policies, and even the governing structure of the United States, that were developed following independence, including parts of the Constitution, were influenced by the significant role of agriculture and the farmers that had led the fight for independence. In fact, the first President, George Washington, considered himself a farmer and experimented with new crops, fertilizers, crop rotations, and livestock breeding at his Mount Vernon, Virginia estate throughout his life.

The Louisiana Purchase in 1803 increased the size of the United States by 828,000 square miles. President Thomas Jefferson sent Meriwether Lewis and William Clark on the Corps of Discovery Expedition (1804–06) to find a passage across the continent and to record the resources of this newly acquired land. Agriculture during this time was labor intensive and difficult. There were some early innovations such as the cotton gin invented in 1793 by Eli Whitney, which made harvesting cotton much more efficient, and the steel plow introduced by John Deere in 1837, which replaced cast iron versions. By the mid-1800s, wheat and corn belts began to develop in the Midwest. In 1870 Illinois, Iowa and Ohio were the top producers of wheat, and the Great Plains began to develop the cattle industry. During the 1800s, the amount of agricultural products arriving from the newly developed farms in the West caused New England farmers to specialize in products such as dairy and

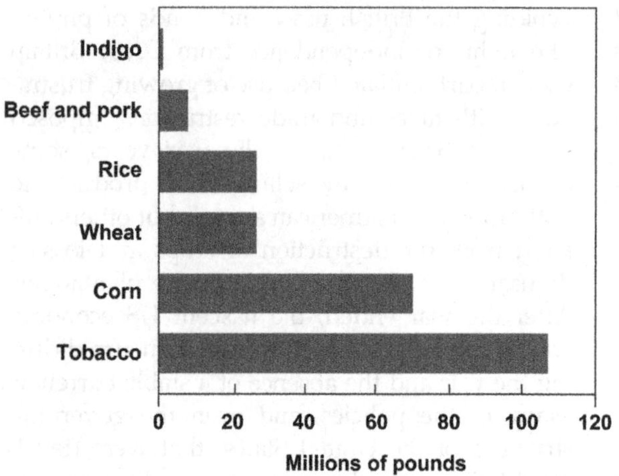

FIGURE 2.2 By the late 1700s colonists were exporting large amounts of tobacco and corn. Wheat and rice were exported, along with somke beef and pork. Indigo was popular for use as a blue dye. *Source: USDA Census of Agriculture (www.usda.gov); Shepherd (1970).*

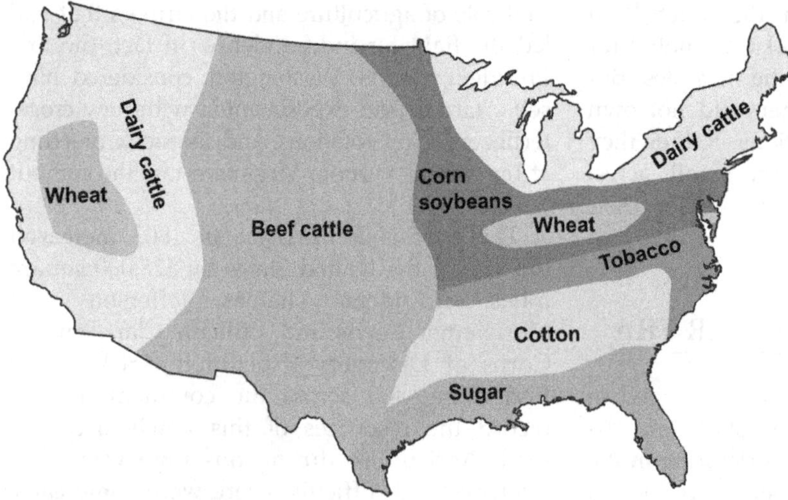

FIGURE 2.3 Regional concentrations of agricultural production in the United States began in the 1800s as farmers moved away from urban centers and began to specialize in crops or livestock. *Source: Illustration by Reannon Overbey. University of South Florida Maps ETC (www.etc.usf.edu); Hurt (1994), Schlebecker (1975).*

cheese. And the fertile land and access to slave labor created a positive environment for southern agriculture to produce cotton, tobacco, sugar cane, and rice (Fig. 2.3).

2.4 CIVIL WAR ERA

The Civil War (1861–65) is generally considered to have been a catalyst for advancing the mechanization of agriculture in the United States. In the years preceding the war, several labor-saving types of farm machinery had been developed but were not widely used. As the war progressed, many farmers and young men left their land to join either the Confederate or Union armies, leaving fewer people to manage the farms. Turning to mechanization allowed farmers to maintain, or even increase, their production despite the loss of laborers. The value of farm machinery on Iowa farms nearly doubled between 1860 and 1865 and Wisconsin was able to increase wheat production mainly because of the mechanical reaper. In the South,

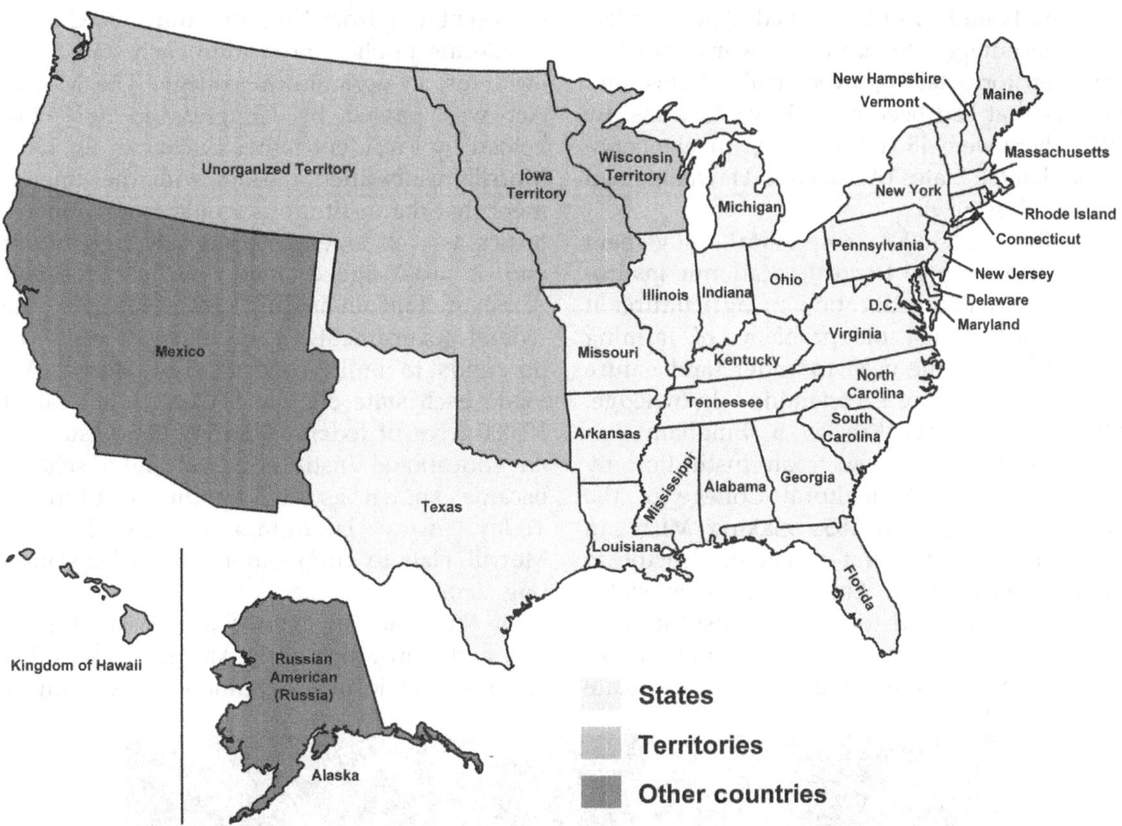

FIGURE 2.4 Prior to 1846 the western region of the United States was held by Mexico, and Russia controlled Alaska. Thereafter these regions were ceded to the United States. *Source: Illustration from User: Golbez CC BY 2.5 (http://creativecommons.org/licenses/by/2.5), via Wikimedia Commons.*

however, mechanization of cotton was delayed due to laws that allowed land owners to continue using African slave laborers. During the secession of the South, Congress was able to pass key laws that enhanced the pace of agriculture advancement, including establishing the United States Department of Agriculture, the Homestead Act, and the Morrill Act. They also were successful in supporting the construction of the transcontinental railroad. These measures were resisted by Southern states because they considered them as taking power away from individual states.

The Homestead Act, signed into law by Abraham Lincoln in 1862, created opportunities for pioneers to obtain tracts of unappropriated public land (160 acres initially, then raised to 320 and 640 acres) to settle on and farm. It is estimated that 270 million acres or 10% of the United States was claimed and settled as a result of this act. The Homestead Act was in effect until it was repealed in 1976, but a provision for homesteading in Alaska continued until 1986. The Guadalupe Hidalgo treaty drew an end to the Mexican–American War (1846–48) and ceded the land that is now California, New Mexico, Arizona, and parts of Utah, Nevada, and Colorado to the United States (Fig. 2.4). Mexico also released all claims on Texas and the Rio Grande river became the southern boundary of the United States. Settlers were encouraged to head westward into this new territory, taking over many farms and ranches that had been owned by Mexicans. By 1912, there were 48 states making up the continental United States (Alaska and Hawaii gained statehood in 1959).

In the mid-1800s, a political movement arose with the goal to create academic institutions to promote education in agriculture. It was apparent that the problems of farming were primarily the need to understand nature and this required scientific knowledge. Michigan governor Kinsley S. Bingham was the first to approve such an institution by establishing the Agricultural College of the State of Michigan in 1855 making Michigan State University the first college of agriculture in the United States. Soon after, other states recognized the need for such an institution to provide education and research about agriculture. In 1853, Justin Smith Morrill, the state representative from Vermont, introduced a bill to allocate public land within each state to be used for an agricultural college. The Morrill Act was passed by Congress in 1859 but vetoed by President James Buchanan. In 1861, Morrill resubmitted the act with the amendment that the institutions would teach military tactics as well as engineering and agriculture and it was signed into law by President Abraham Lincoln in July 1862. However, the federal government, in the midst of war, had no funds to build colleges. The Morrill Act made each state eligible to receive a total of 30,000 acres of federal land credit to establish an educational institution and these schools became known as "land-grant" institutions. Today, many land-grant colleges have a Morrill Hall to commemorate this legislation (Fig. 2.5).

By this time, the Civil War was well underway and a provision of the Morrill Act was that "no state while in a condition of rebellion or

FIGURE 2.5 Morrill Hall on the Iowa State University campus honors the creation of the land-grant college system, and is named for Justin Smith Morrill, who introduced the bill to provide public land for agricultural colleges. *Source: Photo from Iowa State University Relations.*

insurrection against the government of the United States shall be entitled to the benefit of this act," referring to several Southern states that had seceded during the war. In September 1862, Iowa was the first state to accept the terms of the Morrill Act, which provided funding to support their State Agricultural College and Model Farm, which eventually became Iowa State University of Science and Technology. Kansas State University soon followed in 1863. After the Civil War, the Morrill Act was opened to all states and over 17 million acres of land were provided for these institutions. To maintain their status as land-grant institutions, programs in agriculture and engineering, as well as a Reserve Officers' Training Corps, must be continually provided.

Even after the Civil War ended the practice of slavery, some land-grant colleges, particularly in Southern states, remained segregated. A second Morrill Act was passed in 1890 that required each state to allow entry of all people regardless of race to their institutions or to designate a separate land-grant institution. This led to the creation of the historically black colleges and universities (HBCUs or 1890 schools) mostly in the southern United States. There are currently 19 HBCU institutions. Tuskegee Institute, now Tuskegee University, was founded in 1881 by Booker T. Washington, first in a church and then on a plantation that he purchased with his own money. Tuskegee grew and prospered, at first with funding mainly from private sources, and is now a major educational institution. George Washington Carver (Fig. 2.6), the first African American to earn a master's degree from Iowa State College in 1894, had oversight of the Agricultural Experiment Station farms at Tuskegee. With his careful research and innovative approaches, Carver developed new uses for peanuts, sweet potatoes, and soybeans, which diversified agriculture in the South. In collaboration with Booker T. Washington, then president of Tuskegee, they took agricultural education directly to African American farmers to improve farming methods. In 1994, funds were also allocated for land-grant colleges to serve Native Americans. Today there are thirty-five 1994 tribal land-grant colleges and universities recognized by the USDA.

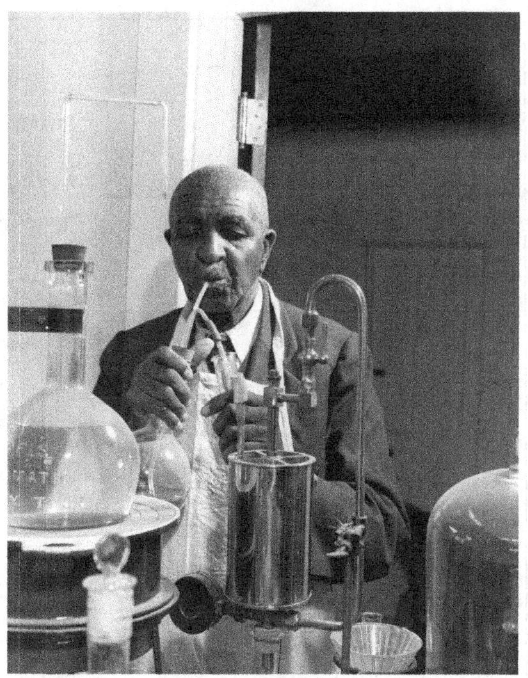

FIGURE 2.6 George Washington Carver made important contributions to agriculture, including development of new ways to use soybeans and peanuts for foods and products while on the faculty at Tuskegee University. *Source: Photo from the USDA History Collection of the National Agricultural Library.*

In the midst of the Civil War, efforts were put forth to build a railroad network to connect the entire US territory from east to west and north to south. Between 1866 and 1916 this monumental feat was accomplished and created new opportunities for agriculture and markets. The development of the railroads is discussed in Chapter 3, Innovations in US Agriculture (Fig. 2.7).

The years following the Civil War were difficult for farmers because of the political

Rail Routes in the West, 1906

FIGURE 2.7 Advances in US agriculture developed rapidly after construction of the railroads by providing fast and efficient transport of goods from rural areas to urban centers. The expansion of railroads from the Midwest to the West Coast, as shown in this map, had been completed by 1906. *Source: From Albert Bushnell Hart, L.L.D. (1919).* The Amercian Nation *(Vol. 25). New York, NY: Harper and Brothers. Downloaded from* Maps ETC, *on the web at http://etc.usf.edu/maps (map #02803). Illustration from the University of South Florida Center for Instructional Technology, www.etc.usf.edu/maps/pages/2800/2803/2803.htm.*

uncertainty, low prices for commodities, high fees charged by railroads, and shortage of credit. The situation was exacerbated by a series of droughts that occurred between 1870 and 1900 throughout the northern prairie and Midwest, grasshopper plagues in 1874–76 in the West, and blizzards in the Great Plains in 1886. Changes in the organization of farming had occurred during the war as well. Farmers were less self-sufficient because they accrued debt from purchasing machinery, which made them dependent on creditors, merchants, and the railroads. Many farms were foreclosed when debts could not be paid. The unrest of this period led farmers to form alliances to promote legislation that would improve their situation. One such group was effective in several midwestern states, mainly Minnesota,

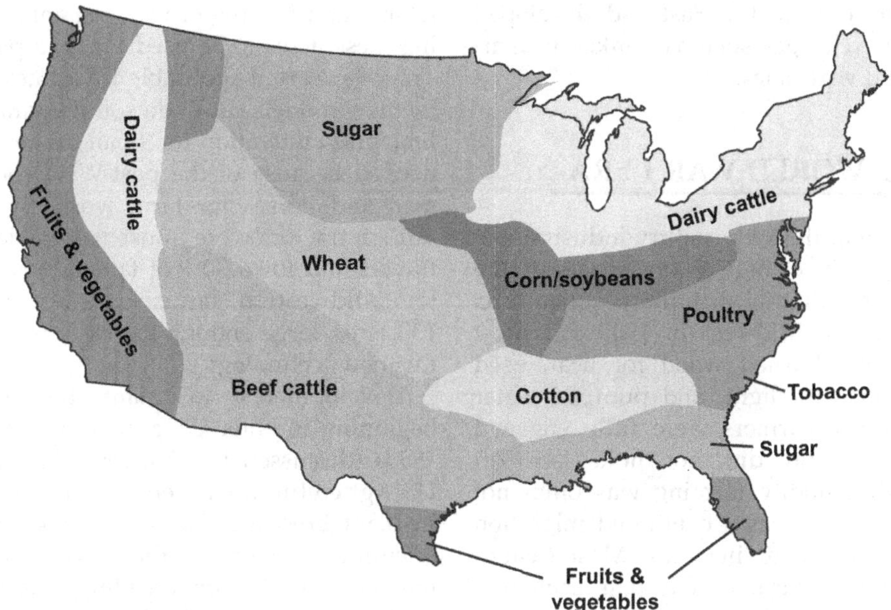

FIGURE 2.8 During the 1900s regional agriculture concentrations began to develop across the United States depending on access to land, transportation routes, and markets. *Source: Illustration by Reannon Overbey. USDA Economic Research Service, Census of Agriculture, www.usda.ers.gov.*

Iowa, Wisconsin, and Illinois, in passing the Granger laws to control prices charged by railroad and grain elevator companies to make them more favorable for farmers. These laws were challenged and eventually overturned and replaced by the Interstate Commerce Act of 1887, which gave oversight of railroads to the federal government rather than individual states.

Advances in agriculture were occurring rapidly during this period. Improved equipment was being developed such as gang plows and steam tractors, silos, and the horse-drawn combine to harvest wheat. The simple invention of barbed wire in 1870 transformed cattle ranching by defining land ownership, and by keeping animals fenced in, eliminating the need for cowboys to round them up. Many of the state land-grant colleges were starting to carry out experimental research on agricultural problems with federal funding provided through the Hatch Act (Section 3.2.1) to support agricultural research stations. Some refer to this period as the "golden age," with a change from self-sufficiency to one-crop agriculture, the adoption of new practices, and use of equipment for profitability. With these changes, regionalization of agricultural production across the United States began to develop (Fig. 2.8). In the Midwest, farming remained largely done by manual labor on small family-based farms. Cotton, tobacco, and rice were the primary crops of the South, grown on small farms (<50 acres) by sharecroppers or tenant farmers, some of whom were African American. In the Northeast, farmers were leaving for employment in urban communities and agriculture shifted to sheep, dairy, eggs, poultry, and vegetables. Agriculture in the Pacific West was specialized to grow citrus and vegetables on large-scale commercial farms. The West Coast provided fertile lands, cheap labor (Mexican, Mexican American, and Filipino) and plenty of water. Western growers relied on the railroad to transport crops to the East and developed cooperative exchanges such as Sunkist to market fruits and vegetables.

2.5 WORLD WAR I ERA

At the turn of the 19th century industrialization was rapidly changing all aspects of life in the United States, including agriculture. A farmer in the Midwest in 1900 did most work by hand, burned wood for heat, used kerosene lamps for light, and pumped water with windmills. Farmers were fully engaged in agriculture and did not hold jobs off the farm. Commodity farming was often not profitable so there was a continued migration to better employment in cities. Most eastern and midwestern farmers felt that schooling meant escape from the farm and encouraged their children to get an education and pursue a life in town that was viewed as better than life on the farm. The farm family was sustained by the food they produced or that could be traded with neighbors. In 1900, 41% of the US population was employed in agriculture but over the next 30 years that number was cut in half. Industrialization was being developed that allowed more work to be done by fewer people. The first production tractor ("traction machine") was manufactured by Hart-Parr in 1902. International Harvester sold its first tractor in 1906 and built 1000 by 1910 but they were too heavy and too expensive except for plowing, threshing, and pulling combines. In 1917, Ford mass produced a lightweight, low-cost, 2-plow tractor, making it more affordable, and the tractor soon became essential farm equipment. Between 1920 and 1950, the number of farm animals used for draft work dropped from 25 million to about 7 million while the number of tractors rose from 1 million to 3.5 million. This decrease in farm-working animals released nearly 70 million acres of cropland that had been needed for the raising and feeding of working animals, allowing these acres to be used for food production. Tractors were a profitable investment because the new models could do soil preparation, drilling, and cultivation for small grains and were needed because of the post-WWI shortages of men and horses for farm work. By 1930, 1.2 million tractors were in use nationwide, and in Illinois and Iowa 30% of famers owned a tractor. Midwestern farms had an average of 130 acres, large enough to justify investment in this new technology.

Developments in plant breeding were beginning to enter US agriculture in the mid-1930s (discussed in Chapter 3: Innovations in US Agriculture). Farmers had always engaged in plant breeding, but not in a systematic or scientific manner. By the early 1900s experimentation with corn breeding was beginning to show promise for increased vigor and higher yields. Corn was the dominant crop of the United States at the time, but yields had been stagnant. Research efforts in land-grant universities focused on corn *hybridization*, the process in which controlled cross-pollination of plants to selectively define the genetic makeup of the seeds, was being refined, leading to crops with higher yield and more uniformity (Fig. 2.9). Between 1935 and 1960 the percentage of acres planted with hybrid corn varieties rose from 0 to near 100%.

Electricity had been provided to urban areas of the United States by the early 1930s but rural areas lagged behind. The challenge of stringing electric lines across wide areas was considered too expensive and risky for private electric companies. President Franklin Roosevelt believed it was necessary for rural areas to have electricity and he pushed for the government to step up and provide service across the United States. These were established as local cooperatives rather than government agencies or private companies. The Rural Electric Administration (REA) began bringing electricity to rural areas in 1935, which created

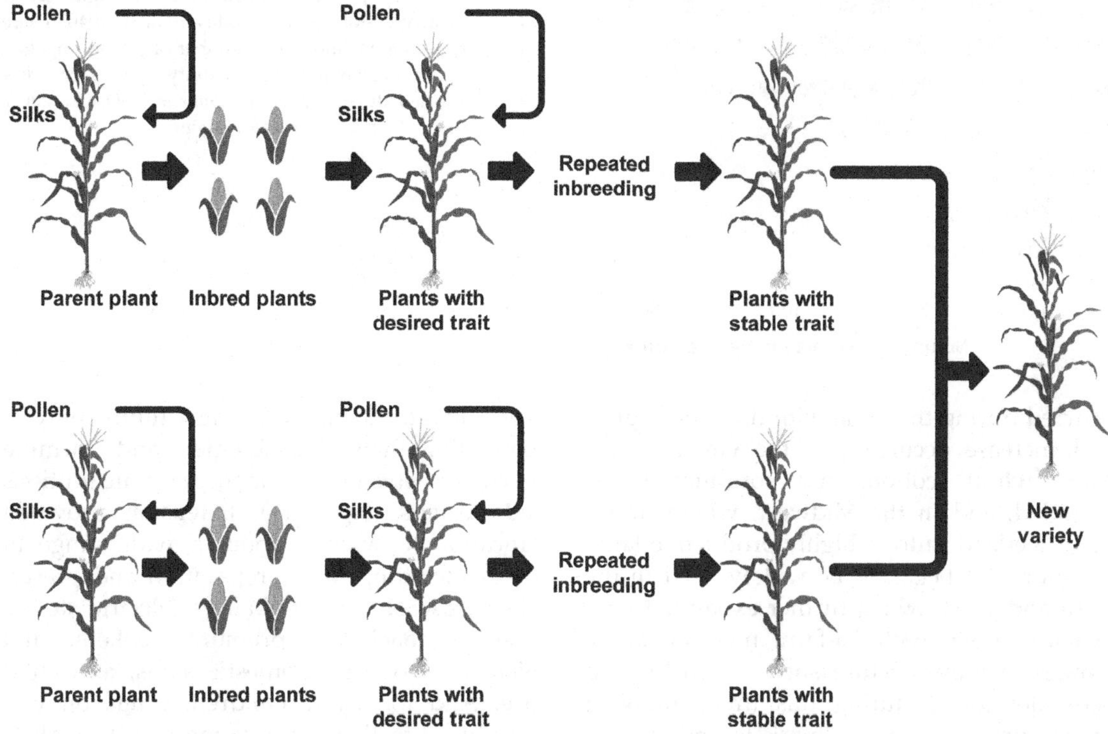

FIGURE 2.9 Hybridization is the process through which desired traits are developed in plants. Hybrid corn was very successful because it made the crop uniform and more efficient to grow, and generated higher yield. *Source: Illustration by Reannon Overbey.*

opportunities for more modern farming systems. Rural electric cooperatives were established by the REA, which encouraged private utility companies to also build infrastructure. The Electric Home and Farm Authority provided loans to farmers that allowed them to purchase appliances through their local power companies and electric cooperatives with low-interest loans. By 1950, more than 75% of farms had electricity. Electricity on the farm made farmers more efficient and aided their conversion away from manual labor. Rural electric cooperatives still operate today in many communities across the United States.

The move to mechanical power profoundly changed the efficiency of farming, leading to fewer workers per farm and larger farm sizes. In 1900, 10.9 million farmers produced food for 76 million people but, by 1950, 7.5 million farmers produced food for 151 million people. That represents an increased efficiency of 1 farmer supporting 7 people to 1 farmer supporting 27 people within a 50-year period. Today that ratio is roughly estimated to be 1 farmer supporting 155 people (Fig. 2.10).

Between 1900 and 1950, the average farm size increased from around 146 to over 215 acres. The number of farms decreased mainly in the North and East, the size of farms increased in the Midwest, and more land was brought into farming in the South and West. A net increase in land used for agriculture

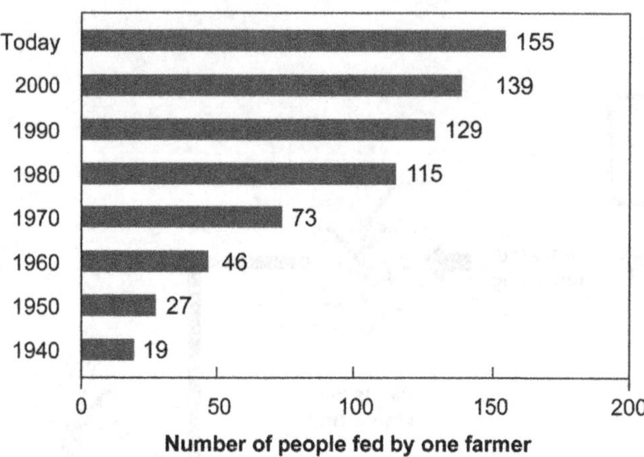

FIGURE 2.10 The productivity of US farmers increased significantly since WWII. The quantity of agricultural products generated by a single farmer is nearly 10-fold greater today than in 1940. There has been a decline in the number of people engaged in farming while productivity per farm has increased during this time. *Source: USDA Economic Research Service, www.usda.ers.gov.*

occurred during this time with the most significant increase occurring in the Great Plains, where wheat, cotton, and sorghum farms expanded, and in the Midwest, where drainage of wetlands added highly productive land. Irrigation also began to be widely used in the South and West, which further expanded land capacity. Along with on-farm mechanization, changes in infrastructure such as hard-paved roads developed during this time, allowing farmers greater access to markets and materials. Rural road paving exploded between 1900 and 1950 from around 387,000 to over 1.65 million miles (discussed in Chapter 3: Innovations in US Agriculture).

Between 1900 and 1920, the total value of farm land rose dramatically from around $20 to $65 billion. Government policy during this time was to encourage farmers to increase production. One of the ways this was supported was with the passing of the Smith-Lever Act Capacity Grant (1914), which established the Cooperative Extension Service. The extension system was intended to engage land-grant institutions in sharing practical knowledge with citizens of their state. Each state created county extension offices that employed extension agents who provided hands-on education programs to farmers, homemakers, and youth. Extension has been credited with assisting the rapid advancement in US agriculture by disseminating new technologies and farming practices between the land-grant institutions and farmers (Fig. 2.11). Extension provided education to women about a wide range of topics including food preparation and preservation, use of new appliances (electric stoves, washing machines, pressure cookers, and microwave ovens), domestic skills, and child care. And for many children, extension was introduced to them at a young age through 4-H (head, heart, hands, and health) programs that taught self-reliance, life skills, and leadership. Cooperative Extension continues to operate and is an agreement between the land-grant institutions and the United States Department of Agriculture (USDA) to provide the public with new information about agriculture and home economics. Land-grant colleges are provided with federal funding, based on a formula that considers the proportion of the population that is rural and engaged in farming within the state, and the states provide matching funds. With this new approach, extension educators were able to take research findings and new agricultural approaches to the farmers directly and thereby disseminate better practices very efficiently.

The years 1910–14 were known as the "golden age" of agriculture because prices for

FIGURE 2.11 Cooperative Extension educators advanced US agricultural practices through direct communication with farmers. In the Midwest, extension educators traveled to communities by railroad and offered classes to farmers on train cars. *Source: Photo from Iowa State University Special Collections.*

wheat, livestock feed, and cotton were high. But as WWI came to an end in 1918, farmers were left with too much product and shrinking markets. Many were in economic trouble having incurred debt from high land prices and investment in equipment. Then in 1920, farm prices collapsed. With low prices for their products, farmers cultivated more and more acres so they could produce enough product to make a living. There was a perceived need for the federal government to regulate production and marketing to control surpluses and increase prices but President Calvin Coolidge vetoed such legislation in 1927 and 1928. President Herbert Hoover approved the Agricultural Marketing Act in 1929 but it had little effect. That year the stock market crashed resulting in foreclosures, riots, strikes, and farm auctions. In 1932, cotton sold for 3¢ per pound (down from 6.8¢ in 1929), wheat for 38¢ a bushel (down from $1), and cattle were $4 per 100 pounds (down from $12). Because the farm sector was a large percentage of the population and farmers were struggling economically, politicians felt the need to act to prevent unrest but there were other problems arising at this time.

2.6 GREAT DEPRESSION ERA

From the early 1900s increased access to new lands, and access to mechanical farming equipment, allowed production to expand to areas of the United States that had not previously been farmed. Large sections of the south-central United States, including the panhandle of Oklahoma and Texas and regions of Kansas, Colorado, and New Mexico, attracted homesteaders who moved in and started farming. Settlers plowed up the native grasses to plant wheat. When WWI began, the increased demand for wheat, the introduction of tractors, and recent good harvests incentivized cultivating more land. Additionally, the United States needed corn and wheat to feed troops during WWI so prices were high. Driven by the promise of a high demand for wheat and the access to land, and facilitated by modern farming equipment, thousands of acres of prairie were rapidly tilled and planted. But conversion of these native prairies to cultivated fields, without consideration of the historic climate of this region, turned out to be disastrous. The Great Plains produced a bumper crop of wheat in 1931, which was met with low prices due to lack of demand, pushing farmers into debt. Then a period of severe drought developed (1934–37) and the exposed topsoil, which was no longer held in place by the native prairie grasses, was swept up in the winds creating the Dust Bowl. Over 300,000 square miles of prairie land in parts of Kansas, Texas, Oklahoma, Colorado, and New Mexico were the focal point of the Dust Bowl. The Dust Bowl has been described as the worst manmade environmental disaster in US history. So much soil was pulled into the air during the dust storms that it was impossible to plant a crop. People and animals were unable to be outside and dirt buried houses and roads. The continual exposure to dirt in the air led to lung damage and thousands of people suffered breathing problems, leading to many deaths. The combination of exposed soil, extreme heat, lack of rain, and strong winds created intense dust storms over

FIGURE 2.12 The Dust Bowl was an economic and environmental disaster caused by aggressive farming combined with severe drought. The center of the Dust Bowl occurred in the southwestern United States around the panhandle of Oklahoma. Clouds of dust periodically engulfed the region, affecting the lives and livelihoods of thousands of people. *Source: Photo from the USDA Natural Resources Conservation Service, www.htpps://photogallery.sc.egov.usda.gov.*

the next 6 years (Fig. 2.12). It is estimated that in a 2-day period 12 million tons of soil from the plains blanketed Chicago, Illinois, and the dust reached as far as New York City and Washington DC. Coupled with the Great Depression and low farm prices, the economic situation in the United States was dire during these years. Many people faced hunger and economic hardship including the loss of their homes and land. As many as 2.5 million people fled from these areas, with many going to California hoping for a new life as depicted in John Steinbeck's novel *The Grapes of Wrath* (1939).

EXPANSION BOX 2.1

IMPACT OF THE DUST BOWL ON US AGRICULTURE POLICY

The causes of the Dust Bowl were evident to scientists and to President Franklin D. Roosevelt (FDR) . Changes were needed to ensure agriculture was in balance with conservation of land and natural resources. FDR created the Soil Erosion Service, which showed farmers improved techniques such as terracing and strip crops, contour plowing, crop rotations, and mulch farming and encouraged planting drought-tolerant crops. The concept of shelter and wind-breaks of trees was introduced by the Forest Service and miles of trees were planted from Bismarck, South Dakota to Amarillo, Texas. Additional trees covering over 18,000 miles were planted in shelterbelts across the plains as part of the Prairie States Forestry Project, which lasted through 1942. Farmers were paid by the government to reduce the amount of their land that was cultivated to provide windbreaks and reduce soil erosion through the Agriculture Adjustment Administration. In addition, a land-purchase program was initiated to buy the most severely eroded land from farmers to remove it from cultivation. These regions today are national grasslands such as Cimarron National Grassland in Kansas and Kiowa National Grassland in New Mexico. The conservation methods were successful and by 1938 the dust was under control.

The lessons of the Dust Bowl are a continual reminder of the consequences of ignoring the environmental impact of agriculture. It created a partnership between farmers and government agencies to work together to develop better agriculture technology and environmental conservation. Today, the Farm Bill retains key elements that were first implemented during the Dust Bowl years, including the Grassland Reserve Program and the Conservation Stewardship Program. The Soil Conservation Service was renamed the Natural Resources Conservation Service in 1994 with an expanded mission to monitor water and air quality, wildlife habitat, and energy production. Droughts have returned to the plains several times since the 1930s, including a severe drought in the 1950s and again in 2012, but another Dust Bowl has not occurred (Fig. 2.13). Hence, the practices that were adopted by farmers to reduce soil erosion have been successful. But soil erosion continues to be a major problem in agriculture and will be discussed further in Chapter 8, Sustainability of the Food System.

The struggle between agriculture and the environment has not ended. Current areas of concern include nitrogen run-off into streams and rivers, airborne particulates and odors

EXPANSION BOX 2.1 (*cont'd*)

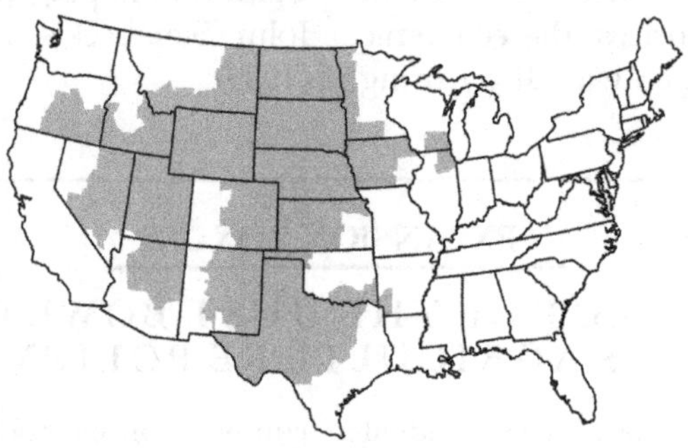

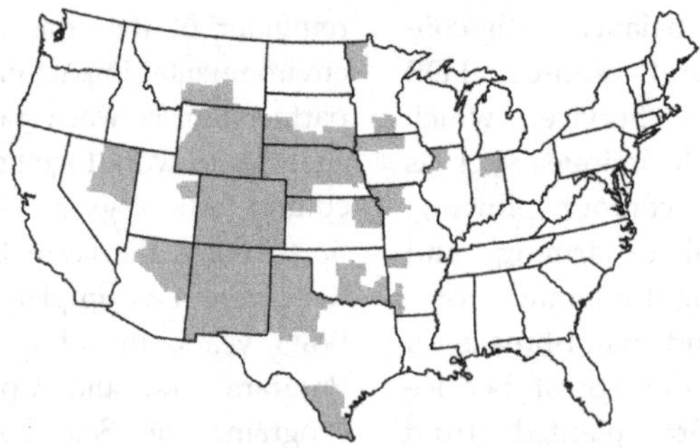

FIGURE 2.13 The regions affected by the Dust Bowl continue to be affected by drought, as shown in highlight, but better land management and government oversight has prevented another disaster. *Source: National Climatic Data Center, www.climate.gov.*

from confinement operations, soil erosion, and pesticide drift. These challenges are not as acute as the Dust Bowl was, but are just as important for the future of food production.

Suggested video: The Dust Bowl, A film by Ken Burns (PBS, 2012).
Suggested reading: The Dust Bowl: An agricultural and social history (Hurt, 1981).

2.6.1 Impact of the Great Depression on Agriculture

The Great Depression began in October 1929, when the stock market crashed, leading to an estimated loss of $30 billion in stock values. US agricultural, financial, and social policy was in many ways shaped by the tumultuous years between 1929 and 1941. Between 1930 and 1932, unemployment rates increased across the country. During that time, net farm income dropped from $6.7 billion in 1929 to a record low of $3 billion in 1932 due to the economic crisis. Farmers were already in difficult financial status following the end of WWI as surpluses were high and prices low due to decrease in demand as productivity was regained in Europe, so their resiliency for another economic downturn was tenuous. President Herbert Hoover was ineffective in stemming the economic tailspin and was soundly defeated in the 1932 presidential election by Franklin Delano Roosevelt In his inaugural speech, FDR promised to take "direct, vigorous action." Within his first 100 days in office, a steady stream of bills was passed, beginning with the Emergency Banking Act of 1933.

Concurrently, with the United States deep in the Great Depression, the Dust Bowl was wreaking havoc on agriculture in the Southwest. Henry A. Wallace was named the Secretary of Agriculture in 1933 and led US agriculture through the Dust Bowl years. He implemented the Soil Conservation and Domestic Allotment Act (1935), which reduced plowing of fields, created windbreaks of trees and shrubs, and oversaw the Soil Erosion Service. A series of programs, referred to as the New Deal, established the Civilian Conservation Corps (CCC), which put young people to work on national projects. The Civil Works Administration, implemented in 1933, created jobs for up to 4 million people to build bridges, schools, hospitals, airports, parks, and playgrounds. FDR signed legislation creating the Works Progress Administration in 1935, which later became the Work Projects Administration (WPA). The WPA employed more than 8.5 million people, including artists, to work on projects across the country. The WPA director Harry L. Hopkins is reported to have said "Give a man a dole and you save his body and destroy his spirit. Give him a job and pay him an assured wage, and you save both body and spirit" (Sherwood, 1948). The Bureau of Reclamation, within the WPA, employed thousands of people to build large dams in 16 western states that provided irrigation water and hydroelectric power, allowing expansion of agriculture in these regions. Among them were the Grand Coulee Dam on the Columbia River (Fig. 2.14) and the Shasta Dam in California's Central Valley. These building projects continue to benefit agriculture today. FDR also signed the Social Security Act of 1935, which ensured workers would be provided benefits after their working years and the Federal Deposit Insurance Corporation (FDIC) to guarantee bank deposits. Despite these efforts, the US economy remained unstable through the next 5 years.

Price supports and commodity programs were put in place based on what farmers had produced before the downturn. Through the 1930s, farmers in the South continued to rely on cotton, rice, and tobacco for cash crops. In the Northeast, there was a rapid decline in farm numbers with an increase in part-time farmers producing dairy and chickens. Midwest farms produced corn, pork, and beef and the Great Plains produced wheat. Chemical fertilizers began to be applied to alleviate depletion of the soils that had occurred during WWI, although they were still scarce and relatively expensive, and the use of pesticides such as DDT began.

The Agricultural Adjustment Act provided subsidies to farmers for not growing crops or animals as a means to manage the amount of product produced. Control of production allowed commodity prices to be maintained consistently and avoid the "boom and bust" cycle of overproduction and low prices or scarce crop production and high prices. An additional benefit was that fields taken out of production were allowed to be used by tenant and sharecropper farmers to raise food for their families, thereby improving their quality of life. The intervention and regulation of agriculture by the federal government was

General view of area from north cliffs. -- June 22, 1942.

FIGURE 2.14 The Grand Coulee Dam was built during the Depression and opened new areas of the United States for agricultural production by providing controlled access to water. This photo was taken in 1942 soon after the main dam's 11 floodgates were completed. *Source: Photo from the US Bureau of Reclamation, www.usbr.gov/grandcoulee/history/construction/gallery/30.html.*

criticized because it helped large-scale farmers more than marginal farmers. These programs provided some security and stability, but did not end rural poverty.

Ironically, during the Depression years, farmers found themselves with surpluses of food products because of the widespread unemployment causing drop in demand. To compensate for low prices, farmers planted more acres leading to excess supply. Farmers were destroying their crops and livestock because they could not sell them or afford to maintain them, while people were going hungry. Children could not pay for their lunches at schools and families had inadequate resources to provide food at home, so the risk of nationwide malnutrition was high. This led the federal government to pass the Commodity Credit Corporation Charter Act in 1933, which gave loans to farmers and allowed them to store their nonperishable commodities when prices were very low. *Commodities* is a term used by the USDA that refers to agricultural products of economic value, such as corn, soybeans, rice, cotton, eggs, cheese, or milk. Because of the dire economic situation, some farmers were unable to repay their loans and were forced to turn over their commodities to the government. The government then had to find ways to distribute these products. To mitigate this discrepancy between product availability and markets, Congress passed Public Law 320 in 1935, which provided the secretary of agriculture funds and authority to purchase commodities from farmers and distribute them to schools, nonprofit summer camps for children, charitable institutions, and needy families. This gave farmers an outlet for their products at a reasonable cost and provided people with food that would have otherwise been wasted. Between 1939 and 1942, there were over 5 million children benefiting from this food program. The government donation of surplus foods to school lunch programs, as well as to other needy groups (low-income elderly; women, infants, and children; and Indian tribal organizations) continues today through the Commodity Supplemental Food Program managed by the USDA. Through these programs, the amounts and price paid for agricultural commodities are balanced with market demand and supply to ensure stabilization of prices.

Another USDA program, the Food Stamp Program, was first introduced in 1939 by Henry Wallace to provide food assistance to the needy. The first administrator of the program, Milo Perkins, was quoted as saying "We got a picture of a gorge, with farm surpluses on one cliff and under-nourished city folks with outstretched hands on the other. We set out to find a practical way to build a bridge across that chasm" (USDA). The process allowed people on relief to purchase orange stamps equal to their normal food expenditures, and for every $1 of orange stamps purchased, 50¢ worth of blue stamps were given. Orange stamps could be used to buy any food and blue stamps could be used to buy food determined to be in surplus (Fig. 2.15). Over 20 million people used the program within the first 4 years, and likely many were spared from starvation and malnutrition. This first Food Stamp Program was discontinued in 1943 because of reduced food surpluses and decline in unemployment due to the economic demands of World War II.

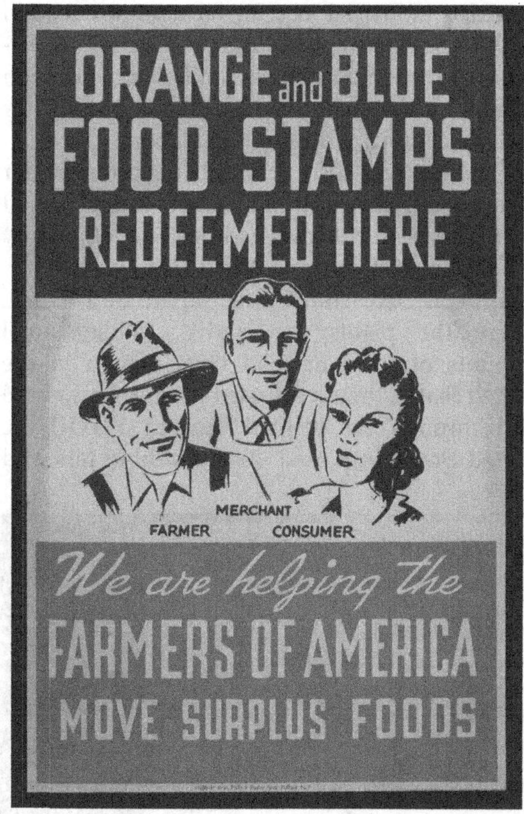

FIGURE 2.15 Food stamps were issued during the Depression and provided a way for the USDA to purchase excess agricultural commodities from farmers and distribute them to people in need of food. This program was discontinued when the United States entered WWII. *Source: Image from the USDA National Agriculture Library Special Collections, www.nal.usda.gov.*

2.7 WORLD WAR II ERA

The entry of the United States into World War II in December 1941 is credited with jump-starting the US economy and ending the Great Depression. Farm incomes reflected the impact of the war effort. The 1944 net income from farming had increased to $13 billion, nearly four times what it was in 1932. Demand for food and materials for the war boosted farm prices, increased the demand for food and fiber, and provided industrial jobs for farmers not needed on the land. WWII ended the problems with surplus production, low prices, and overpopulation in agriculture and provided the stimulus for mechanization and less manual labor.

The war was also primarily responsible for creating new agricultural products in the United States. Soybean cultivars had been developed by USDA scientists during the 1920s with germplasm collected from China, Japan, and Korea, where the plants are native. While small amounts of soybeans had been grown in the United States, they were used mainly for animal feed through the 1940s. Prior to WWII, the United States imported 40% of edible fats and oils from other countries. But during the war, access to these products was disrupted. Ramping up of soybean production began during the war years to provide a domestic source of oil as well as glycerin, which was needed for explosives. Soybean is also an excellent source of high-quality protein and was used as a meat extender in the United States during the rationing periods of WWII and for food aid to Europe following the war. The nutritional value of soybeans was studied by Cornell Extension scientists and recipes were disseminated to encourage consumption of soybeans (Fig. 2.16).

Soybean production continued to expand across the Midwest states in particular where

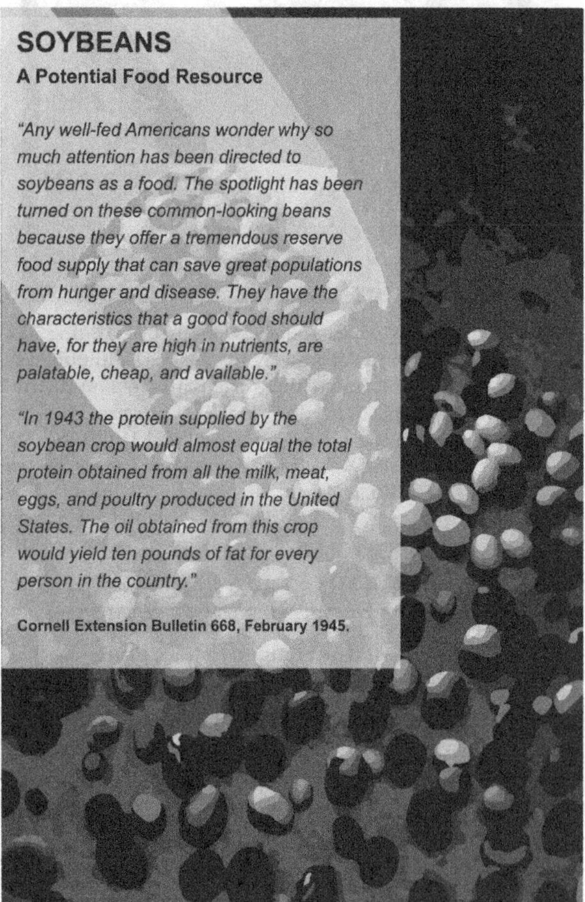

SOYBEANS
A Potential Food Resource
"Any well-fed Americans wonder why so much attention has been directed to soybeans as a food. The spotlight has been turned on these common-looking beans because they offer a tremendous reserve food supply that can save great populations from hunger and disease. They have the characteristics that a good food should have, for they are high in nutrients, are palatable, cheap, and available."
"In 1943 the protein supplied by the soybean crop would almost equal the total protein obtained from all the milk, meat, eggs, and poultry produced in the United States. The oil obtained from this crop would yield ten pounds of fat for every person in the country."
Cornell Extension Bulletin 668, February 1945.

FIGURE 2.16 During WWII food shortages occurred and rationing of foods, including animal products, was necessary. Food and nutrition scientists at Cornell University provided information to consumers about the health benefits of soybeans as an alternative dietary protein source. *Source: Illustration by Reannon Overbey. Cornell Extension Bulletin 668, February 1945.*

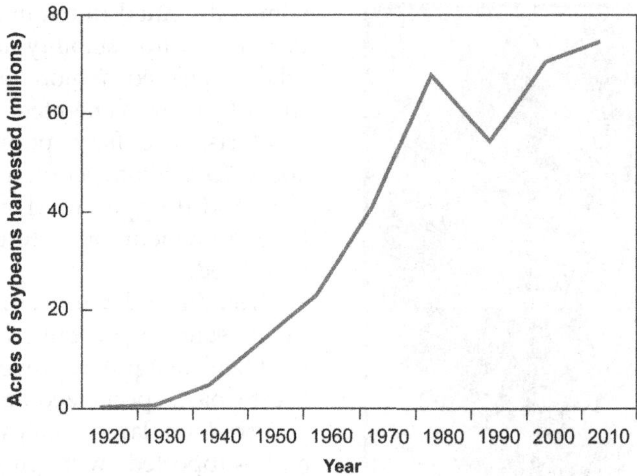

FIGURE 2.17 The number of acres of soybeans planted in the United States increased significantly during WWII to meet the needs for cooking oil, glycerin, and animal feed. The production of soybeans continued to increase after the war, and today soybeans are a major crop for US agriculture. *Source: USDA Economic Research Service, www.usda.ers.gov; Ash and others (2006).*

the soil and climate were most suitable, particularly Iowa, Illinois, Minnesota, Indiana, Ohio, Missouri, and Nebraska. Along with corn, soybeans are used primarily as a source of edible vegetable oil and for animal feed, and increasingly as a source of biofuel and biolubricants. The overall growth in oilseed crop production in the United States, mainly corn and soybean, increased nearly sevenfold between 1948 and 2008, which was much higher growth than any other food crop (Fig. 2.17).

Sugar beet cultivation was also encouraged during WWII to provide a domestic source of sugar when tropical sources of sugar cane were blocked. Farmers were encouraged to plant sugar beets to support the war effort (Fig. 2.18) and California was one of the first states to grow sugar beets in high quantity. Today, sugar beet production and processing occurs in Midwest and Northwest states, accounting for 55% of total domestic sugar production. The majority of sugar cane is grown in Florida and Louisiana.

Five million people left farms during WWII for the military or jobs in cities. But industrialization and technology replaced them. Between 1940 and 1945, net farm income increased and prices rose, yet per capita farm income was still 57% less than nonfarm income. Acreage increased 5% but productivity increased 11% due to use of hybrid seeds, pesticides, insecticides, fertilizer, and mechanization.

2.8 POSTWAR ERA

After World War II ended in 1945, the United States experienced a boom in domestic life. Consumers became a powerful voice in food access and demand. The US economy expanded, the gross national product more than doubled, and government spending to build schools, public buildings, and interstate highways created jobs that fueled the creation of the middle class. Soldiers returning from World War II received subsidized education (the GI Bill) that expanded enrollments at colleges and universities and subsidized mortgages for new housing

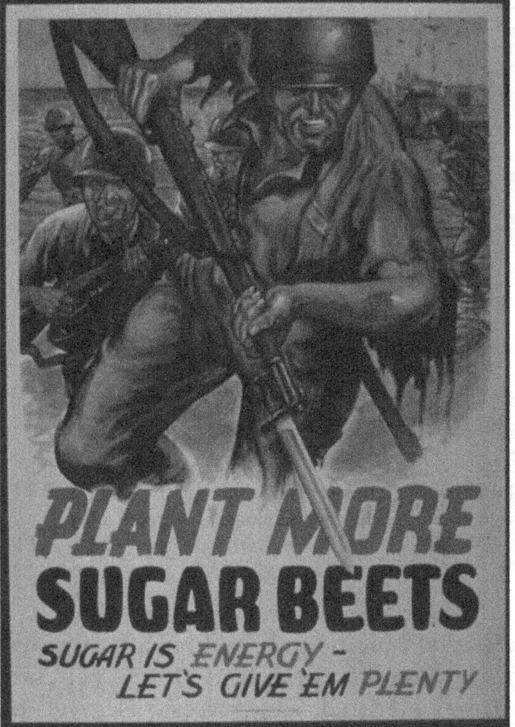

FIGURE 2.18 Farmers were encouraged to plant sugar beets during WWII to provide a domestic source of sugar when imported sugar was not available. Sugar beet production continues to be an important crop for the Upper Midwest and Northwest. *Source: Image from the USDA National Agriculture Library Special Collections, www.nal.usda.gov.*

developments in suburban neighborhoods around the country. This positive economic environment created the "Baby Boom," with birth rates of 24 live births/1000 people in 1950–60. By comparison, the rate was 14 live births/1000 people in 2008 (US CDC data). Baby Boomers make up about 14% of the US population today.

In contrast to the overall economy, the agricultural economy was tumultuous as demand for commodities adjusted down from the high levels that occurred during the war years. The political environment was conflicted about how to provide price support for farmers. Some advocated to retain high fixed price support to ensure stability for farmers, whereas others believed a more free-market approach should be implemented using flexible price supports. The farm population in 1945 was about 25 million people (18% of total population) and dropped to 21 million by 1959 (12%) as employment opportunities in urban areas expanded.

Drought and wind erosion occurred in the plains states again but irrigation, mainly from the Ogallala aquifer, prevented disaster. On the West Coast, specialized agricultural production occurred on large estates that were irrigated and supported with migrant labor. During this time, contract farming became popular, and processing technology to rapidly preserve food (canning and freezing) became integrated with farm production. Consolidation of food companies began to evolve such that large percentages of prune, raisin, wine, orange juice, and other products were produced by a few companies. These corporations were very powerful economically and politically, and heavily influenced marketing, labor, water access, and transportation policies.

Agricultural policies put in place during the New Deal were intended to buffer farmers from financial swings of the market. During the early 1950s commodity prices were high in part because these mandatory price supports continued to be provided. This created problems for the government, including having to purchase large surpluses of certain commodities such as cotton, wheat, and milk, which cost millions of dollars to manage, and loss of export markets due to the high prices. President Dwight Eisenhower requested that Congress pass a Farm Bill that would make farm price supports more flexible (rather than mandated at 90%, which had been implemented since the New Deal), stimulate exports, and promote agriculture research. The political situation at the time was not favorable to these changes and Eisenhower was not able to make any real impact on agriculture policies. Some laws were implemented that allowed surplus commodities to be used for foreign aid programs, but surplus production of some commodities continued to be a problem.

The rapid advances in technology that were occurring during this time in agriculture, including mechanization, hybrid seeds, and selected animal breeds, as well as methods to reduce production and labor costs, made farms more productive. The high production was met with decreased demand leading to lower prices, so overall farm income stagnated between 1950 and 1960. This led to consolidation of farms and greater financial investment by farmers in their operations.

During the 1960s, a period of unprecedented increase in global crop yields began that has been referred to as the Green Revolution. Plant scientists, including Norman Borlaug, who is considered the father of the Green Revolution, created varieties of wheat that were disease resistant and had better characteristics (shorter in height so the plants did not topple over and could be more efficiently harvested), and established agronomic practices that enhanced yields. Improved varieties of rice were also developed. These crops turned out to be very successful and provided food relief for millions of people in India, who otherwise would have succumbed to famine. Overall, the Green Revolution provided people in developing countries with better nutrition, increased incomes, and reduced food costs as well as a more diversified diet. Some parts of the world, notably Africa, did not benefit from the innovations of the Green Revolution, and cultural, sociological, and political issues dampened the benefits in other regions. The impact of the Green Revolution on US agriculture was to further stimulate research in plant breeding and agronomic practices on crops of importance to the United States. This process was further enhanced with the increased availability of war-surplus chemicals such as nitrogen fertilizer (Chapter 3: Innovations in US Agriculture).

The term *agribusiness* was first introduced in 1957 by Harvard business professors John Davis and Ray Goldberg. Recognizing that farming had transitioned from subsistence to commercial status in the United States and that farmers were increasingly connected to business networks, they determined the need to describe these changes with better terminology. In their view, the term agribusiness was defined as "the sum total of all operations involved in the manufacture and distribution of farm supplies; production operations on the farm; and the storage, processing, and distribution of farm commodities and items made from them" (Davis & Goldberg, 1957). The concept of a food system that recognized the interconnectedness of agriculture to a broadening number of businesses, as well as to government and educational institutions, was introduced. With advances in technology, agriculture involved inputs of seeds and animal stocks, transportation, energy, chemicals (fertilizer, pesticides and nutrients), machinery, and feed that were provided by companies. New ways to process agricultural products were being developed by food companies, and a host of other nonfood industries such as fiber, textiles, and chemicals for manufacturing and medicine were on the rise.

2.9 MID-COLD WAR ERA

The 1970s were a complex period for agriculture in US history. Between 1972 and 1974, the USSR began to purchase large quantities of US grain, eventually draining all of the USDA surplus. This sale was engineered by Secretary of Agriculture Earl Butz. The result was a dramatic increase in grain prices, leading Butz to encourage farmers to plant more, famously encouraging farmers to plant "fence row to fence row" and "to get big or get out." Butz was a proactive secretary of agriculture who fostered a belief in US farmers that they should produce as much as possible and sell their surplus globally. His focus was on industrialization of agriculture with little regard for long-term environmental impacts. To accomplish this, farmers increased their debt load to ramp up their production. The farm policies of the time included subsidies and other practices that also encouraged larger operations. But the US economic situation during the mid-1970s became increasingly unstable. President Jimmy Carter imposed an embargo on soybean exports to the USSR to protest their invasion of Afghanistan, which led to a dramatic fall in prices. In addition, the Federal Reserve Board raised interest rates to stem inflation. Prices for farm products remained stagnant and many farmers found themselves with debt loads that were unsustainable. This led to the farm crisis of the 1980s, when many farmers went bankrupt and lost their farms. In addition, banks that were holding farm notes failed because of the high number of defaults.

The 1980s rivaled the Great Depression for the number of farmers who were unable to keep their farms. The Midwest was the hardest hit, leading to unrest and increased activism. Farmers drove their tractors to Washington DC to raise awareness of their plight. Country music artist Willie Nelson led Farm Aid concerts around the country to raise money for farm families. Congress eventually took action in the late 1980s but by then, significant damage had occurred in rural towns where businesses had closed because people left agriculture for jobs elsewhere. The loss of population led to consolidations of school districts and local governments. Impacts were felt by farm machinery suppliers and other agriculture-related industries as well. Rural communities in the Midwest have not yet recovered from the decline of population that occurred during this time.

2.10 PROGRESS IN US AGRICULTURE

Since the 1980s agricultural technology has made significant advances. These are discussed in more detail in Chapter 3, Innovations in US Agriculture. Many factors have influenced the progression of agriculture in the United States and created the system we have today. Key historical events including war and economic and environmental disasters were instrumental in deciding how and where food is produced (Table 2.1). Understanding how US agriculture developed is important to inform future decisions about the food system.

Within the roughly 500 years since the colonization of the United States, agriculture has evolved and changed dramatically. The abundant land and natural resources created opportunities for food production that were initially thought to be unlimited. Industrialization allowed more land to be farmed in less time and increased the amount of food that could be produced per acre, making US farmers the most productive in the world. Infrastructure including railroads, highways, electricity, and dams and irrigation further advanced agriculture production. Research and education played essential roles to facilitate and disseminate technological advances for agriculture. Farming and politics have been closely integrated throughout US history, and regulations, laws, and government policies greatly influenced the formation of our current food system (Table 2.2).

The evolution of agriculture within the states has been influenced by social, economic, and political factors. Each of the 50 states has a history of agriculture production that is interwoven with the culture and people of that state. A comparison of two states, Iowa and

TABLE 2.1 Significant Events in US History and Their Impact on Agriculture

Event and year	Impact
First colonial settlements in Virginia 1607	Individual land ownership and start of agriculture system in the United States
World War I 1914–18	Development of chemical pesticides
Great Depression 1929–41	New Deal policies created farm policies that are still in effect today
Drought 1932–40 Dust Bowl 1933–35	Environmental disaster brought new social policies and federal programs for farmers
World War II 1941–45	Industrialization increased, farmers left farms for military service or work in defense
Farm crisis 1980s	Financial foreclosures of farms and banks, continued migration of farmers to cities

TABLE 2.2 Federal Legislation that Influenced US Agriculture

Legislation and year	Impact
Emancipation Proclamation 1862	Farm work transferred from slaves to tenants
Homestead Act 1862	Encouraged westward expansion and established 160 acre farmsteads
Morrill Act 1862	Land grant and 1890 institutions established
USDA created 1862	Began research and education about agriculture
Reclamation Act 1902	Dams and irrigation projects expanded agriculture production
Smith-Lever 1914	Cooperative Extension Service established
Farm Credit 1933	Supported farmers during the Depression
Agricultural Adjustment Acts 1933, 1938	Programs to control supply and demand and prevent the "boom and bust" cycle
Rural Electrification Act 1935	Increased farm efficiency and reduced manual labor
Soil Conservation Service 1935	Addressed soil erosion that created the Dust Bowl and implemented better land management practices

North Carolina, illustrates the differing progression of agriculture in states of similar land size.

We have learned from our history that natural resources are not unlimited and agriculture can have dramatic negative impacts on the environment. The Dust Bowl was one horrific example of an environmental disaster caused by humans. Other environmental impacts of agriculture are discussed in Chapter 8, Sustainability of the Food System. Agriculture will continue to evolve into the 20th century as new tools and approaches are developed and care must be taken to ensure these are used thoughtfully with ethical and scientific thinking approaches.

EXPANSION BOX 2.2

COMPARISON OF IOWA AND NORTH CAROLINA AGRICULTURAL HISTORY

Over the history of the United States, agriculture has been in a constant state of change. Most dramatically, the percentage of the labor force represented by farmers has steadily declined from 69% in 1840 to 18% in 1940 and is less than 2% today (Fig. 2.19). Mechanization, enhanced understanding of genetics, plant and animal breeding, and use of fertilizers, pesticides, and animal medications have led to increased productivity without an overall expansion in the amount of land used in agriculture. However, how and where we farm has changed across the United States. A comparison of the history of agriculture in two states tells the story of farming in the United States.

Iowa and North Carolina are similar in size. Iowa is 36 million and North Carolina is 31 million acres, but the states differ greatly in population; Iowa has 3.1 million and North Carolina has 9.8 million people. During the peak of agriculture expansion of the 1940–60s these states had about the same amount of farmland (Fig. 2.20). The decline in agricultural land in North Carolina after 1965 was much more significant compared to Iowa in large part due to the pressure of population growth.

The first Europeans settled in North Carolina around 1653. In 1850, there were 869,000 residents in North Carolina with most settled on small farms producing tobacco, cotton, hogs, and cattle. By 1950 there were 4 million people and 288,000 farms. Today, North Carolina has 25,000 farmers and about 8 million acres of farmland, which is about 27% of the total state acreage. The average farm size is about 168 acres with 3% of farms greater than 1000 acres. North Carolina continues to produce about 50% of the nation's tobacco, and is a major contributor to animal food production (Fig. 2.21).

Settlement in Iowa by European Americans began in the early 1800s. In 1850 there were 192,000 residents in Iowa with most settled into diversified farms of small grains, corn, hogs, and cattle. By 1950 there were 2.6 million people and 203,000 farms. Today, Iowa has 50,000 farmers and over 30 million acres are in farmland, which is about 86% of the total state acreage. The average farm size is 345 acres with 9% of farms greater than 1000 acres. Iowa produces about 35% of the nation's hogs and a significant proportion of seed crops.

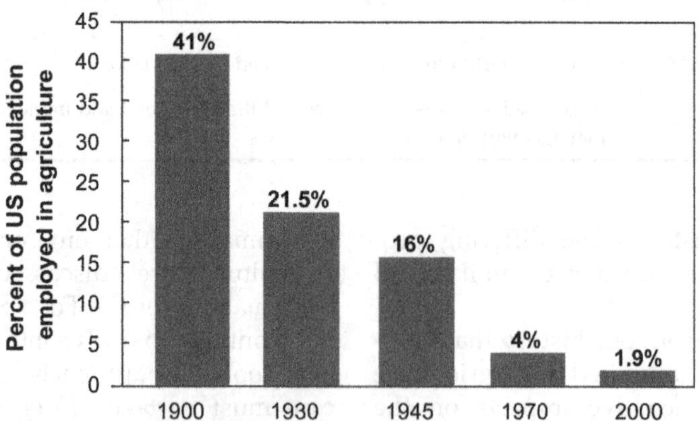

FIGURE 2.19 The percentage of the US population that is engaged in farming has declined significantly since 1900 due to mechanization and consolidation of farming operations. *Source: The 20th century transformation of US agricultural and farm policy, USDA Economic Information Bulletin Number 3, June 2005.*

EXPANSION BOX 2.2 (*cont'd*)

The per capita gross domestic product (GDP) of the two states is similar, $36,773 in Iowa and $35,719 in North Carolina, but North Carolina has a much higher state GDP compared to Iowa (North Carolina $439,672 million vs Iowa $150,512 million). Both states generate agricultural products, but total revenue from agriculture in Iowa is much greater ($35 billion) than in North Carolina ($15 billion). This comparison provides an example of how the societal and economic influences have shaped agricultural production in the United States throughout history. All states continually face pressure to convert agricultural land to commercial and housing uses as the population expands, which requires citizens and local and state legislatures to make difficult decisions. Ensuring adequate agriculture production in the United States depends on a well-educated and informed citizenry.

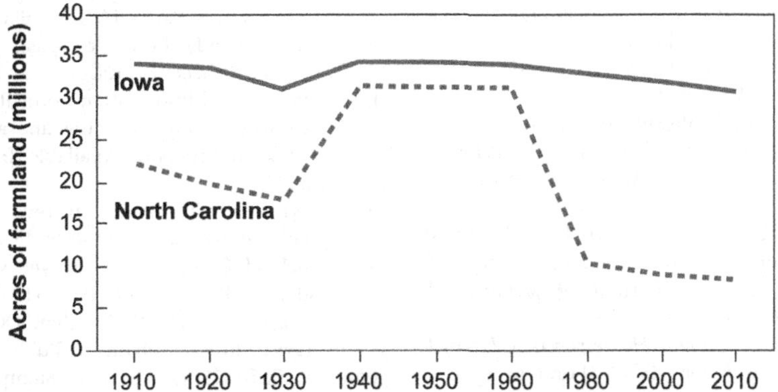

FIGURE 2.20 Iowa and North Carolina have roughly the same amount of total land area. Iowa has maintained the majority of its land for agricultural use whereas North Carolina has become more diversified. *Source: USDA, National Agricultural Statistics, 2007 Census of Agriculture, US Census Bureau 2012, www.agcensus.usda.gov.*

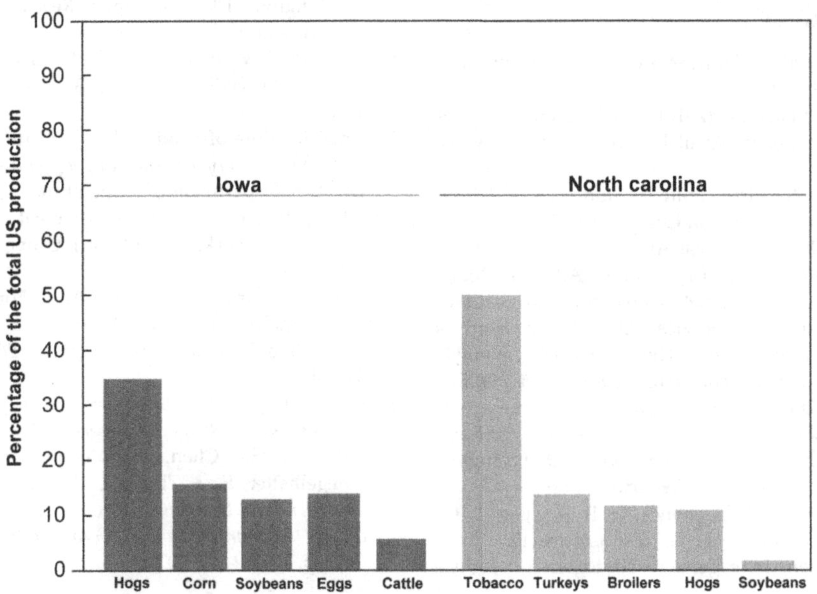

FIGURE 2.21 Both Iowa and North Carolina generate income from agricultural products. Iowa produces mainly row crops of corn and soybeans, which support production of hogs, eggs, and cattle. North Carolina farmers produce about half of the tobacco produced in the United States and contribute to animal food production. *Source: USDA, National Agricultural Statistics, 2007 Census of Agriculture, US Census Bureau 2012, www.agcensus.usda.gov.*

References

Ash, M., Livezey, J., & Dohlman, E. (2006). *Soybean backgrounder. USDA: Electronic outlook report from the Economic Research Service, OCS-2006-01*. Washington, DC: U.S. Department of Agriculture. Available from <www.ers.usda.gov>.

Burns, K., The Dust Bowl: A Film by Ken Burns, 2012, Public Broadcasting Inc. <www/pbs.org/kenburns/dustbowl/>.

Davis, J. H., & Goldberg, R. A. (1957). *A concept of agribusiness*. Boston, MA: Graduate School of Business Administration, Harvard University, 136 p.

Hurt, R. D. (1981). *The Dust Bowl: An agricultural and social history*. Chicago, IL: Nelson-Hall, 214 p.

Hurt, R. D. (1994). *American agriculture: A brief history*. Ames, IA: Iowa State University Press, 424 p.

PBS, 2012. <www.pbs.org/kenburns/dustbowl/>.

Schlebecker, J. T. (1975). *Whereby we thrive: A history of American farming 1607–1972*. Ames, IA: Iowa State University Press, 342 p.

Shepherd, J. F. (1970). Commodity exports from the British North American colonies to overseas areas, 1768–1772: Magnitudes and patterns of trade. *Explorations in Economic History*, *8*(1), 5–76.

Sherwood, R. E. (1948). *The White House papers of Harry L. Hopkins: An intimate history* (Vol. 1London: Eyre & Spottiswoode.

Steinbeck, J. (1939). *The grapes of wrath*. New York, NY: Modern Library, 619 p.

Further Reading

Berdanier, C. D. (2015). Hunger and civil war. *Nutrition Today*, *50*(4), 209–213.

Bureau of Labor Statistics. (2016). Washington, DC: U.S. Department of Labor. Available from <http://www.bls.gov>.

Carson, R. L. (1962). *Silent spring*. Boston, MA and New York, NY: Houghton Mifflin Company, 297 p.

Census of Agriculture Historical Archive (2014). *U.S. farms and farmers. 2012 census of agriculture.*. Albert R. Mann Library, Ithaca, NY: Cornell University and National Agricultural Statistics Service, U.S. Department of Agriculture. Available from <http://agcensus.mannlib.cornell.edu/AgCensus/homepage.do;jsessionid=89EEF62153EA6AAF20CA3F6A3424FFE8>.

Conkin, P. K. (2008). *A revolution down on the farm: The transformation of American agriculture since 1929*. Lexington, KY: The University Press of Kentucky, 240 p.

Dinnes, D. L., Karlen, D. L., Jaynes, D. B., Kaspar, T. C., Hatfield, J. L., Colvin, T. S., & Cambardella, C. A. (2002). Nitrogen management strategies to reduce nitrate leaching in tile-drained Midwestern soils. *Agronomy Journal*, *94*, 153–171.

Duffy, M. (2011). *Estimated costs of crop production in Iowa-2011*. Ames, IA: Iowa State University Extension FM-1712.

Food and Nutrition Service (2013). *Food distribution programs*. Washington, DC: U.S. Department of Agriculture. Available from <http://www.fns.usda.gov>.

Food and Nutrition Service (2015). *Commodity supplemental food program*. Washington, DC: U.S. Department of Agriculture. Available from <http://www.fns.usda.gov/csfp/commodity-supplemental-food-program-csfp>.

Hallberg, M. C. (1992). *Policy for American agriculture: Choices and consequences*. Ames, IA: Iowa State University Press.

Hodgson, E. (1991). Pesticides: past, present and future. *Reviews in Pesticide Toxicology*, *1*, 3–12.

International Food Information Council Foundation. (2009). From farm to fork: Questions and answers about modern food production. Available from <www.foodinsight.org>.

Jensen, M. (1969). The American revolution and American agriculture. *Agricultural History*, *43*(1), 107–124.

Knoblauch, H. C., Law, E. M., Meyer, W. P., Beacher, B. F., Nestler, R. B., & White, B. S. (1962). *State agricultural experiment stations*. Washington, DC: U.S. Department of Agriculture Miscellaneous Publication 904.

Landers, P. S. (2007). The food stamp program: History, nutrition, education and impact. *Journal of the American Dietetic Association*, *107*, 1945–1951.

Lubowski, R. N., Vesterby, M., Bucholtz, S., Baez, A., & Roberts, M. J. (2006). *Major uses of land in the U.S., 2002*. Washington, DC: Economic Research Service, U.S. Department of Agriculture.

Monthly Review (1961). *Recovery forces in the economy*. St. Louis, MO: Federal Reserve Bank of St. Louis, *43*(2), 1–12.

National Institute of Food and Agriculture (2014). *Growing a nation-the story of American agriculture*. Washington, DC: U.S. Department of Agriculture.

Paarlberg, R. (2010). *Food politics: What everyone needs to know*. New York, NY: Oxford University Press, Inc, 218 p.

Pimentel, D., Hurd, L. E., Bellotti, A. C., Forster, M. T., Oka, I. N., Sholes, O. D., & Whitman, R. J. (1973). Food production and the energy crisis. *Science*, *182*(4111), 443–449.

Poppendieck, J. (1986). *Breadlines knee-deep in wheat*. New Brunswick, NJ: Rutgers University Press, 306 p.

Privalle, L. S., Chen, J., Clapper, G., Hunst, P., Spiegelhalter, F., & Zhong, C. X. (2012). *Development of an agricultural biotechnology crop product: testing from discovery to commercialization, . American Chemical Society* (60, pp. 10179–10187).

Rasmussen, W. D. (1965). The Civil War: A catalyst of agricultural revolution. *Agricultural History, 39*(4), 187–195.

Rasmussen, W. D. (1966). The impact of the Civil War on American agriculture: A review. *Agricultural History, 40* (3), 219–221.

Rasmussen, W. D. (1983). The New Deal farm programs: What they were and why they survived. *American Journal of Agricultural Economics, 65*(5), 1158–1162.

Rasmussen, W. D. (1991). The 1890 land-grant colleges and universities: A centennial overview. *Agricultural History, 65*(2), 168–172.

Schapsmeier, E. L., & Schapsmeier, F. H. (1979). Farm policy from FDR to Eisenhower: Southern Democrats and the politics of agriculture. *Agricultural History, 53*(1), 352–371.

U.S. Census Bureau. (2015). Washington, DC: U.S. Department of Commerce. Available from <www.census.gov>.

U.S. Department of Agriculture (2000). Farm resource regions. *Agricultural Information Bulletin No. AIB-760.* Washington, DC: U.S. Department of Agriculture, 2 p.

U.S. Department of Agriculture (2015). *Census of agriculture and National Agricultural Statistical Service 2015.* Washington, DC: U.S. Department of Agriculture. Available from <www.agcensus.usda.gov> and <nass.usda.gov>.

Watson, J. D., & Crick, F. H. C. (1953). Genetic implications of the structure of deoxyribonucleic acid. *Nature, 4361,* 964–967.

CHAPTER

3

Innovations in US Agriculture

3.1 FARM TYPES AND DESIGNATIONS

Since the 1930s, US agricultural policy has focused on producing high-quality, low-cost food. This has been achieved. Americans spend about 6% of their total income on food, the lowest level for any country on Earth. Some developing countries continue to spend as much as 45% of income on food and even countries such as France spend 14%. Comparing 1950 and today, 25% less farmland, 15% fewer inputs, and 78% less labor is used in agriculture while during that time farm production has more than doubled. Some argue that cheap food comes at a cost, however. There is growing concern that US agriculture has become too industrialized, that large farms are stifling rural economies, and that current farming practices have negative impacts on the environment. Americans have moved away from farming as an occupation and most people have no direct experience with agriculture. And for the most part of the last century, people were not very much interested in where food came from or how it was produced. This has recently been changing as consumers have become more interested in food, nutrition, fitness, and health.

A farm is defined by the USDA Economic Research Service as "any operation that sells at least $1000 of agricultural commodities, or that would have sold that amount of product under normal circumstances." Family farms are defined as farms organized as sole proprietorships, partnerships, or corporations where at least 50% of the stock is held by related persons and which do not have hired managers. Farms with less than $350,000 in sales are considered small operations, and those with sales greater than $1 million are considered large. Nonfamily farms are organized as corporations or cooperatives or are operated by hired managers, and include farms held in estates or trusts (Table 3.1).

The number of farms in the United States declined steadily between the mid-1930s and the 1960s (Fig. 3.1). In 1935, there were 6.8 million farms in the United States, which had declined to 3 million by 1960. The number of farms decreased slightly over the next three decades and by 2013, there were 2.1 million farms. Consequently, the size of farms increased. Despite this decrease in farm numbers, farm productivity increased linearly over that same time period while the amount of land farmed has slightly decreased. This is a result of substantial enhancement in agricultural output. For example, since 1948, soybean yield per acre has doubled and corn yields increased fourfold. These increases in production

Understanding Food Systems.
DOI: http://dx.doi.org/10.1016/B978-0-12-804445-2.00003-X

TABLE 3.1 USDA Criteria for Defining Farm Type (2016)

Farm type	Annual gross cash farm income	% of farms	% Total land use	% value of production
Small family farm	<$350,000	89.7	52.1	25.5
Midsize family farm	$350,000–999,999	5.7	21.7	24.8
Large family farm	$1,000,000–4,999,999	2.0	16.2	35.0
Nonfamily farm	Varies	2.7	10.0	14.7

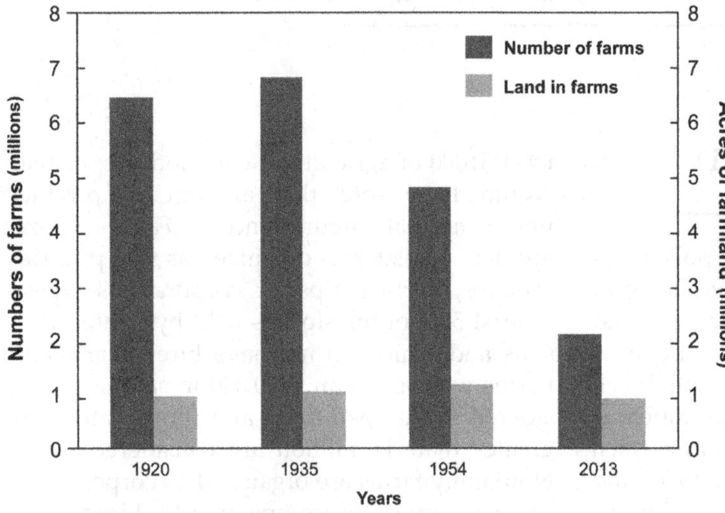

FIGURE 3.1 The number of farms in the United States has decreased significantly since 1920 while the amount of farmed land has only slightly decreased in the same period. *Source: USDA Economic Research Service, www.usda.ers.gov.*

were generated using less land (25% less since 1948) and significantly less labor (78% less since 1948). The US farm structure today is reflective of fewer farms of larger size with substantially higher productivity.

Family-run businesses are the main type of agriculture production in the United States. Most rely primarily on family members to provide the labor for these operations, but may contract with other businesses for needed additional resources. The USDA-ERS estimates that 97.6% of all US farms and 85% of farm production come from family-operated farms. About 90% of all farms in the United States are small family farms and they hold 52% of the agriculture land. However, the majority of production occurs on midsize and large farms, combining to generate 60% of agriculture value of production (Fig. 3.2). Among the large family farm operations, over 3800 farms have sales of at least $5 million and another 42,400 farms have sales of at least $1 million. The major agriculture production in the United States is by a small percentage of farmers with large operations. Farms with more than $1 million in sales produced 59% of US agricultural production in 2012, up from 47% in 2002, indicating increased concentration of farming operations. Large farms are more common when producing row crops such as corn, soybeans, and cotton and in dairy operations, while poultry, beef, and hog operations are dominated by small and midsize producers.

Nonfamily farm operations account for only 2.7% of all farms in the United States and 15%

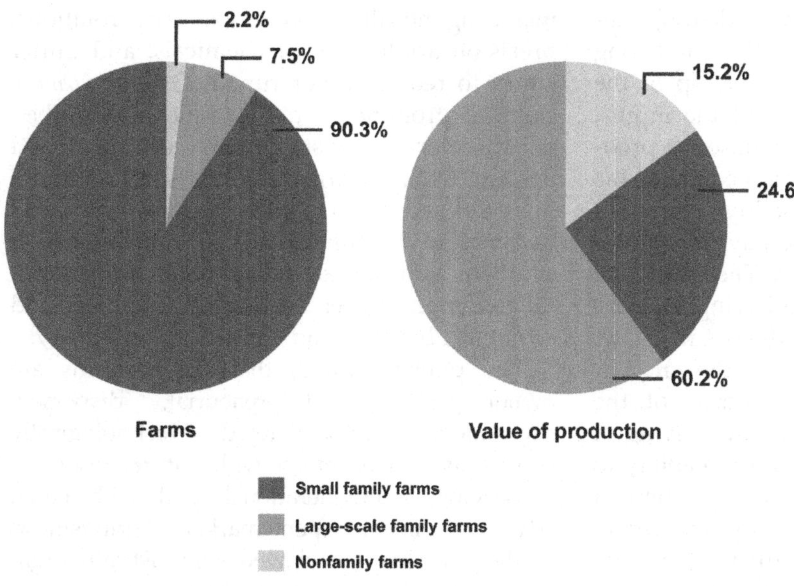

FIGURE 3.2 In the United States almost 90% of farms are classified as small family farms. However, farms classified as large family farms generate about 60% of the total farm production. *Source: USDA Economic Research Service, www.usda.ers.gov; 2004 Agricultural Resource Management Survey, Phase III.*

of farm production. These operations may be owned by corporations, trusts, or partnerships. Corporations that operate farms usually are small groups of tightly held partnerships, often related persons, and usually less than 10 shareholders, who have incorporated for tax benefits and management efficiency. Therefore, agriculture remains predominantly family-based in the United States.

Sharecropping and tenant farming have always been part of US agriculture. According to the 2014 USDA Census of Agriculture, about 39% of all the agricultural land in the United States was rented, and the majority of the landlords, 87%, were not operating the farm. About 44% of the land held by landlords was purchased from nonrelatives, 35% was inherited, 16% purchased from a relative, and 4% purchased at auction. Landlord-owned property was estimated to be worth $1.1 trillion, and landlords also generate income from leasing a variety of rights including mineral, gas and oil, recreational, water, and wind rights. About 63% of rented land is in crop production and 34% in pasture. The renters of farmland may be land owners themselves who are actively farming nearby and lease additional land to increase their profits, or to acquire more pasture to graze their livestock. Other renters may be financially unable to own land, but have a longstanding relationship with the landlord to live on and farm the land. Some landlords are not farmers, may live far from the land they own, and may not have any working knowledge of farm practices or management. A common occurrence is for parents to retire from farming and give the farm to their children, but the children have other career plans and choose to rent or lease the land. Farm management companies provide oversight of rented land to work with both the renters and the owners to ensure the farm is well managed. Some have raised concern that renters of farm or pasture land may trade short-term profits for long-term good in their farming practices leading to environmental damage. There is however limited empirical data to justify this concern.

Contract farms are those that have a legal agreement between a farm operator and another person or firm to produce a specific type, quantity, or quality of agricultural commodity. The

USDA defines these as either marketing contracts or production contracts. For marketing contracts, the farmer retains ownership of the crop or livestock and provides all the inputs, and the contractor agrees to purchase the product at a preset price. In production contracts the contractor owns the commodity, provides needed services and inputs, and pays the farmer a fee for managing the operation. The benefits of contracts to the farmer include having a secure market outlet for their products at a set price, but there may be costs if the crop or livestock do not meet the standards of the contractor. Contractors benefit by having a predictable source of product and the ability to set standards for quality and quantity. Contract farming has increased significantly and represented 40% of agricultural production in 2011. Poultry production has a high rate of contracts, especially among small farms.

In 2013, average farm expenses included tractors and equipment $230,000, diesel fuel $3 per gallon, seed $100 per acre, fertilizer $100 per acre, herbicides and pesticides $50 per acre, labor $11.60 per hour. The average land value in Iowa was $8300 per acre and land rent was over $200 per acre. This demonstrates the high financial infrastructure needed for farming and the barrier it poses to enter farming.

3.1.1 Agriculture Systems

Farmers, ranchers, and other agricultural workers run operations that produce crops, fruits, vegetables, livestock, eggs, and dairy products. This type of farming is considered *conventional farming*. Conventional farmers use a range of inputs to ensure a high-quality crop and healthy animals. Inputs may include USDA- and EPA-approved pesticides and herbicides, and bioengineered seeds. Animals may be treated with antibiotics or growth promotants to enhance growth rate and reduce illness. Conventional farmers use different approaches to limit environmental impacts including no-till, cover crops, crop rotations, precision application of chemicals, and buffer zones to reduce water runoff. *Organic farming* came to prominence in 1990 and is described in Chapter 8, Sustainability of the Food System. Organic farmers follow specific guidelines defined by the USDA to raise crops and livestock in a manner that will limit the use of synthetic substances, develop soil using natural materials, foster cycling of resources and ecological balance, and conserve biodiversity.

The public ideal is that family farms are owner-operator, entrepreneurial, dispersed, diversified, family-centered, technologically progressive, in harmony with nature, resource-conserving, intergenerational, and with equal advantage in an open market. Agribusiness, on the other hand, is viewed as industrially organized, financed from growth, management-centered, capital intensive, standardized production practices, and resource-consumptive, with an advantage in controlled markets. With the majority of Americans not directly involved with farming, a great disconnect has occurred between how farming is actually done and how the public thinks it should be done. Large farm operations, although still family owned, can be viewed as industrial with corporate management driving the business. Today, the expectation is that farms provide not only food and fiber but also biofuels, energy, and ecosystem services (high-quality air and water, carbon sequestration, and wildlife habitat) while staying true to the ideals of small, self-sufficient operations. Farmers today have an increased need to respond to consumer inquiries about how they operate and raise food. This is a fairly new position for farmers who by reputation tend to be independent, private people and don't crave being scrutinized as to how they conduct their farming operations.

3.1.2 Specialized Food Production

Over the years, farming in the United States has become increasingly specialized. In the

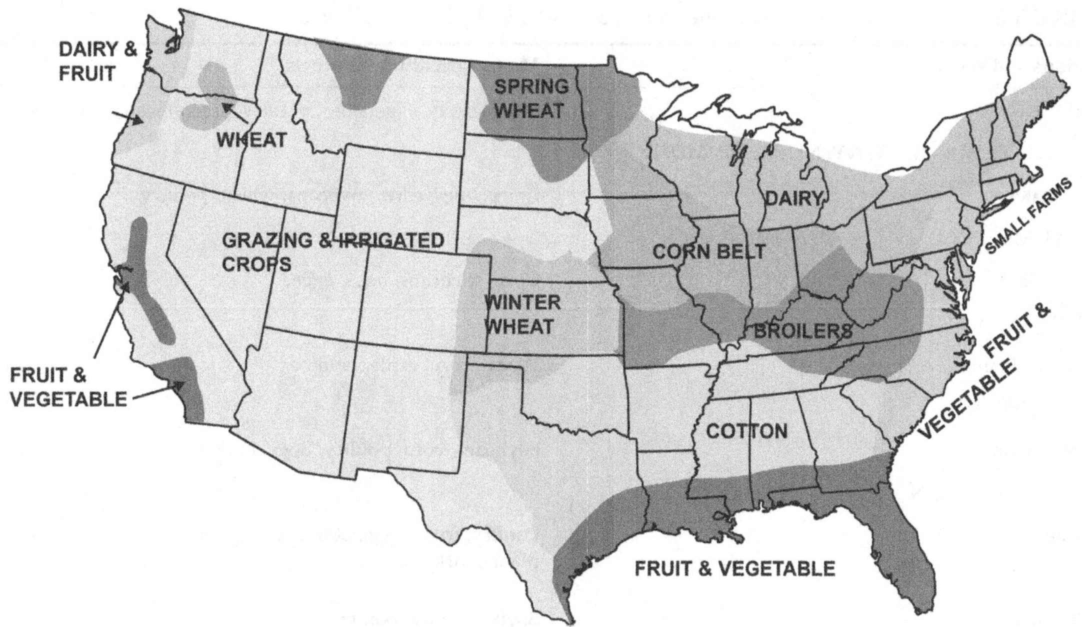

FIGURE 3.3 Where food is produced in the United States is defined by the type and quantity of land, weather, and climate features; access to water and transportation; and market and economic demands. Regional food production has developed to take advantage of efficiencies of scale. *Source: Illustration by Reannon Overbey. USDA Economic Research Service Farm Resource Regions, www.usda.ers.gov.*

early 1900s, most farms had chickens, dairy cows, and hogs and grew corn to provide feed for the animals. Farming has evolved to be focused on a few commodities, 1–2 for small farms and 3–4 for larger farms. Farming has also become more regional across the United States based on climate, markets, and infrastructure. The climate, soils, and terrain of the United States allow for multiple types of agriculture and regions of high productivity for specific crops and animals (Fig. 3.3). The USDA annually monitors agricultural production by county to assess quantity produced, income generated inputs, and other measures. From these assessments, a picture of the range of agriculture across the United States arises. Nationwide, 408 million acres were designated for crop production while grassland and pasture averaged 614 million acres. There has been intensification of production in areas where the climate is most suitable for a particular crop and labor and transportation is most efficient.

The distribution of agricultural products by region of the United States is diverse (Table 3.2). The primary US crops are corn, wheat, and soybeans. These commodities are grown in areas of the country with fertile soil and reliable rainfall. In 2007, the top-ranking states for number of acres in cropland were Iowa, Illinois, North Dakota, Indiana, and Kansas. The states of Iowa, Illinois, Minnesota, Kansas, and Nebraska produce the majority of corn and soybeans, whereas wheat is predominantly grown in North Dakota, Montana, Washington, Kansas, and Oklahoma. The majority of rice is grown in Arkansas, which produces long, medium, and short grain rice. California, Louisiana, Mississippi, Missouri, and Texas also contribute to US rice production. Sugar beets are another major crop in the United States, with Minnesota the leading

TABLE 3.2 Agricultural Production in the United States Varies by Region and State

Regions and states	Major agriculture products
Northeast	Dairy, fruits, vegetables, poultry, greenhouse plants
MN, NH, VT, MA, RI, CT, NY, NJ, PA, DE, MD, DC	
Lake States	Dairy, hogs, corn, soybeans, wheat, poultry
MI, WI, MN	
Corn Belt	Corn, soybeans, hogs, eggs
OH, IN, IL, IA, MO	
Northern Plains	Wheat, corn, cattle, potatoes
ND, SD, NE, KS	
Appalachian	Soybeans, corn, poultry, hogs, horses
VA, WV, NC, KY, TN	
Southeast	Poultry, fruits, vegetables, oranges, tomatoes, greenhouse plants, turkeys
SC, GA, FL, AL	
Delta States	Soybeans, rice, poultry
MS, AR, LA	
Southern Plains	Cattle, cotton, dairy
OK, TX	
Mountain	Cattle, dairy, greenhouse crops, corn, barley, hay
MT, ID, WY, CO, NM, AZ, UT, NV	
Pacific	Fruits, vegetables, dairy, grapes, apples
WA, OR, CA	
Alaska	Greenhouse plants, hay, potatoes
Hawaii	Greenhouse plants, macadamia nuts, sugar cane

state in production followed by Idaho, Montana, and Nebraska.

Vegetable production is classified as either fresh market or processing. Vegetables grown for processing may be canned, frozen, or dehydrated, and are usually varieties that can be mechanically harvested. States in the Upper Midwest (Wisconsin, Minnesota, and Michigan) and the Pacific states (California, Washington, and Oregon) produce the majority of processed vegetables. North Dakota, Idaho, Washington, and Oregon supply the majority of potatoes, and North Carolina, California, and Mississippi are major producers of sweet potatoes.

California and Florida produce the largest variety and quantity of fresh market vegetables. The overall production of vegetables has been increasing in recent years due to higher consumer demand. The use of precision farming techniques such as drip irrigation, plastic mulches, row covers, and high tunnels (Fig. 3.4), along with other advances in farming practice, have made these crops more cost effective to produce. Also, more support for specialty crops

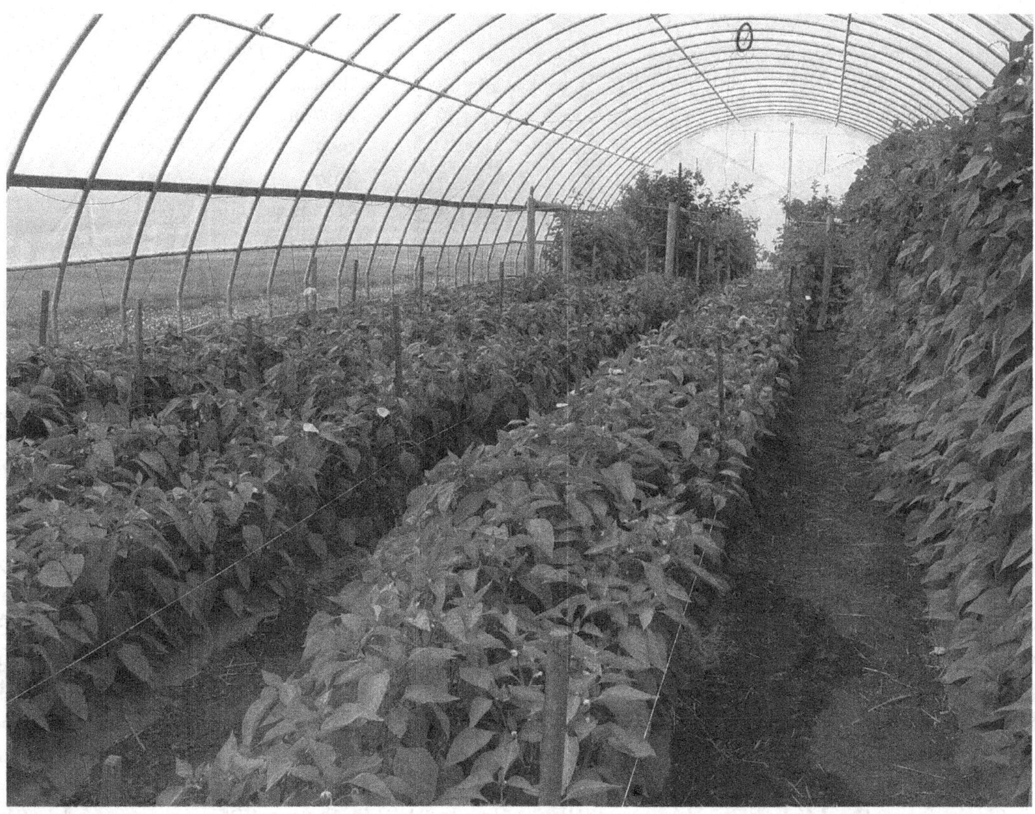

FIGURE 3.4 A current trend in farming is to use high tunnel structures that extend the growing season by keeping crops partially covered allowing control of temperature and water. *Source: Photo from Linda Naeve, Iowa State University Extension and Outreach.*

and financial support programs for farmers in the recent Farm Bills have likely contributed to the growth in vegetable production. Fruit production occurs across the United States and is highly dependent upon weather conditions. Because of its suitable climate, California produces over 50% of the fruits and 25% of the vegetables grown in the United States. Growers in Washington produce 60% of the apples, Florida growers produce 75% of the oranges, and most of the strawberries are produced in California and Florida. Cherries, blueberries, and blackberries are produced in Michigan, Oregon, and Washington; peaches in Georgia; and cranberries in Massachusetts and Oregon.

Animal agriculture is also concentrated in regions. Although Wisconsin is known as the dairy state, California is the largest producer of milk, followed by Wisconsin, Texas, New York, and Idaho. Hog production is highest in the states of Iowa, Minnesota, Illinois, Indiana, Ohio, and North Carolina. The top states for grassland and pasture grazing for raising beef cattle are Texas, New Mexico, Nevada, Montana, and Wyoming. Turkeys are raised throughout the Midwest especially Missouri and Minnesota, and chickens are predominately produced in the Broiler Belt states of Georgia, Arkansas, Alabama, and South Carolina. Top egg-producing states include Iowa and Ohio.

Total land drained in 1985
(1000 acres)

Illinois	9795
Indiana	8085
Iowa	7790
Ohio	7400
Arkansas	7085
Louisiana	7015
Minnesota	6370
Florida	6290
Mississippi	5805
Texas	5760

FIGURE 3.5 Placing drainage tiles allowed farmers in the Midwest to utilize swamp and wet areas for farming. A trenching tool such as this inserts porous tubing that moves water from the fields into drainage ditches. From the 1960s to 1985 large amounts of land in many states had been drained to enhance agriculture production. *Source: USDA Economic Research Service, www.usda.ers.gov; Photo from the AgLeader, Ames, IA.*

California and Texas are the states with highest total farm income, followed by Iowa and North Carolina. Florida ranks fifth in total farm income followed by Minnesota, Nebraska, Georgia, Kansas, and Kentucky.

Access to transportation routes, processing facilities, markets, and labor defines the distribution of agriculture production in the United States. Grain from the Midwest is economically transported by barge down the Mississippi River for global export via the Gulf of Mexico and moved by railroad to cattle-finishing operations in Texas and New Mexico. Processing plants for fruits and vegetables near production sites in California or Florida ensure is efficiently and economically preserved. Migrant workers who pick fruits and vegetables travel defined routes to assist with the harvest.

Human intervention has also played a key role in defining where agriculture production occurs. Two examples of this are the installation of drainage ditches and tiles across the Midwestern states that redirected streams and converted wetlands and swamplands to high-yielding crop land. This practice was encouraged by the federal government with the Swamp Land Acts of 1850 and 1860 and the Flood Control Act of 1944. These laws encouraged draining, tiling, ditching, and channelizing water to convert wetlands to farmland and to reduce flooding. Within the Midwest, the majority of land was modified for drainage (Fig. 3.5). While the theory at the time was that these approaches would improve agriculture production and reduce flooding risk, it has since been found that enhancing the drainage of such a large area may have increased the risk of major floods. Within the past 25 years, periods of high rain and snowfall have occurred leading to significant flooding in areas of the Midwest, especially in 1993 and 2008. In addition, draining of agricultural land concentrates nutrients and chemicals into waterways, leading to algal blooms and damage to aquatic plants and animals.

Another example was the diversion of water for irrigation into the Pacific Northwest. The Yakima River Basin Project, which began in 1905, created dams on the Yakima River to trap water from the Cascade Mountains. And the Columbia Basin Project began in 1951 after the construction of the Grand Coulee Dam. These two projects provided irrigation that allowed extensive expansion of agriculture such that by 2008 the annual value of crops from these areas was $1 billion. While these adaptations of the land and water created opportunities for food production, they and other projects have been criticized for their environmental impacts and potential consequences on natural resources and habitat.

3.1.3 Global Influence on US Agriculture

Agriculture is a major economic driver for many states. In 2016 the value of agricultural production in the US was nearly $429 billion. And 34 states reported net cash farm income of over $1 billion.

Throughout history US agricultural products have been exported around the world. Europe was the primary market in the early decades but in 2014 the top countries receiving US exports were China, Canada, Mexico, and Japan. Bulk commodities of wheat, rice, corn, soybean, cotton, and tobacco have accounted for the majority of agriculture exports, but a wide range of products are sold abroad. The United States has seen an increase of 125% in farm exports over the last century reaching $150 billion in 2014. Top exporting states and products are shown in Table 3.3.

TABLE 3.3 Exported Agricultural Products Generate Significant Revenue for Many States

State	Export product	Million $
California	Fruits (fresh)	3506
Illinois	Soybeans	3069
Iowa	Pork	1962
Kansas	Wheat	1547
Texas	Cotton	1403
California	Dairy	1270
Iowa	Corn	1117
Nebraska	Beef	949
Arkansas	Rice	907
California	Vegetables (fresh)	866
Georgia	Poultry	757
Iowa	Vegetable oils	292

While exporting products is a major source of income for the United States, agricultural products are also imported at an increasing rate. Imports increased from $98 to almost $112 billion between 2011 and 2014. According to the USDA, the amount of food imported to the United States has increased from about 11% to 17% over the past 20 years. Canada is our primary source of imported foods, followed by Mexico. These countries send a wide range of products to US food markets, including meats, grains, and vegetable oils (Canada), and fruits and vegetables (Mexico). But we receive coffee from Brazil, nuts from India, cheese from Italy, fish from Thailand, bananas from Costa Rica, and wine from France. Many countries exchange food and agricultural products with the United States and each other, hence our food system today is truly global. This leads to a wider variety and accessibility to foods but also raises concerns about environmental costs when foods are shipped thousands of miles, and demands increased focus on food security and safety monitoring. People have become accustomed to having a wide variety of foods available year round, which is a significant change from 200 years ago when food was seasonal and regional.

3.2 ROLE OF GOVERNMENT IN AGRICULTURE

Agriculture is both a business and lifestyle for many producers. Farmers and ranchers live and raise their families on the land where they grow crops and manage livestock. The idyllic expectation is that this close personal connection to the farm ensures stewardship for the land and environment. Less than 2% of people in the United States are directly involved in farming, but agriculture is an important component of the US economy. Farmers thereby have a major voice in government. As discussed in Chapter 2, History of US Agriculture and Food Production, agriculture and government have had a close interrelationship throughout history. The secretary of agriculture is a cabinet level officer, appointed by the President and confirmed by the Senate, reports directly to the President and is ninth in the Presidential line of succession. As discussed in Chapter 4, Animals in the Food System, agriculture lobbying groups have been instrumental in defining funding and support programs for farmers. Some would argue that these strong lobbying groups protect farmers from regulations that are needed to protect the environment, give breaks to big farm operations, and do nothing to support small farmers. In some cases, technology may be adopted by farmers before ecological impacts can be known. The Dust Bowl, discussed in Chapter 2, History of US Agriculture and Food Production, is an example of such a situation. Regulations to manage farming practices were eventually implemented following that disaster. The pattern of government responding with regulations after a problem has occurred, or in response to public demand, has been repeated many times in history.

3.2.1 US Department of Agriculture

President Lincoln recognized the value of agriculture to the United States and the need to promote education and best practices to farmers. This led him to request that Congress pass an act to establish a US Department of Agriculture (USDA) in 1862. The act defined the role of the USDA to "acquire and to diffuse among the people of the United States useful information on subjects connected with agriculture...to procure, propagate, and distribute among the people new and valuable seeds and plants." A Commissioner of Agriculture was to be appointed by the President and was tasked with obtaining and preserving information about agriculture, including "by practical and scientific experiments" to test new and valuable seeds and plants. The concurrent passage of the Morrill Act that created land-grant institutions made agriculture a primary focus of US educational systems. The USDA has worked and continues to work closely with land-grant institutions to advance research and education on agricultural issues.

The success of the land-grant institutions was significant and led to two additional funding programs. The first was the Hatch Act of 1887, named for William Hatch, who chaired the House Committee on Agriculture. This act gave each land-grant institution the amount of $15,000 to be used by "agricultural experiment stations." The concept of agricultural experiment stations developed over several years through discussions and debates by scientists involved in agricultural research, university presidents, and legislators. Among these was Wilbur Atwater, who became the first Director of the Office of Experiment Stations. The vision of these leaders was that agricultural research needed to be shared among researchers and made available to farmers, and that consistent funding was needed to ensure the work could be accomplished. A close relationship between fundamental investigation of new science and solutions to practical problems was a unique feature of agriculture and created some intense debate as to who determined what research would be done—scientists, farmers, or government. Through these discussions came the process that continues today, where scientists working at land-grant institutions are independent, but report to the USDA the results of their work and these findings are made publicly available at no cost. In addition, several agricultural experiment stations were created outside of land-grant institutions that employ full-time researchers who report directly to the Secretary of the USDA and carry out research in specific areas. An example of an agricultural experiment station is the USDA Beltsville Agricultural Center, in Beltsville, Maryland, where much research on agriculture, food, and nutrition has been and continues to be conducted. The funding for agricultural experiment stations was

TABLE 3.4 USDA Mission Areas and Agencies that Implement Programs

USDA mission area	USDA agencies
Farm and Foreign Agricultural Services	Farm Service Agency (FSA)
	Foreign Agricultural Service (FAS)
	Risk Management Agency (RMA)
Food, Nutrition, and Consumer Sciences	Center for Nutrition Policy and Promotion (CNPP)
	Food and Nutrition Service (FNS)
Food Safety	Food Safety and Inspection Service (FSIS)
Marketing and Regulatory Programs	Agricultural Marketing Service (AMS)
	Animal and Plant Health Inspection Service (APHIS)
	Grain Inspection, Packers, and Stockyard Administration (GIPSA)
Natural Resources and Environment	Forest Service (FS)
	Natural Resources Conservation Service (NRCS)
Research, Education, and Economics	Research, Education and Economics (REE)
	Agricultural Research Service (ARS)
	Economic Research Service (ERS)
	National Agricultural Library (NAL)
	National Agricultural Statistics Service (NASS)
	National Institute for Food and Agriculture (NIFA)
	Office of the Chief Scientist (OCS)
Rural Development	Rural Development

amended in 1955 to a formula based on factors of each state's rural and farm population to allocate the federal appropriations. The revision also required states to contribute at least equivalent matching funds.

Today, the USDA plays an essential role in agriculture. As shown in Table 3.4, there are seven mission areas of the USDA, touching all aspects of agriculture, natural resources, food and human health. Some of these functions of the USDA will be described in Chapter 7, Nutrition and Food Access. One of the mandates of the USDA has been to collect data about agriculture and to make these data available to the public. The data generated by the USDA is a great resource for farmers, researchers, and the public. In addition to the federal USDA, each state has a department of agriculture with an appointed or elected secretary of agriculture. These offices work in collaboration with the USDA to monitor and assess agriculture issues across the nation and to provide training and regulation at the state level.

3.2.2 Environmental Protection Agency

The Environmental Protection Agency (EPA) was established on December 2, 1970 to consolidate into one agency a variety of federal research, monitoring, standard-setting, and

TABLE 3.5 Pesticide Regulations Issued by the EPA and Their Impact on Agriculture

EPA legislative acts	Impact on agriculture
1972: Insecticide DDT banned	Scientific review of all pesticides
1975: Heptachlor and chlordane banned	Carcinogenic lawn and garden pesticides found in 75% dairy and meat products
1976: Toxic Substances Control Act	Required record-keeping of chemical substance use
1983: Ethylene dibromide banned	Soil fumigant no longer used for nematode control
1985: Approval granted to "Advanced Genetics"	First use of gene-altered bacteria to prevent frost damage on strawberries
1986: Right-to-Know Act	Required reports of storage, use, and release of hazardous chemicals
1989: Proposed to ban Alar (daminozide)	Discontinued use on any crops, previous use as plant growth regulator, mostly on apples
1996: Food Quality Protection Act	Increased standards and regulation of pesticide use on food crops
2004: Methane to Markets International Partnership	Plan for reduction of methane (cows and other ruminants produce 80 million metric tons per year)

enforcement activities to protect human health by safeguarding the environment. The EPA was formed in an era of increased awareness of environmental issues. The first Earth Day was April 22, 1970. The EPA has sought to prevent environmental disasters such as those that occurred at the Cuyahoga River, Ohio in 1969; Love Canal, New York in 1978; Times Beach, Missouri in 1982; and the Exxon Valdez oil spill in Prince William Sound, Alaska in 1989. The Clean Water Act and Clean Air Act are among the significant legislative authorities for the EPA.

Regulations for air, water, soils, and land use, including drainage, pesticides, fertilizers, and their impact on wildlife, are under the purview of the EPA. Farmers must comply with EPA regulations and licensing for pesticide spraying, fertilizer applications, windbreak plantings, field drainage, and invasive species management. In cooperation with the USDA, EPA regulations are communicated to farmers via county National Resource Conservation Service (NRCS) and Farm Service Agency (FSA) offices. The FSA also interprets federal legislation for farmers and manages commodity programs described in the Farm Bill.

EPA regulations have sought to mitigate impacts of agriculture on the environment and protect human health. Based on scientific evidence, some chemicals that were used in agriculture have been banned and legislation has been passed to address specific environmental issues (Table 3.5). Establishment of air quality standards and guidelines for reduction in emissions from leaded gasoline and sulfur oxides of diesel fuel affected what fuels could be used for agricultural equipment. Manufacturers, farmers, and consumers have adjusted to tighter restrictions and all have benefited from cleaner air and less pollution. Farm workers benefit from EPA guidelines for safety regulations and pesticide exposure in farm operations, which ensure fair and equitable working standards with safety training and protections for workers and no pesticide handling by children under 18 years of age.

3.2.3 Animal Plant Health Inspection Service

The Animal and Plant Health Inspection Service (APHIS) is responsible for protecting and promoting US agricultural health, which includes both animal and plant well-being and disease prevention. APHIS administers the Animal Welfare Act, which ensures the proper care and handling of all types of animals, including livestock, wild animals, and pets. In the 2014 Farm Bill, APHIS was charged with preventing the introduction or spread of plant pests and diseases that threaten US agriculture and the environment. APHIS provides funding to strengthen the nation's infrastructure for pest detection and surveillance, identification and mitigation of threats, and to safeguard the production system. APHIS carries out inspections of imported plant and animal products and monitors invasive pests and diseases. Investigation and enforcement of animal welfare reports and the avian influenza outbreak are examples of APHIS responsibility. When animal or plant disasters happen, APHIS provides emergency assistance to manage the situation.

A major role for APHIS is the approval and monitoring of genetically modified organisms (GMOs), in collaboration with the EPA and the USDA. To obtain permission to use a GMO plant, insect, or microorganism, the developer of the GMO must submit a petition to APHIS demonstrating that the GMO does not pose any more risks than the equivalent non-GMO variety. APHIS continually monitors and inspects farmers and biotechnology developers to confirm they are operating within compliance standards.

3.2.4 Government Farm Policy

US farm policy has been designed to ensure domestic productivity, global competitiveness and food security, maintain family farms, reduce risk to crop production, contribute to rural economic activity, provide a safety net for agricultural production, and provide a safe, affordable, abundant food supply. Since the 1938 Agriculture Adjustment Act, farm policy is generated on a 5-year legislative cycle that produces the "Farm Bill." The Agricultural Act of 2014 allocated $489 billion over 5 years for nutrition programs (80% of total budget or $391 billion) and agriculture ($98 billion). Nutrition support programs are discussed in Chapter 7, Nutrition and Food Access. Agriculture programs in the Farm Bill provide for crop insurance, conservation, farm commodity programs, horticulture, research, and bioenergy. The 2014 Farm Bill was notable for making significant changes to how farmers receive financial support. In previous bills, farmers received direct payments for selected crops including corn, wheat, soybeans, cotton, rice, and peanuts to ensure price levels. With many legislators and consumers questioning the reason farmers were being provided with this type of agriculture subsidy, direct payments were eliminated from the 2014 Farm Bill and were replaced with the Price Loss Coverage or Agriculture Risk Coverage programs. These programs take into consideration the historical prices for crops and allow farmers to receive payments when prices dip below a defined standard. There are caps in place for the total amount of funding any one farmer can receive and eligibility is based on gross income. The Farm Bill program has been criticized for providing subsidies for only a few staple commodities that are mainly produced by large family farms (>$250K sales). There has been concern that 58% of commodity payments were given to these large, supposedly economically stable farmers and not helping the smaller, struggling farmers. In an effort to correct that, the 2014 Farm Bill included more support for organic and local farmers and specialty crops; provided provision for beginning farmers, including veterans and minorities to obtain financial support; and expanded access to price support and crop insurance to many more agricultural products.

While farming is a business, it is subject to uncontrollable external factors. An early frost or sudden hail storm can wipe out an entire crop leaving the farmer with all the expenses of producing the crop but low or no income.

TABLE 3.6 Types of Agricultural Subsidies

Direct payments	Crop insurance
Marketing loans	Disaster aid
Price guarantees	Export subsidies
Conservation subsidies	Agricultural research and statistics

Similarly, too much or not enough water at key growing times will impact yields. When corn and soybean prices are low, cattle, hog, and egg producers benefit with lower feed costs, but the corn and soybean farmers suffer economic loss. Farmers have relied on various types of insurance to protect their economic investments, and government subsidies have also been part of that safety net (Table 3.6). A *farm subsidy* is an amount of money provided by the government intended to protect the farmer from financial loss due to fluctuations in market prices. Strong farm lobby groups, especially in agriculture-intensive states, have been successful in getting subsidies and other farmer-friendly legislation passed. Once introduced, assistance and subsidy programs became politically and socially integrated and hard to repeal. Frequently, the Farm Bill approval is delayed for months or years due to political controversy about such federal programs.

EXPANSION BOX 3.1

THE 2014 FARM BILL

The Agriculture Act of 2014 (known as the Farm Bill) was signed by President Barack Obama and covers a 5-year period through 2018. A total of $956 billion (3% of the US federal budget) was allocated for programs (titles) listed in the bill. The majority of funding, 80%, supports nutrition programs including the Supplementary Nutrition Assistance Program (SNAP), Emergency Food Assistance Program, and food and nutrition programs for senior citizens, children, and low-income families and American Indians. The remaining 20% of funding supports the agricultural component including crop insurance (8% of funding), commodity support (5%), conservation (6%), and 1% for trade, credit, rural development, research, forestry, energy, horticulture, and organic agriculture. The complexity of the Farm Bill was noted by President Obama, who said "[i]t's like a Swiss Army knife" because of the wide range of programs that are connected to the Farm Bill (Fig. 3.6).

In the early years of passing Farm Bills (the Agricultural Adjustment Act of 1933 was the first such legislation), the majority of people were farmers and farm programs were seen as a means to ensure that the United States had continued access to abundant, safe, and affordable food. Supporting commodity prices and environmental programs and funding nutrition programs, especially after the tumultuous Great Depression and Dust Bowl era, were considered important and necessary. Today, farmers in the United States represent less than 2% of the population and 0.05% of employed Americans. Farmers are not diverse by gender, race, or ethnicity. Most Americans are very removed from, and have little understanding of, agricultural problems. Farm Bill discussions now include state organizations; national farm groups; commodity associations; conservation, recreation, and rural development organizations; faith-based groups; and other interest groups. Even though the number of farmers is small, the number of conflicting interests is large and these constituents have differing opinions.

EXPANSION BOX 3.1 *(cont'd)*

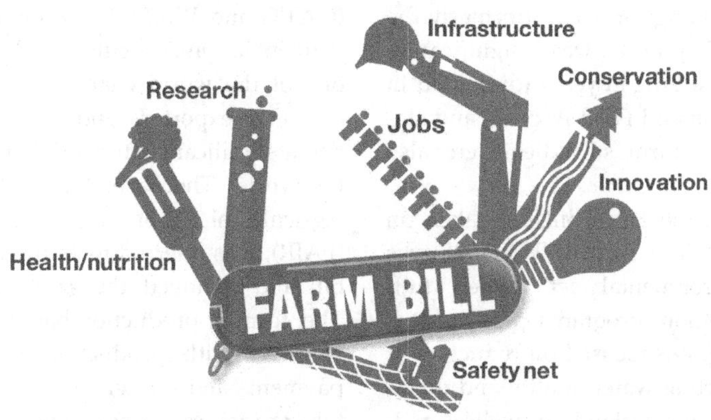

FIGURE 3.6 President Barack Obama referred to the 2014 Farm Bill as a Swiss Army knife because of the many features and components of the bill. *Source: Illustration from www.usda.gov.*

Over the years, legislators from farm-intensive states realized that they had less sway regarding these policies than their urban colleagues in states with higher populations. It became more difficult to garner continued support for farm subsidy programs and other agricultural support. However, those urban states had more people interested in food assistance programs. As a means of satisfying both constituencies, rural and urban, agriculture and food assistance programs were combined under the Farm Bill. Nutrition support programs were first incorporated into the Farm Bill in 1977 when Food Stamps and commodity distributions programs were brought into the legislation. This has created complicated political environments each time the bill is considered for renewal.

The Farm Bill is comprehensive legislation that involves varied interests among producers of different commodities and those with different priorities for farm support, conservation, nutrition, rural development, research, and foreign trade, all of which support President Obama's reference to the Swiss Army knife. Calls to reduce Farm Bill funding come from both sides: those concerned that agriculture subsidies are excessive and unjust, and those concerned that food assistance funding is excessive and unjust. Those who oppose aspects of the agricultural component of the legislation say that too much money is allocated to farmers at taxpayer and consumer expense, that funding benefits the large, wealthy farmers and is unfair to small farmers. Some question the effectiveness of the programs and contend that the Farm Bill does not provide enough environmental protection. Opponents to federal support of nutrition funding argue that the support makes people less interested in working, encourages fraud, and promotes a culture of dependency. Others consider farm policy to be incompatible with current national economic objectives, global trading rules, and federal budgetary or regulatory policies. In fact, in the

EXPANSION BOX 3.1 (*cont'd*)

2014 Farm Bill, funding for SNAP (formerly the Food Stamps Program) was significantly reduced for the first time in years (discussed in Chapter 7: Nutrition and Food Access), and significant changes in farm subsidies were also implemented.

In recent times, the Farm Bill has taken on specific challenges. The Farm Bill of 1985 was called the "environmental act" when funding for conservation programs was added. Conservation programs focused on reduction of soil erosion, protecting water quality and quantity by retirement of farmland, as well as reduction of off-farm impacts of agricultural activities. In the 1990s, farmers were freer but less secure. US farm policy was greatly influenced by world trade agreements such as the General Agreement on Tariffs and Trade (GATT) and World Trade Organization (WTO) Agreement on Agriculture. The United States is one of the world's largest agricultural producers and exporters and thus, US farm policy has a significant effect on farm sectors around the world. The Farm Bill of 1996, the Federal Agricultural Improvement and Reform Act (FAIR), was referred to as the Freedom to Farm bill and removed the safety net for farmers. The former production-based payments were replaced with production flexibility contract payments and farmers received direct payment for crop losses via emergency market loss payments. Energy programs to promote biofuels and cellulosic ethanol production and farm and community renewable energy systems were added to the Farm Bill in 2002. In 2008, farm prices were high, there was a global food crisis,

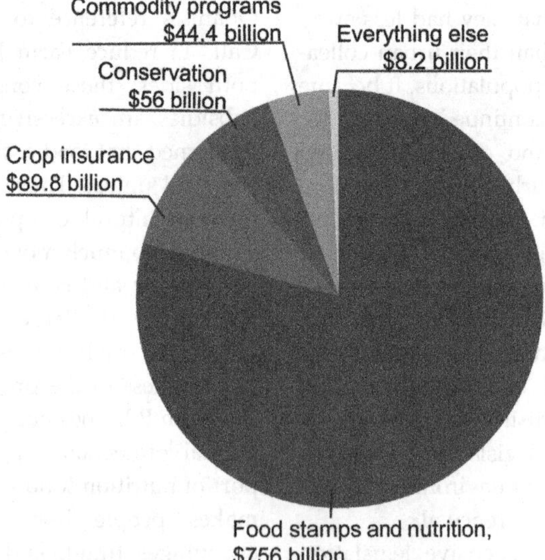

FIGURE 3.7 For the past several decades the majority of funding in the Farm Bill has been allocated for nutrition assistance programs. For example, in the 2014 Farm Bill almost 80% of the funding was for nutrition and food programs while less than 10% was provided as agricultural subsidies. *Source: USDA Economic Research Service, www.usda.ers.gov; Cost Estimates for the Agricultural Act of 2014, January 2014.*

EXPANSION BOX 3.1 (*cont'd*)

TABLE 3.7 Comparison of Major Differences Between the 2008 and 2014 Farm Bills

2008 Farm bill	2014 Farm bill
Direct payments to farmers for major commodity crops (grains, oilseeds, cotton)	Eliminates direct payments to farmers, replaced with crop insurance program
Conservation opportunities for farmers and ranchers	Conservation compliance linked with crop insurance payments
New funding for crop insurance and disaster assistance	
New funding for livestock and poultry	Did not repeal protections for poultry farmers with processing contracts
Established Beginning Farmers and Ranchers Development Program	Increased funding for Beginning Farmers and Ranchers Development Program and support during first 5 years of farming
New funding for fruits, vegetables, specialty crops, and organic agriculture	Doubles funding for specialty crop research initiative and provided new resources for organic farmers
	Tripled funding to farmers market and local food promotion program and new funding for local and regional food systems grants

and food prices were increasing in the United States so the Farm Bill attempted to balance payments to farmers with taxpayer demands for lower funding.

Passing the 2014 Farm Bill was fraught with significant controversy. The economic slump that started in 2007 drove a more conservative climate in Congress. Big spending legislation, such as the Farm Bill, was hotly debated and calls for severe reductions were strong. Eventually, a bill was passed that made some important changes. Nutrition support is the largest part of the Farm Bill (Fig. 3.7) and the decision was made to cut funding for SNAP by \$8 billion (a 4% cut over 10 years). SNAP's basic eligibility guidelines were maintained, but an income deduction that boosted benefits for some households was made more stringent. A comparison of the 2008 and 2014 Farm Bills is shown in Table 3.7. The 2014 bill provided additional SNAP funding for enhanced employment and training activities, increased healthy food options, and expanded antifraud efforts. Other changes addressed concerns about subsidies for farmers, especially when commodity crops were experiencing record high prices. Direct crop payments (subsidies) to farmers, which were unpopular with taxpayers, were replaced with an insurance program. The bill also provided expanded programs for specialty crops, organic farmers, bioenergy, rural development, and beginning farmers and ranchers, in response to calls for increased opportunities for small, local farmers.

Suggested reading: Orden and Zulauf (2015) and Owens (2008).

3.3 ENERGY FOR AGRICULTURE

Farm work was primarily fueled by human and animal power until the 1900s. Wind and water were used in early farming in the United States to drive pumps and grind grain. Major changes in farm work occurred when steam engines followed by diesel and gasoline engines and then electricity entered the picture. These allowed farm work to be mechanized, essentially replacing the need for animals to work the land, and lessening the workload for humans. Use of energy in agriculture comes at a cost. Coal and gasoline are fossil fuels that must be obtained from the earth, refined, processed, and transported. When they are used to generate power, they release greenhouse gases that linger in the atmosphere and contribute to climate change.

3.3.1 Mechanization and Fossil Fuels

The first gasoline-powered tractor was introduced by the Waterloo Gasoline Traction Engine Company in 1893, which eventually became the John Deere Tractor Works. Henry Ford also entered this industry with his low-cost tractor in 1917, which soon accounted for 75% of the gasoline-powered tractor market. Advances in other power equipment followed rapidly with the Massey-Harris corn picker in 1946 and the combine (that picked and shelled corn) in 1954. Mechanical hay balers were introduced in the 1930s and a mechanized irrigation system was developed in the 1940s. The introduction of mechanized equipment made farming less labor intensive and more productive. These machines and implements ran largely on gasoline and diesel, which were cheap and abundant. Additionally, these fossil fuels were the source of many farm chemicals including fertilizers and pesticides. The amount of inputs, especially nitrogen and herbicides/pesticides, ramped up rapidly between 1945 and 1970 (Table 3.8).

TABLE 3.8 Increased Use of Gasoline, Nitrogen, and Chemical Pesticides in Corn Production Over Time

Inputs in corn production	1945	1954	1964	1970
Gasoline (gallons/acre)	15	19	21	22
Nitrogen (pounds/acre)	7	27	58	112
Herbicides + pesticides (pounds/acre)	0	0.4	1.38	2.0

The changes that have occurred over time in accessibility and prices of fossil fuels have impacted and continue to impact agriculture production. Perhaps the most dramatic period of history when fuel prices and availability impacted agriculture was during the oil crises of the 1970s. In 1973, a global spike in oil prices resulted when the Organization of the Petroleum Exporting Countries (OPEC) imposed an embargo on oil in response to US involvement in the Yom Kippur War. The availability of gasoline in the United States reached a crisis level and gas rationing was implemented. Energy-saving programs including a nationwide speed limit of 55 mph and changes in automobile designs were mandated. At that time, interest in developing alternative fuels, including ethanol from corn, developed. In 1978, a federal tax credit program was started that awarded 51¢ per gallon of ethanol used. However, over the next two decades, the price of crude oil stabilized and gas became more available, making ethanol production uneconomical. Since 2008, renewable fuels and natural gas have increased as a percent of total energy consumption in the United States while coal and petroleum have declined (Fig. 3.8). The balance between the cost of crude oil and the costs of ethanol production, has continued and will continue to be the primary determinant for which type of fuel will be used and produced.

3.3.2 Biofuels

In the early 2000s, a renewed interest in ethanol production occurred due to three

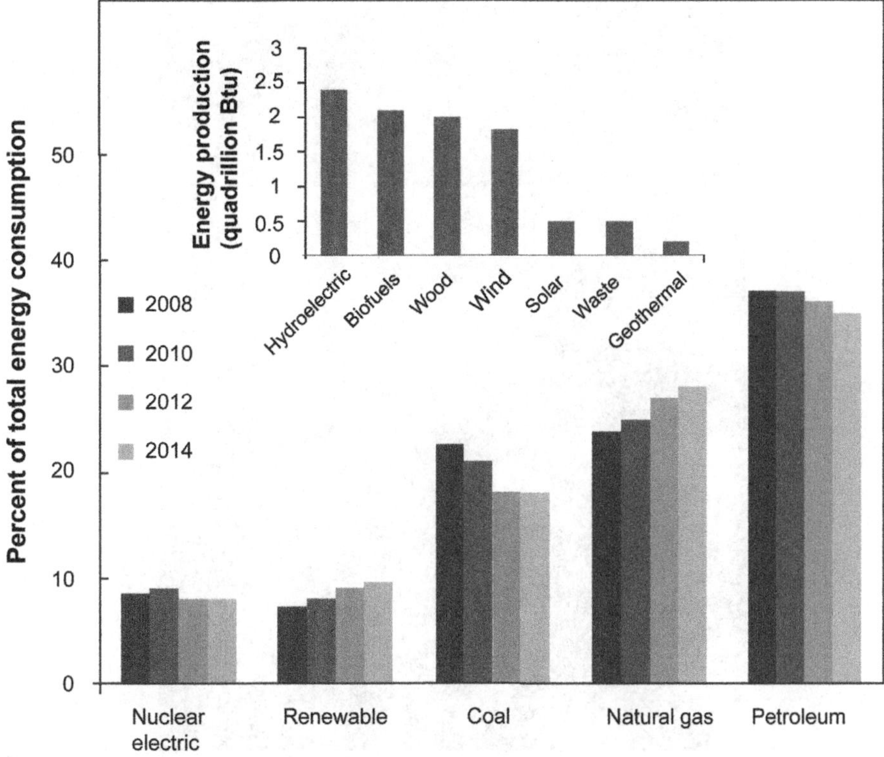

FIGURE 3.8 The use of fossil fuel sources of energy, including coal and petroleum, have decreased within the past decade, while natural gas and renewable fuels have increased. The inset shows the types of renewable fuels and their relative contribution. *Source: US Department of Energy, www.energy.gov.*

factors: crude oil prices surged; a commonly used ingredient in gasoline, methyl tertiary butyl ether (MTBE), was restricted or banned by many states and was replaced with ethanol; and the Energy Policy Act of 2005 and the Renewable Fuel Standard specified using renewable fuels (such as ethanol) in gasoline for automobiles. In addition, the Energy Independence and Security Act of 2007 mandated increasing US use of biofuels to 36 billion gallons by 2022. These factors created a favorable economic environment for ethanol and the industry responded quickly. The production of ethanol in the United States increased from 1.8 to 15 billion gallons between 2000 and 2014. Corn provides a good feedstock for ethanol production because of its high starch content. Starch can be fermented by yeast and bacteria to generate ethanol. But as the demand for ethanol increased, so did the price of corn, limiting the economic return for ethanol producers. This led to pursuit of other feedstocks, particularly those of low cost. The plant material remaining after corn is harvested is called *stover* or *biomass* and is comprised of complex carbohydrates, mainly cellulose (Fig. 3.9). This material is often left on the field to contribute to soil composition as it decays and also to reduce soil erosion, and it is used somewhat for animal feed. Stover is essentially free, except for the cost of recovering it from the field, drying, and transportation. The conversion of cellulosic material to ethanol is more complicated than the conversion of corn starch, but technologies using enzymes and fermentation have been developed

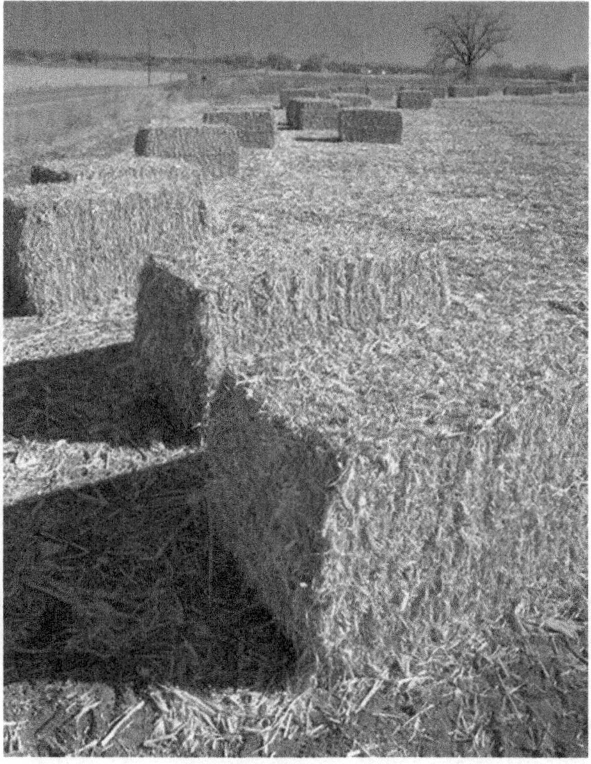

FIGURE 3.9 Stover is the residual plant material left after harvesting corn or other crops. Stover can be converted to ethanol through a process of cellulosic fermentation. Bales of stover are collected after crops are harvested and transported to processing facilities. *Source: Photo from USDA image gallery, www.ars.usda.gov.*

to make it possible and profitable. There is some risk that removal of too much stover from fields has the potential to increase soil erosion, deplete soil organic matter, and increase water contamination. Other sources of feedstock including perennial grasses such as switchgrass and miscanthus are also being explored as more environmentally sustainable options. Cellulosic ethanol is one of the next generation biofuels that also includes biochemical, thermochemical, especially pyrolysis, and other technologies to produce ethanol and petroleum equivalents from renewable sources. Environmental concerns associated with biofuel production include higher economic pressure to grow corn over other crops, reduced incentive to retain buffer zones around waterways in lieu of greater profits from growing crops, and competition between crops for food versus fuel. It is unlikely that any one biofuel process or feedstock will be sufficient or sustainable enough to completely replace fossil fuels but these alternative energy sources allow less fossil fuel to be used.

3.3.3 Wind and Solar Energy

Two other natural sources of energy have become increasingly available for agriculture: wind and solar. The primary limitations to both solar and wind power have been access to low-cost collecting materials and infrastructure systems to deliver and store energy to maintain a constant supply. These challenges have increasingly been overcome in both industries. Solar power capacity in the United States rose from about 44 megawatts (MW) in 1992 to over 22,000 MW in 2013. With advances in technology, the costs of photovoltaic or thermal panels have made it increasingly affordable to install both off-grid and on-grid systems. Solar energy use in agriculture has been traditionally off-grid, for example, using a solar panel to power water pumps for crop irrigation and electric fences or to heat water for dairy operations. However, as more infrastructure has been built, on-grid solar use in agriculture has expanded. California leads all states in using solar energy in agriculture, and with state and federal incentives to reduce fossil fuel use, it is expected that more farmers will turn to solar power. Similar to solar, wind power use in agriculture is on the rise. The cumulative capacity generated from wind was 2500 MW in 2000 and increased to over 45,000 MW by 2011. The majority of wind production has occurred in the Great Plains and Midwest because of the fairly constant wind and access to land for wind turbines (Fig. 3.10). Texas leads all states in wind production followed by Iowa. The installation of large wind turbines on the landscape has raised some complaints about aesthetics, noise, and visibility, but other concerns, such as the potential killing of migratory birds, have been largely avoided with new technology. The USDA Renewable Energy Systems and Energy Efficiency Improvement Program, which was first introduced in the 2002 Farm Bill and was continued as the Rural Energy for America Program (REAP) in the 2008 and 2014 Farm Bills, provides grants and loans to farmers for solar, wind, and other renewable energy projects.

3.4 TRANSPORTATION IN AGRICULTURE

When the majority of people in the United States lived on farms, food distribution was fairly simple. Food was transported from the field directly to homes or local markets. Horse-drawn carts, without refrigeration, traveling on dirt roads limited the distance one could distribute products. As cities and towns developed, moving food via rivers and canals allowed farms to be more distant from consumers. During the 1800s paddle-wheel steamboats moved products along major rivers such as the Mississippi and Missouri, and distribution ports along these rivers grew into cities. Technological advances in transportation came quickly after the Civil War with the introduction of coal-fueled train engines and gasoline-powered cars and trucks. These new means of transportation, and the innovation of refrigeration, changed how and where food was grown and marketed and allowed people to live farther way from farms. The US food system today is dependent on an efficient and complex network of transportation including airplanes and ships that connect to markets around the world, and trains, boats, trucks, and cars that distribute food from California to New York and every small town in between.

3.4.1 Railroads

As the Industrial Revolution began in the United States during the 1800s, a major event that significantly impacted agriculture was the building of the railroad system. Before construction of railroads, products and supplies were moved mainly by water routes or overland by horse and carriage. In 1825, John Stevens invented and built the first railroad track and locomotive in Hoboken, New Jersey and the first steam engine was operational in 1830. Soon thereafter, railroads were built to connect cities along the Eastern Seaboard. Many proposals were made to Congress in the 1800s to fund major railroad construction by eager entrepreneurs. Congress was being pressured to connect the rapidly growing populations on the Pacific coast and to provide a

FIGURE 3.10 Wind turbines have been installed across many states to generate electricity. Fields of wind turbines dot the landscape across the Midwest and Plains states. *Source: Photo from Department of Energy image gallery, www.energy.gov/photos.*

better means to move products across the country and there was general public support for the concept. But the Civil War was draining the government resources, and financial means to fund such a project were unavailable. To overcome this, the Curtis Bill, proposed by Iowa Congressman S.R. Curtis, designated the creation of a Board of Commissioners to oversee the construction and the selling of capital stock of the company ($1000 per share) to raise the funds. This established the railroads as public–private partnerships. The US government granted land rights to the railroad developers that allowed all earth, stone, lumber, or other material within the granted land to be used for construction purposes. Any land that was not used by the railroad could be sold and the profits returned to the railroad company. The first passage of the bill was partially successful, but an amendment made in 1864 expanded the amount of land granted to the railroads and modified the path of the railroad, which launched the project. Two companies were created, the Union Pacific Railroad and the Central Pacific Railroad, as public–private partnerships. The history of the railroads in the United States is fraught with controversies related to the buying and selling of stocks, the engagement of Congressmen in financing projects (and getting rich), and the eventual rise of railroad barons, some notorious, who controlled markets and prices. Cornelius Vanderbilt was one such railroad baron who became very wealthy and had great influence on the US economy, including providing the funding to found Vanderbilt University in Tennessee.

President Abraham Lincoln determined that the first site to begin the Union Pacific Railroad construction was to be a section from the western boundary of Iowa and the eastern boundary of Nebraska, which is today the city of Omaha, to connect to Ogden, Utah with a connection from Ogden to San Francisco. Early on, the construction lagged due to lack of infrastructure, high costs of materials, untrained labor, and skirmishes with Native Americans. History is rife with stories of rowdy railroad crews and rough shanty towns that were hastily built along the route and battles with Native Americans. Thousands of immigrants, including Chinese and Europeans, were recruited to work the construction of the railroads. The impact on Native Americans was dramatic, as many fought and died to resist the crossing of their lands or succumbed to diseases spread by the railroad workers. Between 1866 and 1869, the track was completed from Omaha to Ogden and the connection of the East to the West was widely celebrated as a new era for US commerce. By 1885, four systems connected the East and West Coasts, thus providing fast and economical means to ship products. Over the next decade, additional railroad lines were built to connect major cities across the country from a few thousand miles of track in 1840 to over 250,000 miles by 1916.

The impact of railroads on agriculture was immediate. The ability of cattle ranchers in the Great Plains to move animals quickly by rail, rather than by cowboys pushing cattle drives along rough terrain, created new markets and centralized their industry. Chicago, Kansas City, and Omaha developed huge stock yards around the railroad terminals and the meat packing industry grew. Early experimentation with using ice to cool products in boxcars, called reefers, added a new dimension to food transportation, allowing perishable products to be shipped long distances. The Swift Company and the Armour Company were among the first to successfully ship animal carcasses and cut meat across the United States using ice-cooled boxcars. By the early 1940s, the Pacific Fruit Express Company had developed systems that moved air through the packed ice to create a more efficient cooling environment and better insulation and seals were developed that made shipping perishable produce possible. The ability to move foods, even perishable foods, across long distances changed how food was marketed and produced (Fig. 3.11). Dairies no longer needed to be located close to towns and cities because milk and butter could be shipped in by train. Products could be produced in larger quantities because there were now multiple markets, allowing larger farms producing single products to be profitable. The grain industry expanded with wheat, corn, and soybeans grown in the Midwest accessible to all regions of the United States. Food was no longer region-based and Americans welcomed the variety and accessibility to a wider range of products.

3.4.2 Interstate Road System

Farm-to-market roads are state or county roads that connect farms and ranches to market towns or distribution centers. At the turn of the 19th century, most farmers delivered their products using horse-drawn carriages. The Ford Model-T automobile was introduced in 1908 and rapidly became affordable thanks to an innovative assembly-line production system. Soon after in 1925, the Ford pickup truck

FIGURE 3.11 Innovations in railcars increased the distribution range for many foods during the 1930s. Watermelons grown in Laurel, Delaware are being loaded into traincars for distribution around the United States in 1905. *Source: Photo from the Library of Congress, www.loc.gov/pictures.*

became available. The construction of roads suitable for these new modes of transportation became critical for getting supplies to farms, and transport of farm commodities to markets. In some states, most notably Texas, Missouri, Iowa, North Carolina, Michigan, and Tennessee, farm roads were built during the 1920s and 1930s, but the majority of states developed rural road systems only after WWII. The Office of Road Inquiry was established within the USDA in 1893. The fact that the federal road office resided in the USDA emphasizes the importance of roads to the US farm economy. The American Association of State Highway and Transportation Officials (AASHTO) was formed in 1914 and worked with federal agencies to develop highway legislation, policy, and standards for the US transportation network. The importance of a national system of highways was driven by concerns for national defense and the ability to move soldiers and materials. The Federal-Aid Highway Act of 1944 approved a 64,000-km National System of Interstate Highways and established a federal-aid secondary system of principal secondary and feeder roads. The construction of the interstate highway system began in earnest during the 1950s with the implementation of a series of east–west and north–south highways. A plan that was projected to take 12 years required 35 years to complete. Access to an interstate network of roads created opportunities for delivery of agricultural products by trucks in a more efficient manner than trains. Trucks could deliver directly from door to door whereas train delivery required secondary transport. Railroad owners realized the threat to their industry and lobbied the government to impose stiff highway user taxes on truckers. This was eventually sorted out with legislation that created a tax system to fund the highways. The overall impact of the interstate highway system was to expand access for both the manufacturing and agriculture sectors. But as these roadway systems age, particularly bridges with more cars and trucks using them, there is concern that the US infrastructure is not keeping up with transportation needs of farmers, grain handlers, and freight haulers. Additionally, many rural areas still remain dependent on gravel or unpaved roads. Finding ways to fund road maintenance continues to be a challenge for local, state, and federal governments.

3.5 AGRICULTURAL CHEMICALS

Pests, including weeds, insects, molds, and fungi are a major threat to food crop production. Weeds compete with crops for water, soil nutrients, and sunlight. A field of wheat can be devastated overnight by a swarm of grasshoppers, and some types of molds that grow on plants produce dangerous toxins. Farmers are in a constant battle with pests regardless of where or what types of crops they grow. Agricultural chemicals are part of the arsenal in this battle, but their use has consequences for the environment (Chapter 8: Sustainability of the Food System). Having the right balance of nutrients in the soil is essential for crop health and productivity. Fertilizers have been used to enhance crop production since the 1950s and are necessary for high yields. The proper use and management of agricultural chemicals and understanding their impact on the environment are current areas of much debate and research.

3.5.1 Fertilizer

Early agricultural practices to enhance plant growth included applying human and animal manure, ground bone, and various mineral mixtures to fields. Experimentation during the 1800s led to the understanding that plants require three main nutrients (nitrogen, phosphorus, and potassium), three lesser nutrients (calcium, magnesium, and sulfur), as well as trace amounts of seven other minerals. Nitrogen was found to be one of the most important elements for plant growth and could be provided from saltpeter, a natural compound of sodium nitrate. Large deposits of saltpeter were mined in Chile and Peru during the 1800s and sold to Europe and the United States. But demand for nitrogen for agricultural, industrial, and military use, outpaced these natural sources. A technique to produce nitrogen in the form of ammonia on an industrial scale, the Haber-Bosch process, was developed in Germany in 1913 that provided sufficient amounts of nitrogen for munitions such as TNT and other explosives used during WWI. The expansion of ammonia production in the United States ramped up quickly after the start of WWII. Ammonia production plants were constructed, mainly in the central United States near pipelines of natural gas, the starting material for ammonia production. By the end of the war, the capacity to produce ammonia was significant. With ammonia no longer needed for munitions, the surplus was quickly converted to agricultural use. At first, ammonium nitrate pellets were added directly to the soil, but eventually anhydrous ammonia (Fig. 3.12), which was easier to apply and provided a purer form of nitrogen, was the main form in use. By the 1940s, chemical fertilizers of nitrogen, potassium, and phosphorus were in wide use, resulting in improved crop yields. The extensive use of fertilizers, especially nitrogen, has been controversial ever since. High nitrogen levels from agricultural runoff into water systems causes algal growth and fish kills and can affect human health.

3.5.2 Pesticides

Insects, fungi, plant diseases, and weeds (commonly referred to as pests) are a major concern for farmers because an infestation can

FIGURE 3.12 Anhydrous ammonia has been used as a fertilizer in the United States since WWII. Tanks containing anhydrous ammonia are connected to a distribution system that inserts the nitrogen into the soil. *Source: Photo provided by the AgLeader, Ames, IA.*

quickly destroy a crop or reduce its value. Pesticides, which include chemicals used to kill insects, microorganisms, fungi, rodents, and weeds, are integral components of farming. Early agricultural practices to ward off insects or kill plant diseases involved burning various materials to create smoke. Plant extracts, tree tar, salt, copper, sulfate, and lime were also applied to plants to control disease. One plant extract, pyrethrum derived from the plant *Chrysanthemum cinerariaefolium*, has been in use for over 2000 years and is still an effective pesticide today. Weed management was mainly mechanical through the 1900s, using hoes and harrows or manually pulling weeds. Burning of weeds was also used through the 1940s especially for row crops.

In the United States prior to 1940, chemicals such as sodium chlorate, sulfuric acid, and arsenic, and those derived from coal such as creosote, naphthalene, and petroleum oils, were used to control insects, molds, and mildews. These chemicals were applied in large amounts, lacked specificity, were damaging to crops, and were highly toxic to humans and other animals. During the 1940s synthetic pesticide production increased and a wide range of chemicals, including DDT (dichloro-diphenyl-trichlorethane), parathion, and chlordane were developed. These chemicals targeted specific insects or diseases and were designed to be less toxic to plants, humans, and animals. Among these, DDT became very popular for agricultural as well as community and home use because of its broad-spectrum activity and low toxicity for mammals. DDT spraying to kill mosquitos and other disease-carrying insects led to reduction in malaria, yellow fever, and typhus, which had been common in parts of the United States. Today four main types of chemical pesticides have been developed with abilities to disrupt insect metabolism: organochlorine insecticides, organophosphate pesticides, carbamate pesticides, and pyrethroid pesticides.

Chemical management of weeds in the United States during the 1800s included applying copper and iron sulfates, sulfuric acid, or sodium arsenite to fields. Sodium arsenite was highly effective as an aquatic herbicide through the 1920s and was widely used to treat lakes and waterways. Between 1919 and 1940 petroleum-derived herbicides gained use in the United States, including sodium chlorate, dinitrophenol, ammonium sulfamate, kerosene, and gasoline. A major discovery in the 1930s of the ability of phenoxyacetic compounds to kill broadleaf weeds, but not grasses or corn, began a new era in chemical applications in weed control. The chemistry of 2,4-D ((2,4-dichlorophenoxy) acetic acid) and related compounds, as well as advances in other chemicals, led to an expansion in chemical herbicides. Between 1950 and 1970 the number of herbicides developed and approved for use increased from 15 to 100. The use of tactical herbicides by the US military in Vietnam included Agent Orange, which was a mixture of compounds related to phenoxyacetic acid, along with kerosene and diesel fuel. Agent Orange was effective as a defoliating solution, but contained a byproduct, dioxin, that was highly toxic and carcinogenic. Long-term health effects in soldiers and civilians exposed to Agent Orange have occurred. Phenoxyacetic acid compounds continue to be approved for use as an herbicide by the EPA, although the other compounds used in Agent Orange and dioxin have been banned from agricultural use.

The first legislation concerning pesticides by the federal government was the Federal Insecticide Act of 1910, which arose from concerns about fraudulent or substandard pesticides being sold. The law set standards for the manufacture of pesticide products and inspection of processors. By the mid-1940s, farmers were applying synthetic chemical pesticides to their crops at increasing rates. Growing concerns about oversight of this industry led to the creation of the Federal Insecticide, Fungicide, and Rodenticide Act (FIFRA) in 1947. FIFRA established a requirement that pesticide formulations must be registered with the USDA and that labels on pesticides be accurate but the legislation did not extend to the regulatory process or standards of use. In 1959, a public scare over contaminated cranberries with pesticides led to a series of hearings in Congress about pesticide use and food safety. The Delany House Committee hearings, led by New York Congressman James Delaney, established a provision that the FDA would monitor and define safety limits for pesticides in food. The USDA and FDA continue to monitor pesticide residues in fruits and vegetables and release these reports to the public via their websites.

Systematic government oversight of pesticide use began only after the Nixon administration founded the EPA in 1970. Concern for the environment came to the forefront of attention for Americans with the publication of the book *Silent Spring* (1962) by Rachel Carson. Carson had become alarmed by the heavy use of chemical pesticides in agriculture and other venues to control pests that were having negative effects on wildlife. Her book investigated the effects that chemical pesticides had on human health as well as the environment and brought to light the lack of regulatory oversight of their use. Fueled by the content and implications of *Silent Spring* (1962), environmental activist groups gained followers and funding. The Environmental

Defense Fund was one of the earliest such groups that took on the campaign against DDT. DDT was widely used as a pesticide, not only in agriculture but to control mosquitos in the environment, and insects in buildings and homes. A tipping point came when a connection was made between DDT and bald eagles. Eagles eat small animals that were exposed to DDT in the fields. DDT from these food sources accumulates in the body tissues of the birds over time, alters their calcium metabolism, and causes their eggs to be fragile and break in the nest. The result was a decrease in the population of bald eagles, nearly to the point of extinction. Bald eagles, of course, are a symbol of the United States and so saving this species became a public demand. Concerns about this and other environmental and human health effects brought to light by activists and scientists led to the banning of DDT in the United States in 1972. DDT continues to be used in several countries primarily to prevent malaria.

The Federal Environmental Pesticide Control Act (FEPCA) was enacted in 1972 giving the EPA expanded authority to regulate pesticide sales and use and to certify pesticide applicators. This placed the EPA, rather than the USDA, in the role of regulators of pesticide use in agriculture. The regulation and oversight of pesticide use is the responsibility of the EPA under authority of the FIFRA and the Federal Food, Drug, and Cosmetic Act (FFDCA). The Food Quality and Protection Act of 1996 amended both FIFRA and FFDCA to require the EPA to ensure that pesticides pose "reasonable certainty of no harm" before they can be registered for use in food or feed. Under the law, the EPA must review pesticides at least once every 15 years to determine if they continue to meet the standards of safety for humans, animals, and the environment. The EPA also restricts the use of some high-risk pesticides to certified pesticide applicators. The certification process is managed at the state level and states are allowed to implement pesticide regulations that are more stringent than federal laws.

3.6 PRECISION AGRICULTURE

At the turn of the 21st century, other advances in agriculture technology have enhanced the ability of farmers to monitor their fields and crops and more accurately use fertilizers, pesticides, and soil amendments. Modern farm equipment employs a combination of global positioning systems (GPS) with geographic information systems (GIS), along with monitoring technologies, to record in real-time and space crop yields, plant tissue nitrogen status, soil moisture and temperature, and weed pressure to guide application of fertilizers, irrigation water, and herbicides on a site-specific basis (Fig. 3.13) These tools require substantial technical expertise and generate large amounts of data, thereby requiring sophisticated computer skills and applications. Because of the complexity of these technologies, they have not yet been widely implemented, but hold great potential. Tools are being developed that connect with cell phones and computer tablets that allow farmers to make better decisions in real time on their farms. We are currently in the era of "Big Data," where finding ways to meaningfully interpret and apply information to enhance agricultural outputs is of urgent need. It also has generated new issues in agriculture around the legal ownership of this information, as more private companies are developing and selling their data-collecting technology to farmers. The concerns are that having access to data about farm operations, yields, and costs could be used to define markets or give advantage to others. Regulations and policies about agricultural data ownership are just beginning to be considered.

FIGURE 3.13 Advances in technology have improved agriculture production in the United States. Farming equipment with computer-assisted planting and harvesting technology increases the efficiency and accuracy of farm operations. *Source: Image Provided As Courtesy of John Deere.*

3.6.1 Integrated Pest Management

Integrated pest management (IPM) programs use a wide range of approaches to manage pests and diseases on crops to anticipate and prevent infestations, rather than respond after they have become a problem. Components of IPM include selecting crop varieties that are not susceptible to pests; using competitive crops, insects, or even viruses to combat pests; applying thoughtful crop rotations; and reducing the environmental habitat for pests. A principle of IPM is to evaluate crop conditions and to use economic thresholds to guide use of pesticides. The aim is to minimize the incidence of unnecessary application of pesticides to the environment while maintaining economic return for the farmer. By using a wide range of approaches throughout the year, pests are less likely to develop resistance to treatments and negative effects on the environment are reduced. Some examples of unique tools of IPM include releasing sterile male fruit flies to mate with wild fertile females causing the population to die out, or using natural plant compounds laced with small amounts of insecticide as bait to selectively attract corn root beetles.

3.7 PLANT BREEDING

The great majority of plant foods that we consume today are the result of plant breeding. Breeding and selection techniques allow desirable traits, such as color, size, shape, or tolerance to growing conditions to be developed in a plant. The wide variety of apples in the produce aisle of the grocery store, including those that are red, yellow, pink, or green, were derived from plant breeding. Broccoli serves as an example of a food crop that was not present in nature in the state we currently recognize. It is thought that farmers during the Roman Empire crossed varieties of the wild *Brassica* family to generate variations of the plant. Cultivation of broccoli as a food crop has been attributed to Italian farmers during the 1600s. The crop was probably introduced to Great Britain in the 1700s and brought to the United States by Italian immigrants in the 1800s. Plant breeding continued to be applied over the years to create the short stalk plant with tight flower buds that we consume today. The health-promoting effects of broccoli have made it a popular vegetable in the United States but many consumers would not know that broccoli was created through plant breeding. Similar stories could be told about how plant breeding led to just about all of the fruits and vegetables we enjoy today, including tangelos, seedless watermelon, and purple potatoes.

Tools used to increase the variation in plants have included exposing seeds and plants to harsh chemicals, gamma radiation, or X-rays. These treatments cause damage to DNA, which results in mutations in the plants, so this process is referred to as mutation breeding. Some of the mutations induced in the plants would produce a desired characteristic. Many of the varieties of wheat used today for bread and pasta were derived using mutagenic breeding approaches, as were many of the rice varieties consumed worldwide. This technique is effective in generating mutations in plants but it is not specific. Many changes may occur in the plants in response to exposure to the mutagenic treatments, most of which are not identified or monitored in the resulting crop. No regulations on mutation breeding, or testing of the safety of foods produced using these techniques, have ever been required. Newer approaches to plant breeding are more specific and targeted because of advances in molecular biology.

3.7.1 Hybrid Seed

The genetic code was not fully understood until the breakthrough by Watson, Crick, and Franklin in 1953 that explained the structure of DNA, followed by understanding of how the four-lettered code of DNA could be translated into the 20 amino acids that make up proteins. Three scientists (Marshall Nirenberg, Gobind Khorana, and Robert Holley) are credited with the latter discovery and received the Nobel Prize for their contribution in 1968. As early as 1906, experiments were being done on the inheritance characteristics of corn. Using the technique of inbreeding, in which pollen from a single plant is transferred to the silks of the same plant, identical generations were created. Selection for desired traits within each generation allowed for production of superior characteristics. Cross-breeding was done using selected parents from inbred lines. The first commercial hybrid corn, Burr-Leaming, was released by the Connecticut Agricultural Experiment Station in 1921.

Many state and federal inbreeding hybridization programs began during the 1920s. As interest in hybrids increased, methods to produce larger quantities of seed were needed. A major player in this innovation was Henry Wallace, who founded the Hi-Bred Corn Company in 1926 (later known as Pioneer Hi-Bred), prior to his government service. Based on the principles of cross-breeding corn to produce higher yielding varieties with improved vigor, hybrid crops were substantially more productive and were rapidly adopted by farmers. By 1949, annual sales of Hi-Bred brand seed corn passed $1 million. Farmers were reluctant at first to plant them, but when they were shown the superiority of these seeds, the demand grew quickly. The Extension Service was instrumental in training farmers on the use of hybrid seed. Hybrid crops were so successful that nearly all corn grown in the United States since the 1930s has been hybrid varieties. Use of hybrid varieties has resulted in significant gains in production efficiencies and yields.

3.7.2 Genetic Engineering

Major advances in science began to arise during the early 1980s that brought a return of economic growth in agriculture. One of these was the development of genetic modification technology in bacteria and plants. In 1982, the FDA approved the first medication developed using this new technology. The DNA sequence that encoded the protein insulin was genetically engineered (GE) into *Escherichia coli*, allowing a reliable and safe source of this medication for diabetics. Many people believe a second Green Revolution began when these advances in molecular biology allowed for creation of crops with specific traits. The techniques that led to genetically modified organisms (GMOs) or GE foods were simultaneously reported in 1983 by three research teams including Robert Fraley (Monsanto Company, St. Louis, Missouri), Mary-Dell Chilton (Washington University, St. Louis, Missouri), and Marc Van Montagu (Rijksuniversiteit, Ghent, Belgium). The three shared the 2014 World Food Prize for their discovery of techniques to insert bacterial genes into plants. These techniques followed processes that occurred in nature. By developing these tools, scientists were able to create novel plants that expressed specific traits selected from related or unrelated species.

One process of genetic engineering involves identifying a specific gene sequence that encodes a protein of interest (this may be referred to as a *trait*). The gene sequence is inserted into the cellular DNA of the plant or animal so that it is expressed with the native DNA. The genetic code involves DNA, which is comprised of bases in a unique sequence that transcribes into RNA. Within RNA, codons connect with amino acids and link them together to form a protein. The resulting protein may be an enzyme, or a regulatory or structural protein that implements the desired trait. If the inserted gene is derived from a different organism, for example a bacterial gene inserted into a corn plant, that is called a transgenic modification. Autologous gene transfers are also used in which the gene is derived from the same species.

Tools of genetic engineering include genomic sequencing, in which the entire DNA code of a plant or animal is defined, and gene mapping is done to link DNA sequences with proteins. In addition, scientists must understand the biochemical processes within the plant or animal to identify where and how the proteins will interact to express the trait. The tools to carry out genetic sequencing have advanced rapidly and today there are large public databases of plant and animal genomes, including the human genome, available to researchers. Newer approaches to modify the DNA–RNA–protein sequence are also being developed. For example, gene silencing can be done so that a specific gene is prevented from being expressed in a plant or is turned off for a specific period of time in the plant's development cycle. Gene editing is possible in which the DNA code is selectively edited to change, prevent or increase gene expression. Modifying RNA, which is the translational step in protein synthesis, is also being used. The recent development of clustered regularly-interspaced short palindromic repeats (CRISPR) technology, adopted from a normal bacterial process, allows scientists to selectively silence and edit genes and holds great potential for both human disease and agriculture management.

The first GMO crop commercially released in the United States was the Flavr-Savr tomato in 1994. Researchers at the University of California-Davis in 1982 modified the plant to retain firmness by blocking an enzyme, polygalacturonase, that dissolves pectin in the plant cell wall. The FDA approved the safety of the Flavr-Savr tomato and it was marketed in Davis, California and Chicago, Illinois, where it sold well. In 1996, the product was introduced as a canned tomato paste in the United Kingdom, where it was clearly marked as being derived from GE tomatoes and sales were strong through 1999. However, public fear of the product arose following a British broadcast in 1998 featuring a local physician who claimed that feeding rats genetically modified potatoes resulted in biological effects in the rats caused by the process of genetic engineering. As word of that report spread, sales of the GMO tomato paste fell and the company eventually withdrew the product. The physician who first raised the concern later retracted his study, which was never peer reviewed or published, and independent scientists found no risk of the technology.

During the 1990s the papaya crop in Hawaii was being devastated by the papaya ringspot virus epidemic, which was not being controlled using conventional pesticide approaches. Researchers at the University of Hawaii and Cornell University developed varieties of papaya using genetic engineering that were resistant to the virus. These were approved by the FDA in 1998 and by 1999 GMO papaya represented over 30% of the crop in Hawaii. Today about 75% of papaya grown in Hawaii is the GMO variety.

The power of genetic engineering to address pest management was recognized by seed companies and early research was focused on commercially important crops of cotton, corn, and soybean. Approvals for crops with herbicide-tolerant or insect-resistant traits began in 1996. In 1974 the Monsanto Company released the herbicide glyphosate, commercially known as Roundup. Glyphosate acts by inhibiting an enzyme pathway that is specific for plants (the EPSP enzyme in the shikimate pathway), which causes plant death. Humans and animals do not use these enzymes and so glyphosate was considered a safe herbicide. Scientists had found that *Agrobacterium tumefaciens*, a naturally occurring bacterium, produced an EPSP enzyme that was not affected by glyphosate. Monsanto scientists successfully inserted the *A. tumefaciens* gene into soybeans using genetic modification techniques, resulting in the production of Roundup Ready soybeans in 1996. Because the soybeans were tolerant of glyphosate, fields could be treated with the herbicide after the soybean plants emerged from the ground. The weeds would be killed but not the crop. Varieties of Roundup Ready corn, cotton, spring canola, alfalfa, sugar beets, and winter canola were also developed. By using herbicide-tolerant seeds, farmers could spray their fields at key times to reduce weeds that compete with young seedlings resulting in more efficient crop management. Farmers could thereby use less herbicide and spray fields less frequently, allowing them to have higher yields and lower input costs compared with other varieties of soybean. This technology also meant less physical cultivation to remove weeds, resulting in less soil erosion and fuel consumption.

As the use of these herbicide-tolerant crop varieties increased, glyphosate use expanded rapidly. The USDA estimated that glyphosate accounted for about 15% of all herbicides used in 1996 and 89% in 2006. The most troubling aspect of such widespread use is the development of resistant weeds. When resistance develops, farmers must use more chemicals or switch to other types of herbicides. Glyphosate-resistant weeds have become a significant problem, especially for corn, cotton, and soybean farmers. Recently, the USDA approved new varieties of GM corn and soybeans that are resistant to 2,4-D, which will replace some of the Roundup Ready varieties, as a means to address resistance to glyphosate.

A successful trait that was developed for insect resistance was the Bt variety (e.g.,

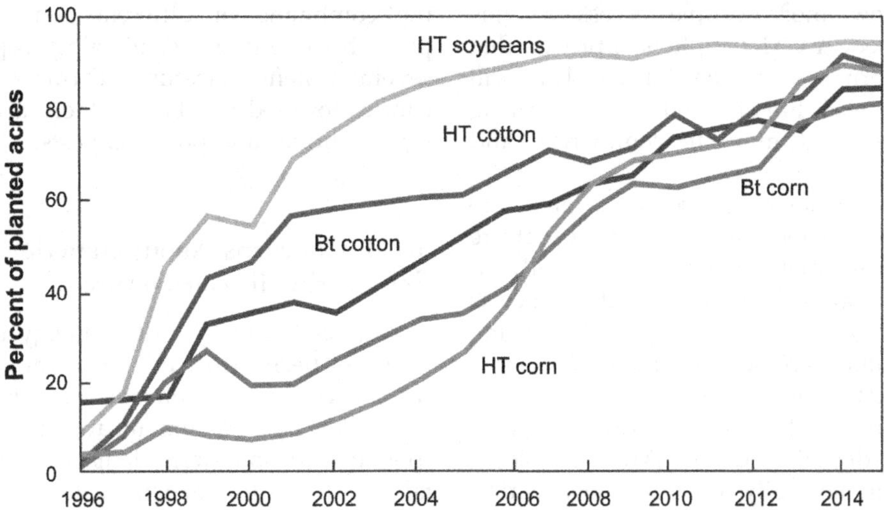

FIGURE 3.14 US farmers began to grow genetically modified crops after their introduction in the mid-1990s. By 2014 nearly all of the corn, cotton, and soybeans grown in the United States contained an herbicide-tolerant (HT) trait or the *Bacillus thuringiensis* (Bt) pesticide. *Source: Fernandez-Cornejo, J and McBride W.D. (2002) Adoption of bioengineered crops. Agricultural Economic Report no. 810, US Department of Agriculture, Economic Research Service.*

Bt corn, Bt cotton, Bt soybeans). *Bacillus thuringiensis* (Bt) is a common soil bacterium that had been known for decades to produce a specific protein (cry protein) that was lethal to certain types of insects. Bt toxin was widely used in the early 1800s in Europe to prevent insect infestations in flour and other products and it has been an approved insecticide in the United States since the 1960s. Bt toxin is currently an approved pesticide for organic farming. Several different Bt proteins are produced by variants of *B. thuringiensis* that are specific for different types of insects. A Bt toxin was found that was highly effective against the corn borer. The corn borer was a significant pest causing financial loss to farmers. The toxin produced by Bt is active only in insects and has no effect on humans because of the specific way it interacts with receptors in the insect's intestine. Insertion of the Bt toxin gene into plants eliminated the need to spray fields with pesticides because the insects would ingest the toxin directly when they consumed parts of the plant. The presence of the Bt toxin in the plant meant that pests were immediately controlled even before evidence of their infestation was noticed by the farmer. The economic benefit of GMO seed to farmers was quickly recognized and the number of acres of row crops planted with GMO traits expanded from almost none in 1996 to 80%–90% by 2014 (Fig. 3.14).

Since the mid-1990s GMO foods and food ingredients have been part of the US food system. These include corn products such as corn oil, corn starch, corn meal, and high-fructose corn syrup, and soybean products such as soybean oil, textured soy protein, and soy flour, as well as many food-grade chemicals that are produced from corn and soybeans. Canola oil, sugar from sugar beets, squash, and papaya are approved GMO products as is alfalfa for animal feed. The FDA approved a GMO animal, a salmon that reaches maturity in half the time of wild salmon. Salmon is an excellent source of omega-3 fatty acids and dietary guidelines recommend increased consumption of salmon. The ability to produce more salmon should allow more people access to this healthful food. A GMO apple is approved that does not turn brown when sliced. This will increase the shelf-life of sliced apples allowing them to be used more in commercial and home meals.

GMO technology is being applied to enhance key nutrients in staple crops that are lacking in the diets of people especially in developing countries. Vitamin A deficiency is the leading cause of blindness in developing countries, and increases the risk of infectious diseases. Beta carotene is a naturally occurring form of vitamin A that is converted to the vitamin during digestion. Using GMO technology, foods that are normally consumed such as rice, bananas, and cassava can be modified to produce beta carotene. These foods are useful to prevent the devastating effects of vitamin A deficiency in children and young women. As more crops are genotyped and biochemical pathways understood, GMO applications have the potential to improve the nutritional and organoleptic qualities of food and to reduce food spoilage and waste. With climate change, alterations in temperatures and rainfall are occurring around the world. Crops that are more suitable for drought, soil salinity (caused by higher sea levels), temperature extremes, and pest resistance (changes in climate directly alter the types and distribution of pests) will be needed to sustain food production. GMO technology has the potential to address these challenges in a more efficient manner than conventional plant and animal breeding.

Some of the economic advantages for farmers and the environment from using GMO seeds include reduced amounts of herbicides and insecticides (less cost for chemicals, less fossil fuel to run sprayers), higher yields (crops are healthier without pests and weeds, more yield on less land), and less need to till the soil between growing seasons (less use of fossil fuels, less topsoil loss, less soil compaction). As the technology advanced, stacking traits, or the combining of different genes into one plant, became possible, allowing crops to have several beneficial features. Farmers had more choices for seed that best fit their farm's growing conditions and potential pests.

3.7.3 Concerns About Genetic Engineering in Agriculture

Application of GMO technology in agriculture has been criticized for a variety of reasons. These include concerns over the safety of the technology to human and animal health (potential for long-term damage), environmental risks (including genetic drift into non-GMO crops, overuse of pesticides, and resistant weeds and pests), economic pressures (high cost of seeds and related chemicals, patent rights regarding seed ownership and favoring large operations over small producers), and ethical concerns (changing nature, creating superspecies, and putting profits ahead of safety). Critics of GMO technology have raised concerns about the impact of this technology in the following ways.

Weed resistance: GMO technology has been criticized by environmental groups because of the risk of developing resistant weeds and pests. Resistance is a feature that most invasive weeds and pests (and microorganisms) develop over time with repeated exposure to the same chemicals. The rise of resistance occurs through natural selection pressure and processes. The genes for resistance may already exist within a small percentage of the population. As the same pressures (exposure to a pesticide) are applied over time, the susceptible weeds and insects will be eliminated while those that are resistant will become dominant. Weeds and pests have become resistant to all types of pesticides and herbicides, including glyphosate and Bt toxin. Resistance is not unique to GMO crops and the history of agriculture includes many situations of having to use different types of pesticides and herbicides to manage infestations. As more fields were planted with crops bearing similar traits, it was inevitable that resistance would arise. Recommended agronomic practices including rotation of crops and use of refuge seeds (non-GMO seeds within a field of GMO seeds) were not well followed. The rapid and widespread adoption of GMO crops, especially corn and soybeans in the Midwest, has likely contributed to development of resistant weeds and pests in a shorter timeframe than would have occurred without the GMO introduction. The USDA tracks the number of resistant weed species and has recorded a steep increase in the number since 1975. The top four crops for herbicide-resistant weed species were wheat (66), corn (59), rice (51), and soybeans (47) in 2014. Note that of these, only corn and soybeans have GMO traits, hence the increase in resistant weeds is not solely due to the introduction of GMO technology. It is possible that the large number of acres of crops all treated with one or two types of chemical herbicides, particularly glyphosate, hastened resistance. Careful crop rotation and use of different types of pest control have always been important in agriculture but are even more so today with the rapid expansion of GMO technology.

Economics: A small number of companies have been at the forefront of GMO technology for agricultural crops. These companies have focused their efforts on crops of economic benefit, mainly corn, soybean, cotton, and alfalfa, and the biotechnology applications have been focused on improved farm efficiencies. These advances may not be viewed as having much direct benefit to consumers. The power of GMO technology to enhance nutritional quality has not yet been fully implemented. GMO technology is patented by the companies that created them. Farmers must follow legal restrictions when using GMO seeds, and may be required to use pesticides and herbicides made by the same company. This has led to concerns of monopolizing agriculture, limiting competition, reducing biodiversity, and excluding small farmers. Examples of big seed companies suing farmers over seed rights or patent abuse have been widely publicized and fuel the belief that these companies are concerned only about profits. Making a profit is the right and expectation for all companies in the United States as is protection of intellectual property. To some, agriculture is viewed differently than other industries, because growing plants or raising animals is considered a process of nature, and raising food may be held to a higher ethical standard than other types of business.

Human and animal health: GMO corn and soybeans have been used in animal feed and human foods since the late 1990s. Animals fed GMO grains throughout their lifespan and over multiple generations have shown no negative changes in their overall health, reproductive status, growth, or well-being. In fact, overall animal health and productivity has improved over the two decades in which GMO crops have been in use. Similarly, humans have been consuming products produced from GMO corn and soybeans, including corn, soybean and canola oil, corn starch, corn sweeteners, soybean protein, as well as dietary fibers and other food ingredients, for this same period with no evidence of negative health effects. Public concerns over GMO foods as being unsafe or potentially harmful to human health have been raised. Some of the concerns are that a higher incidence in food allergies and sensitivities have occurred since GMO foods were introduced, and that rates of autism, obesity, and other chronic diseases may be associated with the application of this technology in food. While it is possible to draw graphs showing correlations between disease incidence rates and introduction of GMO to the food system, there has been no credible scientific evidence showing a causal relationship between these factors. Major health organizations around the world have reviewed the scientific evidence and have concluded that there is no risk to human health from consuming GMO foods. These include the American Medical Association, the American Academy of Pediatrics, the Center for Science in the Public Interest, the European Commission, and the World Health Organization. The National Academies of Sciences, Engineering, and Medicine completed an extensive review of GE crops and released a report in 2016 that concluded GMO foods were safe for humans and animals and the technology provided important benefits for agriculture and the environment.

GMO labeling of food: Many consumers remain confused and concerned about the use of GMO technology in foods. For some, there may be uncertainty about long-term impacts of consuming products with selected genetic traits, or for others the concern may be with the wider use of a few chemicals such as glyphosate. This inspired a movement to mandate labeling of GMO ingredients on food packages. Since 1992, the FDA policy on bioengineered foods has been that there is "…no basis for concluding that bioengineered foods differ from other foods in any meaningful or uniform way, or that, as a class, foods developed by the new techniques present any different or greater safety concern than foods developed by traditional plant breeding." Therefore, the FDA has not required GMO foods or ingredients to be listed on the package label. A voluntary provision that food manufacturers may label their products as either not containing ingredients produced using bioengineering or that they do contain such ingredients has been in place since 1992. A movement to mandate GMO labeling on all foods gained national attention in the mid-2000s. The effort has been well funded and effective in using social media to raise awareness about their cause. However, many consumers are not able to accurately express what GMO is or why they might be concerned about GMO in foods. Several states tried to pass legislation to require GMO labeling within their own borders. Such state-by-state labeling regulations would be nearly impossible for food manufacturers to comply with and they mobilized to defeat the bills. While most of the state bills failed, Vermont approved a GMO labeling law that went into effect in July 2016. Under pressure from the food industry and concerns from the FDA that state-based food labeling laws would undermine a national uniformity of labeling, Congress proposed a compromise bill that would require manufacturers to provide information on food packages about GMO ingredients using one of three routes: direct wording or symbols on the package, information provided via a QR code, or a website link. The bill was signed into law by President Obama in the summer of 2016. Several arguments against mandatory GMO labeling of foods include the potential increase in food costs that would result from tracking, verifying, and monitoring the wide range of ingredients used in foods; the need for more FDA inspectors and monitoring tools (some consider the FDA to have insufficient time and resources to deal with the current food security issues); and potential confusion for consumers about GMO safety (labeling infers some risk). As is commonly observed, some food manufacturers have recognized the consumer interest in non-GMO foods and see a marketing opportunity. As a result, foods that never did contain GMO ingredients, such as oatmeal or wheat flour, are being marketed with non-GMO labels.

Regulatory oversight: In 1986, the government created the Coordinated Framework for the Regulation of Biotechnology to oversee the introduction of GMO crops. Three agencies were tasked with providing the regulatory process: the USDA APHIS, the EPA, and the Department of Health and Human Services Food and Drug Administration (DHHS-FDA). The focus of regulation is to ensure that new GMO products would not negatively impact the environment (EPA) or existing agriculture (APHIS) and that they were safe for consumers as part of the food system (FDA). Before a GMO food enters the food system it must be shown to:

- be nutritionally identical to the non-GMO variety
- pose no human or animal health risk
- have a trait that is digestible and does not persist in the body
- have a trait that poses no allergenic potential
- have a trait that poses no risk to the environment

EXPANSION BOX 3.2

ETHICS OF GMO FOOD

The history of agriculture in the United States is described by continual adaptation to economic stress and political and social change by using mechanization, science, and technology. The general trend in US agriculture has been more and more food production by fewer and fewer people. Economists would call this increased efficiency and utilitarians would likely agree that the benefits of increased amounts of food to feed more people outweigh any harm to the public or the environment. The use of GMO technology is credited with some of the increase in food production and is a technology expected to be important in future agricultural developments.

However, much of the dissatisfaction with the current food system is related to adoption of genetic modification technology. An anti-GMO organization would contend that the costs of environmental damage; injustice and inequities to minorities, rural communities, and small farmers; and violation of individual rights are valid reasons to question current agricultural methods. There is considerable controversy and skepticism about the ownership, decision-making and use of GM crops.

Who owns technology, specifically genetic modification of plants, and who determines how this technology will be used? Who decides what plants will be used for crop production? How can ethics help us decide? Table 3.9 describes various positions regarding ownership and use of GMO in agriculture.

The companies that have invested in the development of GMO seeds and pesticides believe that they have ownership of the seeds and have the right to profit from their use (recent legal cases have upheld this view). They also state that there are many benefits to the use of GMO crops, among them the requirement for less herbicide use and less cultivation (so less fuel and less soil erosion) than non-GMO crops. Utilitarians view the increased production using fewer resources as providing the greatest good for the greatest number, which makes it a beneficial technology.

Crop farmers also profit from GMO technology and believe they are providing safe and wholesome grains to the benefit of society with less damage to the environment. They also state that they are in the best position to decide the appropriate seeds, plants, and technology to apply to food production. Farmers are continuing in the rights perspective of the independent citizen-farmer for whom individual property rights are of the utmost importance.

Another group may argue that control of plants belongs to everyone in the community and that everyone has a right to food. Control of seeds by a few corporations violates the rights of small farmers, organic farmers, and farmers in developing countries who do not want GMO seeds or cannot afford the GMO seeds and pesticides. Consumers who demand food companies to provide labeling to identify

EXPANSION BOX 3.2 (*cont'd*)

TABLE 3.9 Ethical Perspectives on GMO Use in Agriculture

Stakeholder	Seed and chemical companies and corporations	Farmers	Advocates of small farmers in developing countries	Environmentalists
Favored Policy	Regulated market with government oversight	Individual property rights and responsibilities	Public ownership for the common good	Favor diversity and government protection
Primary Value	Plants and seeds are assets to be used	Ownership and control of own property	Individuals have a right to grow crops for food	Intrinsic value of plants and land and virtue of its protectors
Ethical Principle	Greatest good for greatest number	Noninterference	Fairness and equity	Ecological integrity
Philosophy	Utilitarianism	Rights	Egalitarianism	The land ethic

GMO and non-GMO foods probably agree with this view. Equal access to and distribution of resources, and doing no harm, are important ethical principles of beneficence.

The ethical view of environmentalists has become more prominent in recent decades as Americans have gone from taming the wilderness to appreciating the natural world for its own value. This group views the use of GMO crops, and maybe all agricultural monocultures, as a threat to the diversity of the ecosystem. This ethic is also linked to building strong moral character or virtue of individuals.

Recognition of various viewpoints, each based on sound and rational ethical values and principles, should aid discussion of complex dilemmas, such as the use of GMO technology in the food system. The policies and programs that stakeholders support are based on their individual and collective philosophies. Difficult problems can be resolved when there is respect, scientific thinking, civil discourse, debate, and compromise.

Suggested reading: Comstock (2008) and Thompson (2007).
Suggested websites: www.GMOAnswers.com *and* www.GeneticLiteracyProject.org

References

Carson, R. L. (1962). *Silent spring*. Boston, MA and New York, NY: Houghton Mifflin Company, 297 p.

Comstock, G. L. (Ed.), (2008). *Life science ethics* Ames, IA: Iowa State University Press and Blackwell Publishing Co, 380 p.

Fernandez-Cornejo, J and McBride W.D. (2002) Adoption of bioengineered crops. Agricultural Economic Report no. 810, US Department of Agriculture, Economic Research Service.

Orden, D., & Zulauf, C. (2015). Political economy of the 2014 Farm Bill. *American Journal of Agricultural Economics*, *97*(5), 1298–1311.

Owens, J. T. (Ed.), (2008). *The farm bill and its far-ranging impact* New York, NY: Nova Science Publications, 265 p.

Thompson, P. B. (2007). *Food biotechnology in ethical perspective. The International Library of Environmental, Agricultural and Food Ethics*. Dordrecht, The Netherlands: Springer, 335 p.

Further Reading

Beckman, J., Borchers, A., & Jones, C. A. (2013). *Agriculture's supply and demand for energy and energy products. ERS Economic Information Bulletin-112.* Washington, DC: Economic Research Service, U.S. Department of Agriculture.

Brase, T. (2006). *Precision agriculture.* Clifton Park, NY: Thomson and Delmar Learning, 224 p.

Brookes, G., & Barfoot, P. (2005). GM crops: The global economic and environmental impact—the first nine years 1996–2004. *AgBioForum, 8*(2&3), 187–196.

Committee on Genetically Engineered Crops: Past Experience and Future Prospects; Board on Agriculture and Natural Resources; Division on Earth and Life Studies; National Academies of Science, Engineering and Medicine (2016). *Genetically engineered crops: Experiences and prospects.* Washington, DC: The National Academies Press, 388 p.

Dimitri, C., Effland, A., & Conklin, N. (2012). *The 20th century transformation of U.S. agriculture and farm policy. ERS Economic Information Bulletin-3, June 2005.* Washington, DC: U.S. Department of Agriculture. Available from <http://www.ers.usda.gov/publications/eib-economic-information-bulletin/eib3.aspx>.

Economic Research Service. (2015). *Effects of trade on the U.S. economy-2014.* Available from <http://www.ers.usda.gov/data-products/agricultural-trade-multipliers/2014-data-overview.aspx>.

Environmental Protection Agency. (2015). *EPA history.* Available from <https://www.epa.gov/aboutepa/epa-history>.

Federal Highway Administration. (2015). *General highway history.* U.S. Department of Transportation. Available from <http://www.fhwa.dot.gov/highwayhistory/history_misc.cfm>.

Fernandez-Cornejo, J., Wechsler, S., Livingston, M., & Mitchell, L. (2014). *Genetically engineered crops in the United States. ERS Report Number 162.* Washington, DC: Economic Research Service, U.S. Department of Agriculture.

Gonsalves, D. (2004). Premier papaya plantations rescued through science and teamwork. *Agriculture Research, 52* (1), 2.

Hallberg, M. C. (1992). *Policy for American agriculture: Choices and consequences.* Ames, IA: Iowa State University Press.

Harper, C. L., & Le Beau, B. F. (2003). *Food, society and environment.* Upper Saddle River, NJ: Prentice Hall, Pearson Education, Inc, 260 p.

Heckman, J. (2005). A history of organic farming: Transitions from Sir Albert Howard's *War in the Soil* to USDA national organic program. *Renewable Agriculture and Food Systems, 21*(3), 143–150.

Hill, J. (2014). *The WPA: Putting America to work.* Detroit, MI: Omnigraphics, 246 p.

Hodgson, E. (1991). Pesticides: Past, present and future. *Reviews in Pesticide Toxicology, 1*, 3–12.

Institute of Medicine; Committee to Review the Health Effects in Vietnam Veterans of Exposure to Herbicides; Division of Health Promotion and Disease Prevention (1994). *Veterans and Agent Orange: Health effects of herbicides used in Vietnam.* Washington, DC: National Academy Press.

International Food Information Council Foundation. (2009). *From farm to fork: Questions and answers about modern food production.* Available from <www.foodinsight.org>.

Kinkela, D. (2013). *DDT and the American century: Global health, environmental politics, and the pesticide that changed the world.* Chapel Hill, NC: The University of North Carolina Press, 256 p.

Lu, Y. C., Daughtry, C., Hart, G., & Watkins, B. (1997). The current state of precision farming. *Food Reviews International, 13*(2), 141–162.

Matthews, G. A. (Ed.), (2006). *Pesticides: Health, safety and the environment* Oxford, UK: Blackwell Publishing Ltd.

Nandula, V. K. (Ed.), (2010). *Glyphosate resistance in crops and weeds: History, development and management* Hoboken, NJ: John Wiley & Sons, Inc, 321 p.

Nickerson, C., Morehart, M., Kuethe, T., Beckman, J., Ifft, J., & Williams, R. (2012). *Trends in U.S. farmland values and ownership. Economic Information Bulletin No. EIB-92.* Washington, DC: Economic Research Service, U.S. Department of Agriculture, 55 p.

Nicolia, A., Manzo, A., Veronesi, F., & Rosellini, D. (2014). An overview of the last 10 years of genetically engineered crop safety research. *Critical Reviews in Biotechnology, 34*(1), 77–88.

Prohens, J., & Nuez, F. (Eds.), (2008). *Vegetables I: Asteraceae, brassicaceae, chenopodicaceae and cucurbitaceae* Valencia: Springer, 426 p.

Ricroch, A., Chopra, S., & Fleischer, S. (Eds.), (2014). *Plant biotechnology: Experience and future prospects* New York, NY: Springer Science + Business, 291 p.

Schimmelpfennig, D., & Ebel, R. (2011). *On the doorstep of the information age: Recent adoption of precision agriculture. ERS Bulletin Number 80.* Washington, DC: Economic Research Service, U.S. Department of Agriculture.

Taylor, W. J. R. (1973). The federal Food and Drug Administration in the United States: Born and bred in crises. *International Journal of Clinical Pharmacology, 7* 156–162.

Timmons, F. L. (2005). A history of weed control in the United States and Canada. *Weed Science, 53*(6), 748–761.

U.S. Department of Agriculture. (1980). *Report and recommendations on organic farming.* Washington, DC: National Agricultural Library, U.S. Department of Agriculture. Available from <http://pubs.nal.usda.gov/report-and-recommendations-organic-farming-usda-1980>.

Van Eenennaam, A. L., & Young, A. E. (2014). Prevalence and impacts of genetically engineered feedstuffs on livestock populations. *Journal of Animal Science*, *92*(10), 4255–4278.

Weingroff, R.F. (2015). *Moving the goods: As the interstate era begins.* Highway History, Federal Highway Administration, U.S. Department of Transportation (updated November 18, 2015; cited 2015). Available from <http://www.fhwa.dot.gov/infrastructure/freight.cfm>.

White, T. K., & Hoppe, R. A. (2012). *Changing farm structure and the distribution of farm payments and federal crop insurance. ERS Economic Information Bulletin No. 91*. Washington, DC: Economic Research Service, U.S. Department of Agriculture.

Zobeck, T. M., & Schillinger, W. F. (Eds.), (2010). *Soil and water conservation advances in the United States. Soil Science Society of America Special Publication 60* Madison, WI: Soil Science Society of America, Inc.

CHAPTER

4

Animals in the Food System

4.1 ANIMAL FOODS IN HUMAN DIETS

The consumption of animal products, such as meat, milk, and eggs, may have positively influenced the development of early humans by providing a consistent and high-quality diet. There is evidence that early humans were consuming meat as early as 2.5 million years ago. When fire was used to cook food the quality of the diet improved by increasing the digestibility of both animal and plant foods. Fire that provided heat to dry and smoke to preserve foods allowed food to be stored for later use. Animals were harnessed for labor, and contributed hides, feathers, and wool for clothing. As human civilizations developed they continued to assert control over their environment by domesticating food sources. There is fossil evidence of domestication of animals beginning over 12,000 years ago. Domestication of pigs and cattle likely began about 10,000 years ago, and goats and sheep about 8,000 years ago in the Fertile Crescent (now the Middle East). Using DNA mapping of current species, along with archeological findings, it is possible to identify how animal domestication progressed concurrently with human civilizations. From these records it is evident that animal foods have been well integrated in many cultures. Animal-source food products account for over half of the value of US agriculture, according to the USDA, exceeding $100 billion/year. These products are classified as aquaculture, cattle and beef, dairy, hogs and pork, poultry and eggs, and sheep, lamb, and mutton. In the American diet, the primary animal foods are beef, chicken, pork, milk (and dairy products), and eggs. Fish and seafood are also consumed and domestication has recently occurred for these foods.

4.2 HISTORY OF THE CATTLE INDUSTRY

Bison are the only cattle native to the United States and were a source of food, clothing, and shelter for Native Americans. Domestic cattle were brought to the United States in the early 1500s by various expeditions including the Spanish, Mexican, and French. Cattle are in the family Bovidae and therefore are called bovines. There are dozens of breeds of cattle around the world, derived from crossing and selection over many generations. The most common breeds of beef cattle in the United States are Angus, Hereford, Gelbvieh, Limousin, and Simmental, but a large number of specialty breeds have been developed.

Understanding Food Systems.
DOI: http://dx.doi.org/10.1016/B978-0-12-804445-2.00004-1

Angus and Angus crosses may account for as much as 60% of the American commercial cattle population.

Cattle are ruminants, meaning they have a four-compartment gastrointestinal system that allows the breakdown of grasses and plants that are not digestible by humans or nonruminant animals (Fig. 4.1). This physiology allows ruminants to survive on foodstuffs with low nutrient content. Ruminants graze on forage (grass, hay, or plant material), which is mixed with saliva that contains sodium, potassium, phosphate, bicarbonate, and urea. The mixture, or cud, is passed via the esophagus into the first stomach compartment, the rumen. The cud cycles back and forth between the rumen and the second stomach compartment, the reticulum. The muscles lining the reticulum contract and cause the cud to be regurgitated back into the mouth to be further chewed. This is referred to as "chewing the cud" and ruminants spend about one-third of their time in this ruminating process. In a mature cow the capacity of the reticulum and rumen is about 40 gallons. As breakdown of the cud occurs, liquid and small particles pass into the third compartment, the omasum. The fourth compartment, the abomasum, is referred to as the "true stomach." In the abomasum acids, buffers and enzymes are secreted to fully digest the remnants of the cud. The digested material is released into the small intestine where digestion continues and nutrients are absorbed into the body. Ruminants are able to thrive on forages because of the chemical factory within their digestive system. The reticulum and rumen are populated by a variety of microorganisms including bacteria, protozoa, and fungi that work in the anaerobic (without oxygen) environment to break down plant cell walls and synthesize a wide range of compounds. Cellulose and other complex carbohydrates are broken down to simple carbohydrates, protein is synthesized from nitrogen

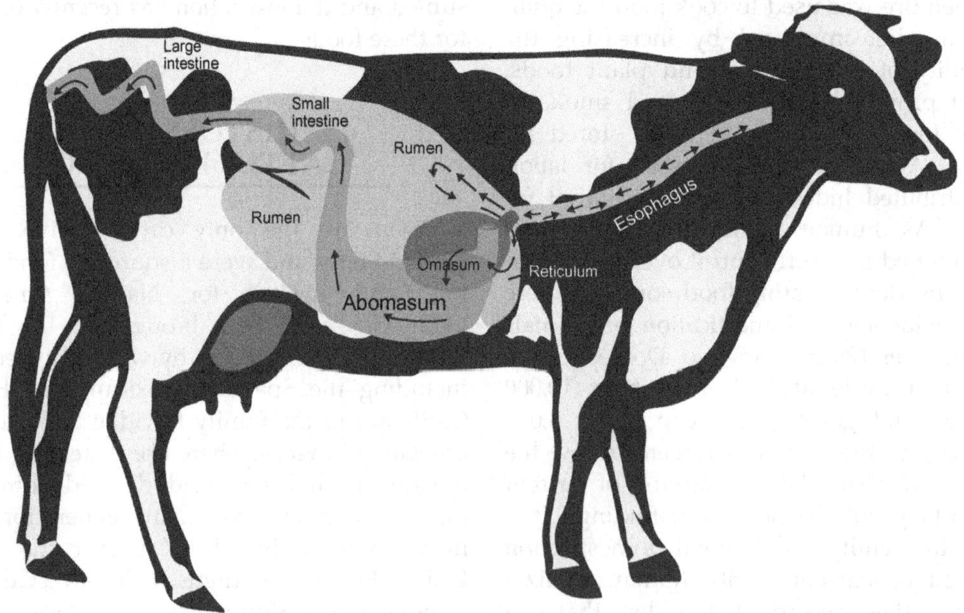

FIGURE 4.1 Cows are ruminants and have a four-compartment stomach that allows them to digest cellulosic material such as hay and grasses. *Source: Illustration by Reannon Overbey.*

present in the plant materials, and volatile fatty acids such as acetate (a precursor of fatty acids), propionate (a precursor of glucose), and butyrate are generated, as are B vitamins and vitamin K. Gases are also generated including carbon dioxide, methane, and hydrogen sulfide, which are expelled via the mouth and anus. The microbial population of the rumen is a symbiotic relationship; the bacteria utilize the rumen nutrient composition to grow and reproduce, and these microbes are digested to provide a source of nitrogen- and carbon-containing compounds for use by the animal. In this manner ruminants turn low nutrient quality grasses and plants into muscle and milk that provide high-quality food for humans. It is estimated that about 60% of the land mass is comprised of land that is not suitable for cultivating crops but is appropriate for ruminant animal grazing. Grazing cattle on this land provides a food source from otherwise unproductive land. Cattle also provide a source of fat (tallow), hides for leather, bone meal and manure for fertilizer, and other beneficial materials.

The American colonists raised cattle to provide meat for consumption and for export trade. Exporting meat to Britain from the United States began in the late 1800s. By the early 1900s three companies—Armour, Swift, and Morris—dominated the export of beef (chilled) and cattle (live) market. Most of the meat was shipped "cooled" but not frozen. The trip across the Atlantic Ocean from the United States to Great Britain took about 8–10 weeks. Once the ships reached ports the meat had to be quickly distributed to markets. The quality of chilled meats was probably more favorable during the winter months, but the lack of refrigeration on such a long trip certainly caused significant spoilage. Moving live cattle on ships was also done, but had major limitations due to the challenges of keeping the animals healthy on the long voyage. Concerns about transmitting diseases carried by the US cattle into Great Britain eventually led the British to impose quarantines on imported animals that added time and handling costs to the shippers.

Prior to the Civil War the cattle industry in the United States was growing slowly, mainly in the South and Great Plains where cattle were being raised. The war left the South in an economic depression and the freeing of slaves meant that farms were unattended and many animals were released to the wild. Ranchers, especially in Texas, eventually realized that they could round up these wild and semiwild cattle and move them across the country to markets in the North to sell at higher prices than could be obtained in the South. These cattle drives along famous routes, such as the Chisholm Trail, were economical because the cattle would consume grasses found along the way for food. The cowboy became an American icon, illustrative of the tough, independent adventurers of the new frontier. This trade was also responsible for establishing new towns and trading posts across the Southwest. Because of the ready access to open grazing land, cattle ranching expanded across several states including Kansas, Missouri, Colorado, Montana, Wyoming, and the Dakotas, which were built on the breeding stock that was brought along with the cattle drives. The post–Civil War expansion of the railroad network across the United States facilitated the construction of stockyards and feedlots in key locations near railroad hubs that transported the animals and meat to both coasts. The economic impact of stockyards and feedlots in the central Midwest was substantial and aided the population growth of cities such as Kansas City, Chicago, St. Louis, and Omaha.

4.2.1 Meat Safety Regulations

As the cattle industry grew in the United States, veterinarians, ranchers, and meat packers became concerned about the lack of regulation to prevent the spread of diseases in livestock. This prompted passage of the Bureau of Animal Industry Act and formation of the Bureau of Animal Industry (BAI) under the USDA in 1884. BAI's primary responsibility was to prevent diseased animals from being used in food, and to inspect and certify live animals for export markets. BAI's inspection authority did not extend into meat packing plants. This changed when Upton Sinclair published his novel, *The Jungle* (1906), about the lives of immigrants working in the meat packing industry in Chicago. Sinclair intended to bring to light the deplorable working conditions for immigrants in meat packing plants. But his descriptions of filthy plants, spoiled and diseased meat, and overall unsanitary practices raised the public's concerns about their food. President Roosevelt commissioned a team of inspectors to investigate these reports, believing at first that Sinclair was merely pushing his Socialist agenda. The inspectors largely confirmed the ugly practices that Sinclair had written about, which led Congress to pass the Federal Meat Inspection Act (FMIA) in 1906 (Table 4.1). This law was passed on the same day as the Pure Food and Drug Act, which are two milestones for food safety regulations in the United States. FMIA had four sanitary requirements for the meat industry: (1) mandatory inspection of all livestock before slaughter; (2) mandatory postmortem inspection; (3) explicit sanitary standards for slaughterhouses; and (4) USDA authority to inspect and monitor slaughter and processing operations. The quality and safety of meat produced in the United States was significantly improved following the implementation of these federal regulations and inspection programs. Within the first year of the law being passed the USDA had hired 2200 meat inspectors. A meat research center was opened in 1910 at the USDA experiment research station in Beltsville, Maryland to develop methods to test and evaluate meat for safety and quality. In 1967 state inspection programs were approved for meat processors selling only within that state, but products sold across state lines or for export continue to require federal inspection.

Since 1981 the Food Safety and Inspection Service (FSIS) within the USDA has had the responsibility for the safety of meat, poultry, and egg products. FSIS operates primarily under the authorities provided by the Federal Meat Inspection Act (1906), the Poultry Products Inspection Act (1957) and the Egg Products Inspection Act (EPIA) (1970; Table 4.1). FSIS inspectors are responsible for the inspection of all poultry and livestock throughout harvest and processing. This includes all beef, pork, chicken, turkey, lamb, and veal as well as some wild game, and all liquid, frozen, and dried egg products. Foods such as ham, sausages, stews, soups, pizzas, and frozen dinners that contain 2% or more cooked poultry or 3% raw meat also require FSIS inspection. The main goals of FSIS are to define and ensure food safety standards throughout the processing, handling, and packaging of meat and egg products; to analyze products for microbial or chemical adulterates; and to provide public education to improve food-handling practices.

In 1993 ground beef sold in the Pacific Northwest was the source of a major foodborne illness that caused 400 illnesses and 4 deaths. The bacteria that caused the outbreak was *Escherichia coli* 0157:H7, which is a particularly dangerous pathogen because it can cause permanent kidney and liver damage (hemolytic uremia). The public became alarmed by this outbreak and demanded improved food safety oversight by the government. Until this time, FSIS inspectors relied on sight, touch, and smell to detect adulteration of products, which was recognized as being inadequate after this incident. To correct this, the Pathogen Reduction/Hazard Analysis Critical Control Points (HACCP) Systems approach

TABLE 4.1 Historical Events in Meat and Egg Safety and Inspection

Date	Government action	Impact
1906	Pass the Federal Meat Inspection Act (FMIA)	Prohibited sale of adulterated meat, standards for sanitation in meat processing facilities, USDA inspection
1917	Pass the Farm Products Inspection Act	Provided for voluntary inspection and grading of eggs
1926	Establish the USDA Federal Poultry Inspection Service	Provided for voluntary inspection and grading of poultry processors
1946	Pass the Agricultural Marketing Act	Replaced Farm Products Inspection Act and gave authority to USDA to define quality standards of eggs
1957	Pass the Poultry Products Inspection Act	Defined standards for sanitation in poultry processing facilities and gave USDA oversight of inspection of poultry
1958	Pass the Humane Methods of Slaughter Act	Standardized humane slaughter protocol
1967	Pass the Wholesome Meat Act	Required state inspection of meat and poultry processors to be equivalent to federal standards
1968	Pass the Wholesome Poultry Act	
1970	Pass the Egg Products Inspection Act (EPIA)	Set standards for sanitation of liquid, frozen, and dried egg production facilities
1971	Establish the Animal and Plant Health Service (APHIS)	Transferred all meat and poultry inspection to APHIS
1981	Establish the Food Safety and Inspection Service (FSIS)	Transferred all meat and poultry inspection to FSIS
1995	Define egg inspection oversight	Gave FSIS responsibility for liquid, frozen and dried egg products and USDA and FDA responsibility for shell eggs
1996	Pass Pathogen Reduction/HACCP Systems Rule	Defined measurable checkpoints for monitoring sanitation and food processing
2002	Pass the Animal Health Protection Act	Gave APHIS authority to monitor and address infectious diseases of livestock
2010	Passed Food Safety and Modernization Act (FSMA)	Enhanced regulations for food safety, stronger recall authority for FDA, sanitation standards for farms and small producers
2012	Implement FDA defined "Egg Rule"	Provided regulations for egg producers to prevent food-borne pathogens

was developed and became a rule in 1996. This process moved responsibility to ensure the safety of food products from the government to the manufacturer. HACCP plans require food manufacturers to define specific steps in their processing facility that would be monitored and evaluated for contamination, temperature control, or other measures and to retain these records for inspection. The regulation specifically defined that processors needed to implement practices to reduce *Salmonella* contamination and to test for *E. coli*. By 2000, all food manufacturers, not just meat and egg producers, were required to have HACCP plans in place. The Food Safety and Modernization Act (FSMA) was passed in 2010, which increased the range of HACCP programs, strengthened the FDA's recall authority, enhanced traceability systems for food products, defined sanitation standards for farms, and redesigned assessment of safety of imported foods. HACCP and FSMA are discussed in more detail in Chapter 6, Food Processing.

4.2.2 The Cattle Industry Today

The cattle industry in the United States involves several specialized types of operations. Cow-calf ranchers maintain a breeding herd and oversee the birth of calves. In some facilities animals are bred using artificial insemination procedures wherein sperm from a commercial source is used to impregnate females. This allows breeders to be very specific in maintaining the genotype of their herds and enhances the efficiency of the operation. Calves spend several months, about 7 on average, suckling with their mothers and grazing on grass in pastures. Between 6 and 10 months of age the calves are weaned and continue to be grass-fed in pastures, or provided hay or silage. Silage is defined by the USDA as "any crop that is harvested green and preserved in a succulent condition by partial fermentation in a more-or-less airtight container such as a silo" (USDA National Agricultural Library). Silage is higher in nutrients than hay and it is used with hay to enrich rations, especially in the winter. Cattle are sold at livestock auction markets when they reach maturity and are transferred to feedyards. Feedyards are also referred to as animal feeding operations (AFOs), which are defined by the EPA as facilities that confine animals for at least 45 days/year with no grass or other vegetation in the confinement area. If the number of animals in an AFO exceeds 1000 the facility would be classified as a concentrated animal feeding operation (CAFO). Cattle spend about 6 months being fed a balanced diet under the supervision of veterinarians in the feedyards. The purpose of the feedyard is to increase the growth rate by feeding a more nutritious diet, to closely monitor the animals' health and growth, and to ensure the quality of the meat. The diet fed to the animals in the feedyards is a mixture of forage and nutrient-dense ingredients including corn, soybeans, and other grains. Because cattle are ruminants, these higher nutrient feeds can cause problems with their digestive system and must be balanced with the right mix of cellulosic material (hay or straw). If the feed is not balanced well, microbial overgrowth can occur in the rumen, which causes metabolic problems for the animal. These problems, such as acidosis, which is a lowering of the pH of the blood, will inhibit animal growth and can be fatal. Animal nutritionists are typically part of the management team at commercial feedlots to ensure the diets are well designed and to prevent such problems.

Cattle raised for meat are usually given steroid hormones to increase the amount of lean muscle. These act similarly to endogenous hormones produced by the animal, and are dissipated before the animal is processed for meat. A further discussion of growth-promoting hormones in beef cattle is provided in Section 4.9. When the cattle reach market weight, typically around 1200–1400 pounds or 18–22 months of age, they are sent to a processing facility where they are slaughtered and butchered.

Some producers do not send their cattle to feedyards, but allow them to continue grazing in pastures until they reach market weight. There is an increasing market for grass-fed meat as consumers have become more concerned about hormones in meat, the use of CAFOs, or they prefer the taste of grass-fed meat. Some research has shown that grass-fed meat has a better balance of fatty acids, particularly more omega-3 fats, than meat from animals raised in feedyards. The amount of omega-3 fats varies widely and the overall nutritional benefit to humans is very small due to the low total amount of these fatty acids present in meat. Grass-fed animals require more land per animal, and more time to reach market weight than feedyard-fed animals, therefore the costs are higher for the producer and the consumer. At this time, there are no federal standards for a grass-fed beef label so consumers must determine on their own how the animals were raised. Animals raised under the USDA Organic Standard would be considered grass-fed.

The United States has been a major exporter of beef since the 1900s. In 2014 the beef and veal (meat from calves) exported from the United States was over 2.5 trillion pounds carcass weight and Japan was the top recipient. However, the United States actually imports more beef and veal than it exports, almost 3.0 trillion pounds carcass weight in 2014. Australia, Canada, and New Zealand are the top countries that supply beef and veal to the United States.

Since the 1970s the total amount of red meat (beef and pork) consumed has steadily decreased in the United States (Fig. 4.2). The decline in beef consumption began when medical and nutrition researchers reported that beef, with its high amount of saturated fat and cholesterol, increased the risk of cardiovascular disease. Chicken was lower in saturated fat and was recommended to replace beef in the diet. The beef industry responded to the health concerns about beef and since the 1970s the amount of saturated fat in beef (marbling) has been reduced significantly and today some cuts of beef have about the same amount of fat as chicken. The role of red meat in health and disease is discussed in Chapter 7, Nutrition and Food Access. Americans consume ground beef at a higher rate than cuts of meat. This preference for ground beef was true for all income levels and race/ethnicities. Ground

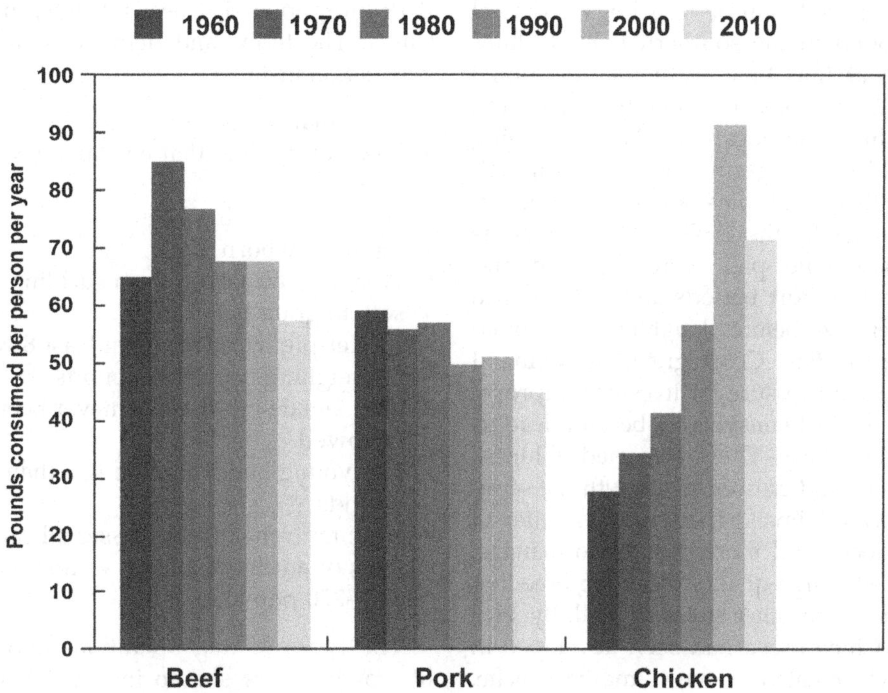

FIGURE 4.2 Americans decreased their consumption of beef and pork between 1960 and 2010, and increased their consumption of chicken. *Source: USDA Economic Research Service, www.usda.ers.gov.*

beef also held the dominant place in type of beef consumed away from home and in restaurants, mainly in the form of hamburgers.

4.3 HISTORY OF THE HOG INDUSTRY

According to fossil records, wild pigs roamed the earth 40 million years ago, and pigs were domesticated 7000 years ago in eastern Asia. Pigs are in the family Suidae, which includes wild boars and warthogs and are collectively referred to as swine. The National Pork Board tracks the start of the pig industry in the United States to Hernando de Soto, a Spanish explorer, who supposedly brought 13 pigs to Florida in 1539. The expedition took the pigs with them across the South as they made their way to Mexico. Many of the pigs escaped or were stolen by Native Americans, leading to a population of wild hogs and feral pigs still found in the southern United States today. Pigs at that time were the traditional foraging type, adapted to range wooded areas for food. In some locations, large roaming herds would ravage grain fields and crops. (It is notable that feral pigs are considered an invasive species by the USDA today.) In some regions, these wild pigs were captured and held in lots for short periods and fed grain to increase their size before slaughter. As populations along the East Coast grew, the demand for pork created pressure for livestock improvement. Farmers in Pennsylvania became leaders in swine breeding. They imported Chinese breeding stock and crossed them with the semi-wild American animals. These new varieties of pigs were docile and more suitable to manage, and produced a high-quality meat. Pig breeding became popular in other states as well. By 1816 the Poland China breed had been developed in Ohio, which created a new market niche. Within 20 years Cincinnati, Ohio become the largest pork processing city, earning the nickname Porkopolis. Large packing plants for both hogs and cattle were developing around railroad terminals in cities such as Chicago, Illinois; Kansas City, Kansas; St. Joseph, Missouri; and Sioux City, Iowa.

4.3.1 The Hog Industry Today

In states where corn and soybeans are a major crop, including Iowa, Illinois, Minnesota, Nebraska, Indiana, and Missouri, hog production expanded after WWII. The economic synergy of raising the feedstock for the hogs (corn and soybeans) and production of fertilizer for the crop (manure) in near proximity aided this expansion. Research in swine genetics, feed, and housing led to more efficient animal production and healthier, leaner animals. The main purebred swine breeds include Yorkshire, Duroc, Hampshire, Landrace, Berkshire, Spotted, Chester White, and Poland China. The terms and definitions used in hog production include:

- Boar: male hog
- Sow: female hog that has farrowed at least once
- Farrowing: giving birth
- Piglet: newborn pig
- Weaning: transition from suckling on the sow to grain-based diets
- Feeder pig: weaned pig at least 8 weeks old or weighing up to 100 pounds
- Gilt: female pig that has never been farrowed
- Pig: young hog weighing less than 120 pounds
- Hog: term that encompasses all categories of pigs; or adults of market weight, i.e., 240–270 pounds

The types of hog operations and stages of life for hogs are shown in Fig. 4.3. Hog breeders provide careful monitoring of genetics within purebred lines and aid producers in

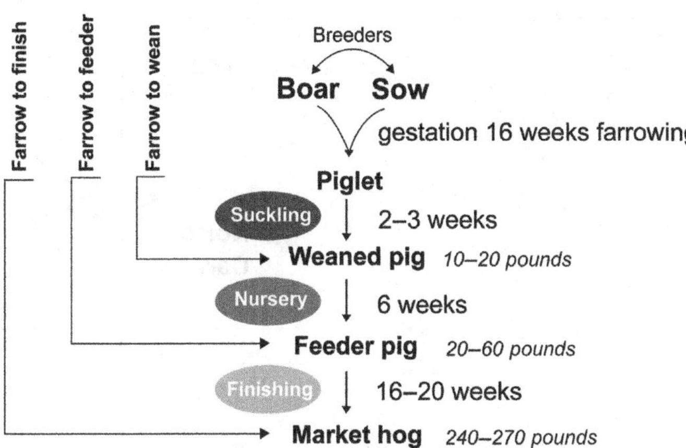

FIGURE 4.3 The stages of hog production from conception to market are shown. Specialization of operations has become more common to increase efficiencies and reduce costs. *Source: Illustration by Reannon Overbey.*

management of their breeding programs. The gestational period for a sow is about 16 weeks, which allows about 2, and the average litter size is 8–10 piglets. It is important that piglets suckle on the sow to obtain immunity and nutrients for the first 2–3 weeks of life. Thereafter they are separated from the sow and weaned onto specialized diets to enhance growth. These feeder pigs are continued on balanced diets until they reach market weight of about 240–270 pounds.

There has been a decrease in the number of operations that manage hog production from farrow to finish over the past decade and an increase in specialized operations. Focusing on a limited period of hog production allows facilities to be more efficient in their operations, utilize advances in production technology, and increase in size to optimize economic value. Specialization has led to concentration of hog production to fewer and larger operations. Since 1992 the number of farms raising hogs declined by 70% while inventory remained stable. The number of hog operations in 1992 was 240,000, dropping to about 60,000 in 2004. Operations with more than 5000 head represented only about 25% of total operations in 1992 and about 90% in 2014. And the USDA estimates that in 2002 nearly half of the total US inventory was owned by operations of more than 50,000 hogs. Iowa has held the spot of number 1 in hog production for many years, in part due to ready access to corn and soybeans for feed (Fig. 4.4). Recently, growth of hog production has occurred in North Carolina, which is now the second largest producer of hogs. Several factors contributed to the rapid growth in hog production in North Carolina, including development of genetic breeds that were more efficient in feed utilization and by utilizing housing systems that reduced disease and improved growth. Political and economic policies that support and encourage larger hog operations have also contributed to the expansion of this industry in North Carolina.

A major contributor to the growth in North Carolina and in other states has been the application of production contract agreements in which the hog owner (or contractor) engages a producer (a grower) to care for the pigs in the producer's facility. The producer is paid a fee for the service of raising the pigs, and the contractors provide inputs (feed and medications), technical assistance and market options. In some cases, meat packers act as contractors. For example, Smithfield Foods was the largest hog contractor in 2004 with over 800,000

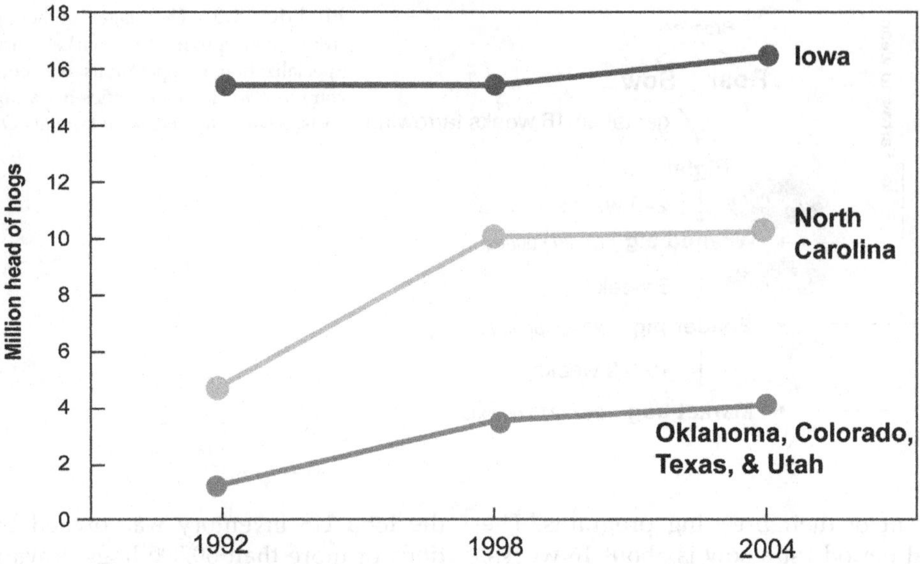

FIGURE 4.4 Iowa produces more hogs than all other states. North Carolina has increased hog production in recent years. *Source: USDA Economic Research Service, www.usda.ers.gov; Key and McBride (2007).*

sows. Production contract operations reduce the economic variability and risk of raising hogs, but take away some autonomy from farmers.

4.3.2 Hog Housing Systems

Pigs are nonruminant animals and have a digestive tract that is quite similar to humans. They are able to consume an omnivorous diet and were often fed scraps and garbage by farmers in the early years of US agriculture. But progress in understanding swine nutrition has led to more efficient growth and higher meat quality. Pig feeding systems have been well researched and defined to optimize growth and animal well-being. At birth piglets suckle from the sow to obtain colostrum that provides immune factors. They are then weaned to diets that are balanced for each stage of growth. The main ingredients include corn, barley, milo, oats, and wheat, along with soybean meal, vitamins, and minerals. With a focus on reducing the amount of fat in pork, producers have used selective breeding and dietary management to develop hogs that are 75% leaner than in the 1950s.

Pigs are raised in confinement barns, hoop barns, or open pastures. Confinement housing operations are often large, with thousands of hogs per unit, and are the predominant way hogs are housed in the United States. Confinement barns provide group or individual spaces for animals with a grated floor construction that allows manure to fall through to be collected and removed. Confinement barns have heat and humidity controls to reduce stress on the animals and can be cleaned thoroughly to prevent disease. In confinements, hog producers and workers are able to monitor animal behavior, feeding, and health efficiently and safely. The downsides of large confinement facilities are that when many animals are housed in one location, the large amount of manure must be managed to prevent environmental and water contamination and strong odors from the facility can reduce quality of life for people living or working nearby.

Pregnant sows raised in confinement barns may be placed in gestational stalls, which have bars to keep the sow separated from her piglets, but allow the piglets access to the sow for nursing (Fig. 4.5). Use of stalls was implemented primarily to protect the piglets from being hurt by the sow. Sows can weigh as much as 300–350 pounds and piglets can become trapped under the sow and suffocate. The stalls also make it easier for workers to inspect the piglets and the sow and control food and medical care. Gestational stalls have become unpopular with consumers in recent years because they are viewed as cruel or unethical due to the small space provided for the sow. To increase efficiency in the confinement facility, the stalls are narrow and do not provide room for the sow to walk or turn around. Research has found that sows spend the majority of their time while nursing piglets laying on their sides and are not unduly stressed by being confined. Farmers who use gestational stalls argue that their animals are well cared for and this housing system is appropriate, while animal activist groups and some consumers view them as inhumane. Ethical considerations of animal housing are discussed in Section 4.11. In response to these criticisms by activist groups and consumers, most major hog producers are moving to other types of housing that give sows more room to move, while still protecting the piglets and workers.

Hoop barns are used to house pigs and are typically open on two sides, with dirt floors, and fenced areas for animals to be outside. Management of temperature is not possible

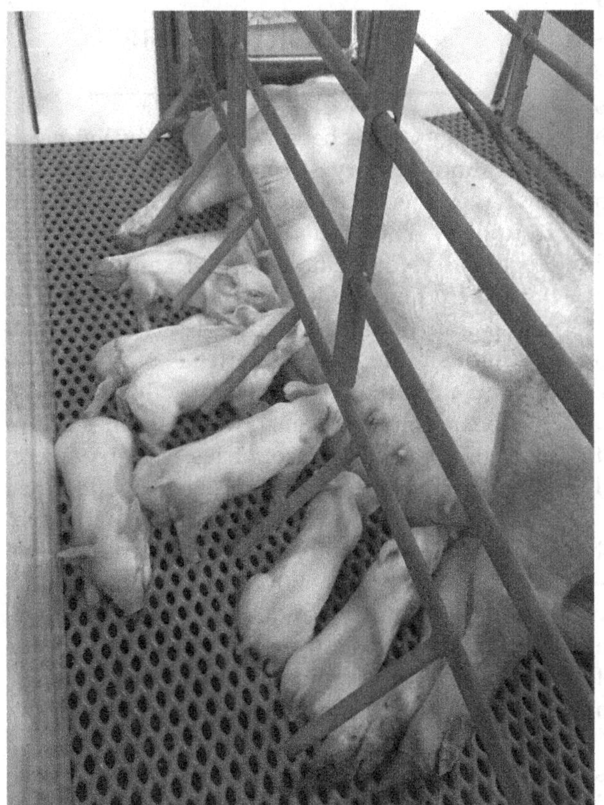

FIGURE 4.5 Gestational stalls are used to restrain sows during the time that they are nursing piglets. The stalls protect the piglets from being suffocated by the sows and allow workers to safely access and monitor the piglets. *Source: Photo provided by the National Pork Board and the Pork Checkoff, Des Moines, IA.*

and there is increased risk of disease with exposure to the elements and closer contact with manure (removed with the bedding by tractor or shoveling). Pasture-raised pigs roam in open fields and root and forage more naturally. This is a low-cost system, but pigs are exposed to the weather, predators, and higher risk of parasites and infections and take more time to manage. Generally, fewer pigs can be managed when using hoop barns or pasture raising compared to confinement housing systems, which results in higher costs for the farmer and consumer.

4.3.3 Pork Products and Consumption

The consumption of pork has remained fairly constant in the United States since 1970, around 52 pounds/capita (Fig. 4.2). Pigs provide a good source of meat, as well as a wide range of other products. Lard, fat obtained from the pig, was widely used in cooking through WWII, but has fallen out of common use. Pig fatback is the layer of subcutaneous fat under the skin on the back. Fatback may be rendered (heated) to make lard, made into salt pork, or used in sausage. Pork belly cuts are cured using salt and smoked to make bacon. Salting and smoking were means of extending the shelf-life of meat prior to refrigeration but continue to be used to make specialized meat products. Ham is made from pork legs by coating it with salt then exposing to smoke, or curing with sugar and salt. The hams are allowed to age for several months or even years during which time enzymatic and chemical changes occur in the meat giving it a distinct flavor and color. Unique processes for making ham are treasured by many cultures. Ham, lunchmeat, sausage, and bacon are the most commonly consumed types of pork in the United States, followed by fresh pork, including pork chops, ribs, and roasts.

In addition to providing human food, hogs are a source of pharmaceutical and industrial coproducts. According to the Pork Producers Association, over 20 drugs and pharmaceuticals are derived from hogs including blood compounds (albumen, plasma, fibrin), hormones (cortisone, estrogens, insulin, oxytocin, thyroxin), and enzymes (trypsin, lipase, pepsin). Heart valves obtained from hogs are suitable for replacement use in humans because of the similarity in anatomy. Currently, research is being conducted to develop strains of pigs that can provide tissues for human implants without generating an immune response that would cause rejection. Hogs are a source of many other products used in adhesives, glass, lubricants, cosmetics, crayons, insulation, upholstery, and leather goods. What is another name for a football? The pigskin, of course!

4.4 HISTORY OF THE EGG INDUSTRY

Bird eggs have been consumed by humans throughout history. Domesticated poultry or fowl were recorded in India and Asia from around 6000 BC and in Europe from 600 BC. Historical records indicate that early civilizations including the Romans and Asians had a variety of poultry breeds. It is likely that Columbus brought some varieties of poultry to the US colonies in the late 1400s. And European colonists transported varieties common in their native countries. Poultry were often carried on sea voyages to provide eggs and meat for the crew. This allowed wide dissemination of poultry to many parts of the world. The British enjoyed developing exotic breeds of poultry, and formed the Poultry Club of Great Britain in 1877 to record, show, and protect these animals. Hundreds of varieties of poultry were developed with differing size, coloring, and feather types.

In the colonial period and as America was being settled, chickens were raised on most farms, were generally allowed to roam freely, and the eggs were consumed locally. Women and children were usually responsible for gathering eggs and feeding the flocks (Fig. 4.6). Mortality was high due to diseases and predators, pecking orders caused some chickens to dominate and fight the weaker ones, and egg production was often sporadic. Larger flocks with about 80 birds were common in the 1900s as eggs became a source of farm income. During the 1920s USDA scientists developed two new varieties of chickens, the Rhode Island Red and the Single-Comb White Leghorn (Fig. 4.7), which produced large, well-shaped eggs. Leghorns are the predominant variety of laying hens being used today in the United States. Concurrently, caging systems, improved feed quality, and monitoring of breeding led to substantially higher egg production. Egg producers in California first demonstrated that caging systems improved animal health, reduced mortality, and increased egg production. The use of housing systems soon spread across the country and fostered technological advances such as conveyer belts that collected eggs as they were laid and removal of manure from below the cages.

4.4.1 Laying Hens and Egg Production

Egg production starts in the hatchery. Males mate with females and the female produces a fertile egg within 23–32 hours. Fertile eggs are collected from many hundred hens, placed in racks within a temperature controlled incubator, and held for 18 days. The eggs are then transferred to hatching baskets where on day 21 the baby chicks hatch. The chicks are separated by sex and vaccinated. The female pullets are transferred to a rearing house

FIGURE 4.6 Mrs. George W. Ferguson feeding chickens in Ida County, Iowa. This photo was most likely taken in 1934. *Source: Photo taken by A.M. Pete Wettach, used with permission from Iowa Public Television.*

FIGURE 4.7 The White Leghorn chicken breed was brought to the United States in the early 1800s. During the 1930s USDA scientists enhanced the egg-laying capacity of the breed, and today it is the standard for the egg industry. *Source: Photo provided by the American Egg Board.*

where they are fed and monitored for 18 weeks. Pullets are then moved to a layer house, provided a diet high in calcium, and exposed to 14–16 hours of light per day to stimulate egg production. The hens produce unfertilized eggs because they are not exposed to roosters. After about 60–65 weeks, the hens are rested (molted) and egg production halted. This allows the birds to rebuild their body reserves. Hens are returned to production for typically three laying cycles. At each cycle the number of eggs is reduced and quality becomes poorer.

Terms used in poultry production include:

- Hen: adult female chicken, turkey, or duck
- Rooster or cock: adult male chicken
- Pullet: female chicken that has not laid an egg
- Chick: newborn chicken
- Layer: female chicken actively laying eggs
- Rock Cornish game hen or Cornish game hen: chicken marketed for meat at less than 5 weeks of age
- Broiler or fryer: chicken marketed for meat at less than 10 weeks of age
- Roaster: chicken marketed for meat at 8–12 weeks of age
- Capon: castrated male chicken marketed for meat at less than 16 weeks of age

Housing systems for laying hens have been a topic of public debate within the past few years. To promote efficiency, and provide better access for workers to monitor animals, battery cage systems with several rows of cages each holding several hens have been used (Fig. 4.8). These systems include conveyer belts to collect the eggs and to remove manure, which reduce labor costs. Feed and water are distributed automatically and measured for each cage. Using these systems has allowed very large facilities, with many thousand hens, to operate. Recent large-scale food-borne illnesses associated with eggs that were produced in such facilities raised consumer concerns about the use of battery housing systems. Animal rights groups have also called for discontinuing the use of battery cages on the premise that hens are unable to move freely when confined in cages. This has led to increased consumer demand for eggs produced from cage-free housing. The social, political, and ethical aspects of laying hen housing is discussed in more detail in Section 4.12.

FIGURE 4.8 Battery cages for laying hens allow eggs to be captured on a conveyer belt to reduce contact with manure. Food and water are delivered to and monitored for each cage automatically. *Source: Photo provided by the American Egg Board.*

4.4.2 The Egg Industry Today

By the 1960s large-scale egg production had become profitable and flock size increased. There are about 340 million laying hens in the United States today, each producing on average 274 eggs/year, totaling over 80 billion eggs. The top egg-producing state is Iowa, which generates twice as many eggs (15 billion) as the next highest state, Ohio (8 billion). Consolidation of egg production has been rapid, with the American Egg Board estimating there were 59 egg-producing companies that had over 1 million hens and 16 companies with over 5 million hens in 2013.

Eggs produced for direct consumption are called shell eggs. In the United States most eggs are white shelled, but brown shell eggs are also produced. The difference between white and brown eggs is the breed of bird that produced the egg; there are no significant differences in nutritional quality or taste. Voluntary egg grading was established in the 1917 Farm Products Inspection Act (Table 4.1), and has continued under the auspices of the Agricultural Marketing Service (AMS). Eggs are classified by grade and weight. Grades AA, Grade A, and Grade B are defined based on the condition of the egg white, yolk, and shells. Grades AA and A are the highest quality eggs and sold in retail stores. Grade B are used to make egg products. Egg weights are based on net weight per dozen eggs:

- Jumbo = 30 ounces
- Extra large = 27 ounces
- Large = 24 ounces
- Medium = 21 ounces
- Small = 18 ounces
- Peewee = 15 ounces

A substantial proportion of eggs, especially those that don't meet high standards, are cracked, processed, pasteurized, and sold as liquid or powdered egg for the food industry. Smaller amounts are separated and sold as egg whites and yolks (liquid and dried). Egg consumption in the United States was about 380 eggs per person per year in 1950, but by 1990 intake decreased and has stayed within 250–260 (Fig. 4.9). Egg consumption, as measured by the USDA, includes eggs that are used in a wide range of food products such as breads and bakery goods and prepared meals. The decline in egg consumption was associated with recommendations in the Dietary Guidelines to avoid foods containing cholesterol because of the concern that dietary cholesterol increased the risk of cardiovascular disease. This relationship, and the change in that recommendation in the most current Dietary Guidelines, is discussed in more detail in Chapter 7, Nutrition and Food Access.

4.5 HISTORY OF THE POULTRY INDUSTRY

Chickens provide eggs, but are also a source of meat. Colonialists and early settlers considered chicken to be a special occasion food and it was often reserved for the Sunday meal or celebrations. Soon after WWII researchers in the USDA and at land-grant colleges began to explore new ways to raise and process chicken, and to develop integrated egg and meat production systems. Because chickens have a fairly short reproductive period and lifespan, experimentation results could be achieved quickly

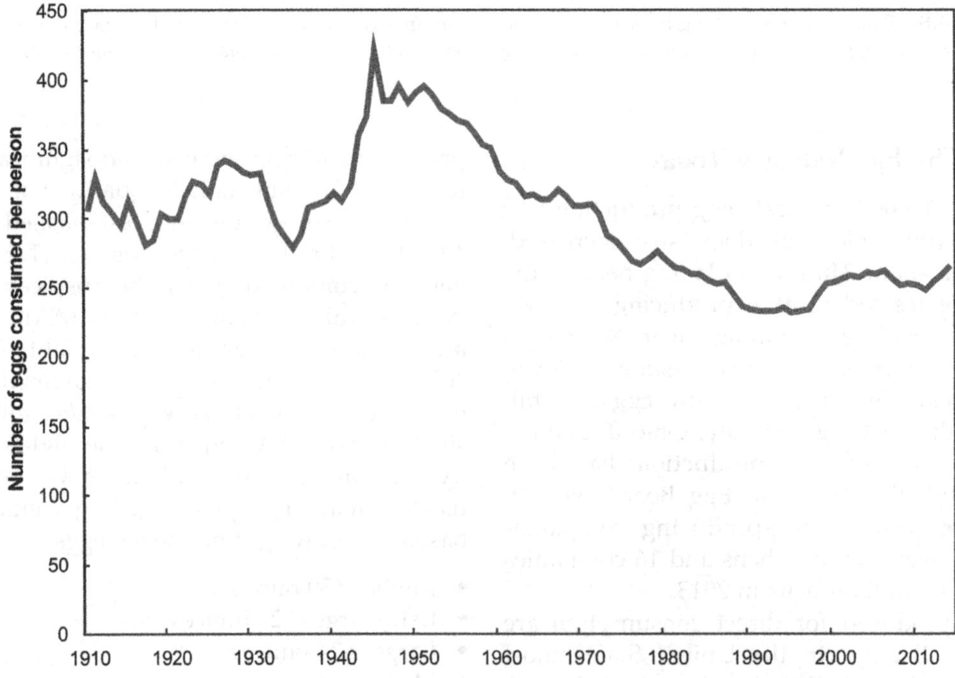

FIGURE 4.9 The consumption of eggs in the United States increased between 1940 and 1950 but decreased thereafter. Recommendations to reduce cholesterol intake in the Dietary Guidelines were likely responsible for consumers eating fewer eggs. *Source: USDA Economic Research Service, www.usda.ers.gov.*

and an understanding of chicken genetics, nutrient requirements, and reproductive traits developed rapidly. Improvements in poultry husbandry led to better quality and lower production costs. Through the Cooperative Extension Service these practices were disseminated to farmers and by the 1950s the poultry industry was thriving. Between 1935 and 1955 the average market weight of chickens increased from 2.8 to 3.1 pounds, while the time to achieve this market weight decreased from 112 to 73 days. This improved the economics of chicken production dramatically. USDA scientists also began development of a smaller, meatier turkey and by the late 1940s, 8–10 pound turkeys were commonly available in stores.

By the mid-1960s poultry production had adapted to technological innovations including using electricity to maintain temperature and ventilation of chicken housing and electric feeders to reduce labor costs. Confinement housing and brooding hatcheries allowed larger-scale production of chickens. These larger populations were at increased risk of disease and several major infections wiped out entire stocks. To address this problem, the USDA BAI was charged with development of inspection and testing protocols that were implemented by the industry during the 1950s. But perhaps the most effective, and now controversial, disease management effort was the FDA approval in 1951 of penicillin and chlortetracycline as feed additives for the poultry industry. These antibiotics were effective in reducing disease in the flocks, inexpensive and easy to mix into feed, and promoted growth, all of which enhanced profits. The use of antibiotics in poultry production ramped up rapidly and by the 1970s nearly 100% of poultry raised in the United States received antibiotics in their feed. Concerns about the potential development of antibiotic-resistant bacteria and the impact on human health have made the industry reevaluate this practice, and pressure from consumers has led major food buyers such as McDonald's to demand elimination of antibiotic use by their suppliers. The development of antibiotic-resistant bacteria is discussed in more detail in Section 4.10.1.

4.5.1 The Poultry Industry Today

Advances in genetic science were being applied to the poultry industry during the past 50 years with a major focus on increasing breast meat yield. Poultry breeding became more concentrated in a few companies that controlled most of the stock available for the entire industry. This intensive breeding approach was very successful and breast meat yield, as well as other enhanced production characteristics, were achieved. An unintended consequence of this rapid selection was that, in some breeds, animals exhibit metabolic and reproductive problems and reduced ability to resist disease. Finding a balance between growth rate, breast meat size, and animal health is now a major focus for the industry.

Since the mid-1980s more pounds of chicken were consumed per capita than either beef or pork (Fig. 4.2). During the past decade, the southeastern United States became the dominant location for broiler chicken production and has been designated the "Broiler Belt." Georgia, Alabama, and Arkansas are the top broiler-producing states. The United States is the largest producer of poultry meat (about 40 billion pounds in 2010) and exports about 18% of production ($3 billion).

4.5.2 Poultry Production Regulations

Federal oversight of the poultry and egg industry developed slowly in the United States. The 1906 FMIA law, described previously for beef, did not include the inspection of poultry or eggs because at the time most poultry was purchased directly from farmers and processing facilities were not common. An outbreak of avian influenza in New York City in 1920 caused some local governments to implement their own inspections of poultry farms but no federal oversight was required. At this time, a few states were using limited regulations to monitor eggs based on their appearance to prevent sales of cracked or damaged eggs. During WWII, the US military required eggs and poultry to be purchased only from processors that were USDA inspected. After the war this standard led to passing the Poultry Products Inspection Act in 1957, which gave USDA the authority to monitor and inspect poultry but did not specifically identify egg regulations (Table 4.1). It wasn't until 1970 that Congress passed the EPIA (as an amendment to the Poultry Products Inspection Act). The EPIA created two distinct categories of eggs: shell eggs and egg products. Under the EPIA, the Poultry Division of the USDA's AMS was given the authority to inspect egg products for safety. FDA retained responsibility for eggs for human consumption, which was largely interpreted as shell egg safety. The wording of the EPIA was ambiguous as to the strict reporting and oversight authority for egg producers and regulation of the egg industry became complicated and confusing. The rapid growth in egg-producing facilities during the 1990s and the apparent lack of systematic inspection raised concerns about food-borne illness associated with egg products. In 1995, an attempt was made to clarify responsibilities for regulating the egg industry. USDA-FSIS was given authority to inspect egg products and egg-producing facilities, and FDA retained authority for the inspection and safety of shell eggs, egg substitutes, and imitation eggs. These government agencies function differently, in that USDA-FSIS conducts continuous daily inspections of food processing facilities for which they have regulatory authority (meat, poultry, and egg products), whereas the FDA tends to inspect facilities randomly or when a safety concern is raised. Two additional USDA agencies have authority for other aspects of the egg industry. The AMS conducts grading of eggs and provides audits of egg-laying barns to determine if industry standards are met. The Animal and Plant Health Inspection Service (APHIS) provides voluntary testing for bacterial contamination in hens and eggs. This complex regulatory environment, with no clear reporting lines, led to lax oversight of the egg industry.

A primary bacterial pathogen that occurs in laying hens is *Salmonella enteritidis* (SE). SE can cause serious and possibly fatal infections in very young children, the elderly, or persons with compromised immunity. In healthy people, SE causes diarrhea, fever, and abdominal pain. The bacteria spreads from the hen to eggs via fecal contamination so cleaning the shells was considered to be an adequate mitigation protocol. SE can also be transferred internally from the hen directly into the eggs. With the bacteria inside the eggs, decontaminating the shells is ineffective and SE detection requires continual monitoring of the hens for infection. In 2010 a major outbreak of SE from contaminated eggs occurred in the United States with over 1900 illness in 11 states. This led to a massive recall of over 500 million shell eggs nationwide. In reviewing the causes of this contamination, the USDA and FDA discovered that the regulatory process to oversee shell egg safety was dysfunctional. To address the problem, the USDA, including FSIS, AMS, and APHIS, and the FDA conducted an audit of the contamination outbreak, including the actions taken (or not taken) by each agency. From this assessment, plans were developed to better define and integrate the responsibilities of each agency. The resulting agreement made FSIS the lead food safety agency with the responsibility to coordinate with FDA, AMS, and APHIS to ensure the safety of the entire egg production process. This new structure to oversee the egg industry went into effect in 2012.

4.6 THE HISTORY OF THE DAIRY INDUSTRY

Milk was likely being consumed and "processed" by humans as early as 9000 years ago based on archeological evidence of milk proteins found in pottery fragments, historical texts, and cave drawings. Processing of milk included fermentation with bacteria, and coagulation by enzymes to produce cheeses, even though the chemistry of these foods was not understood. Fermentation, in which bacteria multiply in the food product and generate by-products such as acids, alcohol, enzymes, and flavor compounds, produces changes in the original food. Milk fermentation includes coagulation of milk proteins (solids) and separation of the whey (liquid), as well as flavor and color changes. Many ancient cultures have some form of traditional fermented milk products such as dahi, kefir, yogurt, and cheese. Depending on the type of milk (cow, sheep, goat, or horse) and the type and mixture of fermenting bacteria, a wide range of dairy foods with many flavors and textures can be produced.

An historical, but unproven legend to explain the origins of cheese making suggests that an Arabian merchant put sheep or goat milk into a pouch he had fashioned from a young sheep or goat stomach to carry on his journey. The pouch contained an enzyme (called rennet) naturally present in the stomach of young ruminant animals to help them digest milk. On the journey the milk was warmed by the sun and exposed to the rennet, leading to coagulation of the milk proteins. When the merchant opened the pouch he found the milk had become solid. Rennet, which is recognized today as the enzyme chymotrypsinogen (or chymosin), is found in the stomach of young ruminant animals and causes the coagulation of milk proteins. Most of the rennet used in cheese production today is produced using recombinant technology (genetic engineering involving bacterial production) and not collected from animals.

The importance of dairy cattle to human nutrition is prominent throughout history. Cow's milk contains about 88% water, 3.4% protein, 4.8% carbohydrate, and 3.5% fat. The proteins, mainly caseins, are of high quality, meaning they provide essential amino acids needed by humans. Milk provides adequate amounts of eight essential vitamins and minerals (especially calcium and potassium needed for bone development). Milk can be obtained fresh daily with access to a cow, sheep, or goat (or water buffalo, camels, or horses in some parts of the world) and therefore is a safe and consistent source of nutrition. Ruminant animals can thrive on low-nutrient grasses and forages that are not useful for human nutrition and therefore do not compete for resources. Domestication of these milk-producing animals was therefore an advantage for human development. Historical records and genetic mapping support a wide distribution of animal herding practices coinciding with human civilizations.

Dairy cattle were not native to the United States and are thought to have been brought to the continent with the earliest explorers beginning with the Spanish Conquistadors. The Europeans are credited with fostering dairy farming in the colonies in the 1600s, and by the 1700s cattle were plentiful along the East Coast. Traditions of dairy farming were well established by the Dutch, Scandinavian, German, and Scottish immigrants who brought these skills to the United States.

Consumption of milk and production of butter and cheese for home use was the main type of dairy farming through the mid-1800s, but enough excess was produced that external butter and cheese markets were needed. A complex web of exporting and importing regulations between the colonies and Europe, which eventually led to the Revolutionary War, kept butter and cheese from being traded with England. But the Caribbean Islands and British West Indies were open markets and dairy food exports became a source of income for the colonists. Butter and cheese were also traded locally and some farmers became well known for their quality products. By the mid-1800s New England, New York, Pennsylvania, and New Jersey were producing the majority of dairy products. The distribution of these products was facilitated by the opening of canals that connected the Great Lakes (Eric Canal in 1825) and the Midwest (Ohio Canal in 1832) to meet the demand from the growing city populations. But within 50 years, the north central regions of the United States including Wisconsin, Iowa, and Minnesota had overtaken the eastern states and became the dominant dairying states. This change was in part facilitated by the railroad system granting a fast and economical way to deliver product, but also because of the ability to grow corn in the Midwest. Having a local source of corn made hog production economical, and the skim milk leftover from butter production was a good food source for hogs. This diversification of farming, allowing the flexibility to switch between corn, hogs, butter, and cheese production depending on the markets and weather proved to be a very successful model.

4.6.1 Butter and Cheese

Butter is produced when cream or unhomogenized whole-fat milk is churned so that the fat particles (butterfat) separate from the liquid (buttermilk). By law, butter must contain 80% butterfat. There is actually a fair amount of water in butter (about 16% by weight) and a trace amount of protein and lactose. Lactose is a type of sugar (disaccharide) and is one reason for the term "sweet cream" butter. Cultured butter is made from cream that has been fermented by lactic acid bacteria (cultured), which metabolizes the lactose and releases lactic acid. Clarified butter is made by heating and then cooling to drive off the water and separating the proteins, leaving the pure butterfat. Butter has been made and consumed throughout history because of its ability to be transported and stored. In the United States, most families made their own butter using wooden churns through the 1900s. Some of the butter was consumed by the family and some used for trade.

By the early 1830s dairy farmers were using cooperatives to pool their cream and milk to ensure adequate butter and cheese production. Eventually, this led to the development of factories to make these products and specialization of the industry into producers and processors. Cheese factories were not immediately embraced by everyone because there were many hurdles. Delivery of milk or cream to the factory was challenging as refrigeration was not yet available, how farmers were paid for their milk was not uniform, and there were those that feared loss of the art and individualism of butter and cheese making. These latter arguments are heard today as there is a desire to return to artisan types of dairy foods. But cheese factories were financially successful and led to advances in technology for production that enhanced the quality and quantity of product. By 1900 factories were producing almost all of the cheese consumed in the United States. Factory-produced butter trailed by about 30 years, however, mainly because butter as a more perishable product was not as easy to store and ship, and there was insufficient technology to assist in large-scale separation of the cream. The invention of the mechanical centrifuge separator and the Babcock device (named for the inventor Dr. Stephen M. Babcock, a professor at the University of Wisconsin) for measuring butterfat changed that dynamic and led to development of creameries where butter and skim milk cheeses were produced concurrently (Fig. 4.10).

FIGURE 4.10 The University of Wisconsin offered courses in dairy science from the early 1900s. In this photo, dairy students are working with cream separators at the University of Wisconsin Dairy School sometime between 1900 and 1910. *Source: Photo provided by the University of Wisconsin-Madison Archives.*

EXPANSION BOX 4.1

HISTORY OF MARGARINE

Animal fats, including butter, tallow (beef fat), and lard (pig fat), are comprised predominately of saturated fats whereas vegetable oils contain polyunsaturated fats (Fig. 4.11). For this reason, vegetable oils are liquid and animal fats are solid at room temperature. In the late 1860s, in response to a competition offered by the French government to create a substitute for butter to address shortages, the chemist Hippolyte Mège-Mouriès invented a product by mixing beef fat with vegetable oils, which he called oleomargarine. And in the United States Henry Bradley in New York filed a patent in 1871 to create margarine from animal fat and vegetable oils. Margarine became popular due to its lower cost, and thereby created a problem for the dairy industry, which became concerned about loss of profits and competition. Following years of lobbying and state-based regulations, the first act of Congress specifically related to a food, the Federal Margarine Act of 1886, was passed. The act imposed a tax of 2¢ per pound on margarine and enforced an annual license fee on margarine producers. This had a devastating effect on the margarine industry and many went bankrupt. But a small loophole in the law was discovered

EXPANSION BOX 4.1 (*cont'd*)

that saved the industry. In 1894 the Supreme Court ruled that uncolored margarine could be easily distinguished from butter and therefore would not confuse consumers. The court allowed the margarine to be sold uncolored, and the manufacturers included packets of yellow coloring that consumers mixed in at home. The dairy industry reacted by adding the natural color annatto to make its butter even more yellow and pushed for state laws that required margarine to be dyed pink, black, or red.

Another invention in the 1890s, by the American chemist James Boyce, led to the process of hydrogenation, in which hydrogen is mixed with oils in the presence of a nickel catalyst to create more solid or "hydrogenated vegetable oils." This process became the dominant way margarine was produced during the Great Depression when animal fat was in short supply. Acceptance of margarine by consumers grew during this time and through WWII. By the late 1940s as production of corn and soybean oil was increasing and available for the margarine market, lobbying efforts began to free margarine from the restrictions of the 1886 act. In 1950 President Harry Truman signed the new Margarine Act into law and by 1955 every state except Minnesota and Wisconsin had repealed their antimargarine laws. These states eventually also relented, but not for nearly 10 more years—Minnesota in 1963 and Wisconsin in 1967. Since the 1990s consumption of margarine has decreased as consumers began avoiding fat in their diet and health concerns about *trans* fat become known (*trans* fats are discussed in Chapter 7: Nutrition and Food Access).

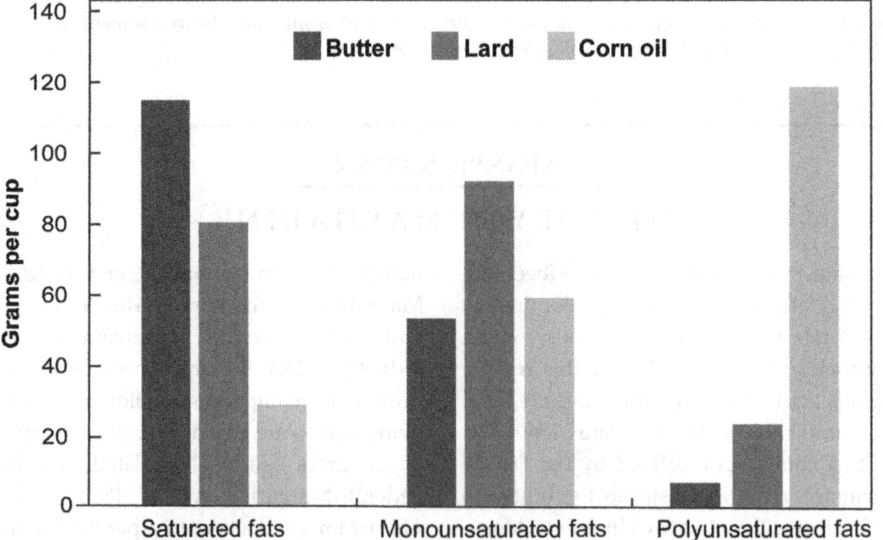

FIGURE 4.11 The amount of saturated fatty acids is higher in animal sources of fat, such as butter and lard, compared to plant sources of fat, such as corn oil. Plant oils are high in polyunsaturated fatty acids. This difference in fatty acid composition makes animal fats solid and plant fats liquid at room temperature. *Source: USDA Food Composition Database, www.ndb.nal.usda.gov.*

4.6.2 Sanitation and Safety of Milk

Fluid milk was a much more difficult commodity to market than either butter or cheese. In the years before pasteurization and refrigeration, milk could be held for only a few days before spoiling. Up until the 1800s dairy farmers were close enough to populations to provide fresh milk daily, which reduced the potential for spoilage. But as cities expanded, such as New York and Boston, farmland was taken over for other industries and housing. Widespread milk adulteration was common during this period, mainly by adding water or taking out fat to use in other products, to increase profits. Milk was carried in open bulk containers, delivered door to door where it was poured into containers with no concern for sanitation.

During this time, distilleries began operating in the cities to make alcoholic beverages. The by-product of this process was a low-quality grain slop. Distillery owners discovered that they could increase their profits by feeding cows on the slop and selling the milk. This feed source was poor quality for the cows, and their milk, called "swill milk" was of very low nutritional quality. Poor families living in the cities, who could not afford fresh farm milk, turned to this source of milk instead. To make matters worse, because swill milk was often not as thick and white in color as farm milk, producers added starch, flour, chalk, eggs, and colorants such as annatto to make the milk look better. The sanitary conditions of these city dairies were very poor and diseased cows were common, posing additional health risks for consumers of their milk. Outbreaks of illness from contaminated milk including cholera, tuberculosis, and scarlet fever were common, especially in children. Regulations for milk quality and sanitation were largely only within states or cities (Table 4.2). These issues were brought to the public's attention in a book called *An historical, scientific and practical essay on milk as an article of human sustenance* by Robert Milham Hartley (1842). Hartley was a staunch prohibitionist with a primary intent to close down the distilleries. But his description of the deplorable conditions of the distillery cows and the dangerously contaminated swill milk eventually led the New York Academy of Medicine to investigate the situation. They concluded that there was sufficient concern about the safety of swill milk for children. The City of New York eventually passed an ordinance in 1862 to discourage the practice (Table 4.2) but it continued until railroads became available to allow milk to be delivered from the countryside into the cities in sufficient quantity and at low cost to drive the distillery dairies out of business.

A lack of understanding of sanitary practices on the farm, transporting milk in open containers, and lack of refrigeration increased the potential for milk contamination. Milton Rosenau was an early pioneer in public health and a promoter of milk pasteurization. Louis Pasteur, a French chemist, had determined that heat treatment prevented spoilage of wines around 1862 and his pasteurization process was widely used in Europe. The practice of heat treating milk was being done in the United States in the early 1900s, but the resulting cooked flavor made the product unpopular. Rosenau developed a low-temperature process (140°F for 20 minutes) that was effective in preserving milk without the negative effects on flavor. Advocates of pasteurized milk were numerous in the late 1800s as an understanding of bacteriology and disease transmission was evolving. But misplaced fears that heated milk was unsafe for infants kept pasteurized milk from being widely accepted. Infant mortality rates were very high during this time, mainly from infectious diseases including those spread through milk.

Pasteurizing milk to reduce infant mortality became a life calling for Nathan Straus, who was the head of R.H. Macy's department store

TABLE 4.2 Historical Legislation to Promote Milk Safety

Year	Level of oversight	Legislation
1856	Massachusetts Act	Prohibited milk adulteration
1859	City of Boston regulation	Prevented distillery slop for feeding cattle
1862	City of New York	Prevented distillery slop for feeding cattle
1871	City of Washington, DC	Prohibited milk adulteration
1879	State of Illinois	Prohibited milk adulteration
1882	Massachusetts	Created Foods and Drug Act
1895	Minnesota	Created dairy inspection law
1908	City of Chicago	Created an ordinance requiring pasteurization of milk
1906	Federal, law	First Food and Drugs Act
1906	Federal, milk standard commission	Created Milk Grade System: A, B, C
1924	Federal, US Public Health Service	Defined Standard Milk Ordinance
1927	Federal, FDA and US Public Health Service	Created Grade "A" Pasteurization Milk Ordinance

in New York. He opened infant milk depots in 1893, which consisted of a milk pasteurizing and bottling plant with a tent pavilion. Mothers would bring their children to the depots to listen to lectures about child care and feeding given by physicians, and purchase pasteurized milk. These depots became popular and had a positive impact on reducing infant mortality in New York. This gave Straus the motivation to invite mayors of other cities to open similar depots to reduce infant deaths and illness and by the early 1900s there were milk depots operating in most of the big cities along the East Coast, as well as in Cleveland, Chicago, and St. Louis.

Pasteurization was found to be effective in making milk safe, but advances in technology were needed before it could be implemented on the commercial scale. Land-grant colleges, mainly Cornell, Wisconsin, and Michigan State, began offering courses in dairy production soon after they opened. The first dairy program was started by the University of Wisconsin in 1891, and with funding from the Hatch Act research on dairy products began in earnest. As was the land-grant mission, the research was done in close collaboration with the dairy farmers and industry. During the early 1900s USDA scientists also began to collect data on dairy herd milk production and developed management plans to improve breeding quality. This led to a significant increase in yearly milk production, making milk less expensive and more available. Technologies to evaluate dairy composition, bacterial load, flavors, and color were developed. Systems to rapidly heat and cool milk, allowing pasteurization without producing cooked flavor, and mechanical bottling lines were created. Glass milk bottles, first produced by a beer bottle manufacturer, were introduced in 1879 and opened a new era in milk delivery, processing, and sanitation. Resistance to mandatory pasteurization was slowly being overcome, but the 1906 Pure Food and Drug Act created a legal argument for pasteurization. Because bacteria were considered contaminants, milk that had more than 50 million bacteria per milliliter could be prevented from being sold for infants and children. A national standard of milk quality was developed by the Milk Standard Commission under the direction of the USDA. The classification system for milk defined in the 1906 National Standard Grades for Milk included:

- Grade "A" milk: for infants and children, pasteurized to have less than 50,000 bacteria/mL
- Grade "B" milk: for adults and cooking, may not be pasteurized but bacteria count could not be excessive
- Grade "C" milk: for cooking only with no bacterial count requirement

By 1920 most large cities required pasteurization of milk and used the grading system, but rural communities lagged behind and contaminated milk continued to be a problem for another decade. The United States Public Health Service (USPHS) developed the Standard Milk Ordinance in 1924 to provide guidance for milk producers to ensure safety and quality. This was followed in 1927 with the Grade "A" Pasteurized Milk Ordinance to be regulated by the FDA and USPHS. Today, the regulations for Grade "A" milk have become much more stringent, making dairy one of the most highly regulated foods. Grade "A" milk must be used for fluid milk and all soft manufactured dairy products (including sour cream, yogurt, cream cheese, cottage cheese, and ice cream). Milk not meeting Grade "A" standards, even in a single dimension (including presence of bacteria or animal cells above the defined thresholds) is termed Grade "B" or manufacturing grade. Grade "B" milk can only be used for hard cheeses, butter, and milk powder. Grade "C" milk is no longer used.

By the 1930s milk bottling plants, tanker trucks, and refrigerated railroad cars had entered the dairy industry. But home delivery, usually daily, was still the way most consumers obtained their milk and dairy products. Customers returned their empty glass bottles to the dairy where they were washed and reused. This practice continued through the mid-1950s until grocery stores became more common. Paper cartons for milk were developed in the 1930s but were not widely used until after WWII, and plastic jugs came into use in the late 1960s.

4.6.3 The Dairy Industry Today

Most of the milk produced in the United States today comes from large dairies with many hundred cows. In 2014 there were 440 fluid milk processing plants that produced on average 115 million pounds of milk per year. Automated milking systems and other innovations allowed the size of dairy farms to increase by reducing manual labor. Milk collected from the dairy is transported by truck to a processing facility where each load is tested for contaminants. The FDA Pasteurized Milk Ordinance for Grade "A" Milk defines the required sanitation, handling, and testing of all milk that is sold. The presence of excessive bacterial loads or antibiotic, or chemical residues would require the milk to be discarded. At the processing plant, the milk undergoes a standardization process in which the cream is separated from the skim milk by centrifugation. The milk and cream are pasteurized according to defined time and temperature standards (e.g., 161°F for 15 seconds). Ultrahigh temperature (UHT) pasteurization, which involves heating milk to 270°F for 2 seconds and then packaging in a sterilized container, creates a product that does not need refrigeration. This type of processing is widely used in Canada, South America, Europe, and Asia, and occasionally in the United States for special products. The cream is then added back to the skim milk using a process that reduces the size of the fat particles (homogenization) so that it does not separate and rise to the top of the container. Milk is then packaged into containers and shipped to the market. Four milk products of differing fat content are commonly available:

- Whole milk = 7.93 g fat per cup (3.5% milkfat)
- 2% milk = 4.83 g fat per cup (2% milkfat)
- 1% milk = 2.37 g fat per cup (1% milkfat)
- Skim or nonfat milk = 0.20 g fat per cup (no cream is added)

Since the 1970s, consumers have decreased their overall intake of fluid milk but increased cheese and yogurt (Fig. 4.12). Consumption of whole milk has declined and skim milk increased, perhaps because of recommendations in the Dietary Guidelines to reduce saturated fat intake. Ironically, whole milk does not meet the standard for a high-fat food. A serving of whole milk provides about 8 g of fat which, is 12% of the Daily Value (65 g) for fat. To meet the definition of "high" the serving would need to provide 20% or more of the Daily Value.

The USDA found that people are less likely to consume milk with lunch and dinner today than they were 60 years ago. The consumption of milk by children has decreased from about 1.7 cups/day in 1977 to 1.2 cups/day in 2007. The recommended intake is 2 cups/day for children 2–3 years, 2.5 cups for children 4–8 years, and 3 cups for children older than 8. Intakes of milk below the recommended levels is a concern for growing children because milk is an excellent source of both calcium and vitamin D, which support bone development. Further discussion of food sources of nutrients can be found in Chapter 7, Nutrition and Food Access.

Milk, cheese, and butter are the primary dairy foods consumed in the United States.

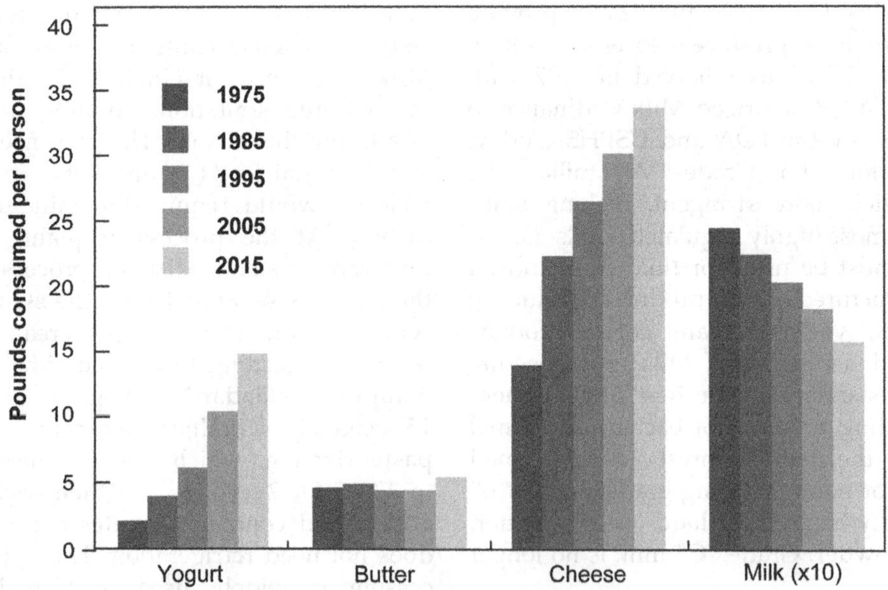

FIGURE 4.12 Americans decreased their consumption of fluid milk between 1975 and 2015 and increased their consumption of yogurt and cheese. Butter consumption decreased slightly after 1985, and then increased in 2015. *Source: USDA Economic Research Service, www.usda.ers.gov.*

Cottage cheese was introduced as a good protein source during WWII because it was a cheese product that did not require aging. After the war it retained its popularity and became a household staple. There have been many other innovations in the dairy industry including novelty ice creams, flavored milks, many types of cheeses, Greek yogurt, and fermented dairy beverages. The by-product of cheese production, whey, is also widely used as a food ingredient to provide nutritional and functional value.

Yogurt and cheese consumption have both increased, reflecting changes in eating habits and preferences. Prior to 1955 yogurt was rarely consumed by Americans. Yogurt gained popularity as a health food based on the long history of use in other countries such as India and Switzerland. Yogurt is made by fermenting milk with two or more types of bacteria (usually *Lactobacillus* and *Streptococcus*). The bacteria metabolize some of the lactose in the milk and produce lactic acid. The lower pH generated by the acid causes the milk proteins to coagulate, forming a creamy product. By using different strains of bacteria, and different setting protocols, a range of yogurts with different flavors and textures are possible. Greek yogurt, which has become very popular, is made by removing more of the whey (liquid) to concentrate the proteins. The introduction of flavorings, fruits, and sweeteners and marketing the "yogurt sundae" have increased yogurt acceptance by children and adults. Cheese has always been part of the American diet, but the growth in cheese consumption is mainly linked with the popularity of pizza and cheeseburgers. In 2014 the consumption of mozzarella cheese was 14.18 pounds and cheddar cheese was 13.47 pounds per person per year, which are each almost 10-fold greater than any other type of cheese.

4.7 ROLE OF ANIMAL FOODS IN HUMAN HEALTH

Animal foods provide high nutrient density (the ratio of nutrients to calories) to the human diet and contribute some essential nutrients that are difficult to obtain from plant foods. These include amino acids, calcium, iron, choline, and trace minerals. An omnivorous diet, which includes all types of foods, is typical for most Americans. Lacto-ovo vegetarians consume dairy and eggs but not meat, pescetarians consume fish and seafood but not meat, and vegetarians or vegans consume no animal foods. Each of these dietary patterns can support a healthy lifestyle if attention is paid to balancing nutrient intake with requirements at each stage of the lifecycle.

Animal foods have been criticized over the years as contributors to chronic diseases. Dairy products, eggs, and meats, particularly red meat, are the main sources of cholesterol and saturated fat in the human diet. During the early 1950s researchers began publishing studies that related dietary fat intake to increased risk of heart disease. During the 1950s it was recognized that cholesterol was present in atherosclerotic plaque material. Researchers were also finding that experimental animals fed high cholesterol and fat diets developed cardiovascular lesions. Although not universally accepted, many scientists and health professionals thought there was sufficient evidence to caution the public about this association between fat, cholesterol and cardiovascular disease. The first Dietary Guidelines, published in 1980, proposed that total dietary fat and a higher ratio of saturated to unsaturated fat may be factors in the pathogenesis of atherosclerosis. Every version of the Dietary Guidelines since then has included guidance on reducing fat, saturated fat, and cholesterol. The message that animal foods, because of their cholesterol and saturated fat content, would increase the risk of heart disease became ingrained in public thinking. In response, consumers largely reduced red meat, as well as whole milk, butter, and egg consumption over the next two decades, switching to more chicken, skim milk, and margarine. With further study, researchers found that the cholesterol from foods was only a minor contributor to cholesterol levels in the body in most people, and that high blood cholesterol was more commonly due to endogenous overproduction or improper clearance. Cholesterol-lowering drugs became widely available to treat patients with metabolic derangements in cholesterol. Simultaneously, in response to the concerns about fat in red meat, the beef and pork industry focused on raising leaner animals with less marbling. These selective breeding approaches resulted in some beef and pork cuts in grocery stores that meet the USDA recommendations for lean meat relative to calories, total fat, saturated fat, and cholesterol, and in fact many have fat profiles similar to chicken (Fig. 4.13). This lean beef and pork provide a healthy balance of protein and are a good source of iron and other minerals.

Science continues to evolve regarding the role of dietary factors and heart disease. Recent studies have suggested that the types of saturated fats in whole milk may actually be protective of heart disease. And it has been suggested that consuming butter and whole-fat dairy foods, including cheese, may be beneficial to overall health. While diet remains an important factor in reducing risk of heart disease, the recommendations have moved toward maintaining a balance in the types of fats (e.g., saturated and unsaturated, omega-6, and omega-3), and reducing overall intake of fat and simple carbohydrates, especially sugar. Lifestyle factors such as lack of physical activity, smoking, and obesity are major contributors to heart disease risk.

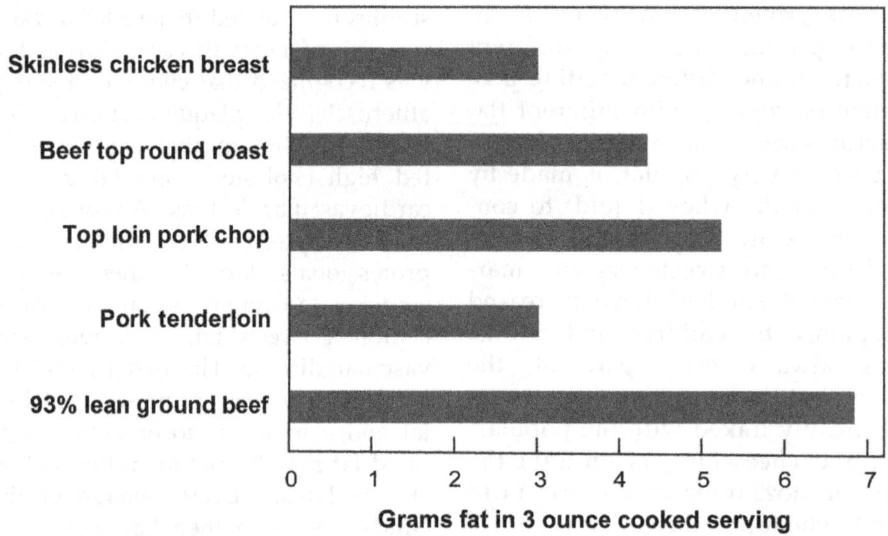

FIGURE 4.13 Selected cuts of beef and pork have similar fat content as skinless chicken breast. Selective breeding and advances in animal feeding practices have been effective in producing leaner beef cattle and hogs to meet consumer demand for lower fat foods. *Source: USDA Food Composition Database, www.ndb.nal.usda.gov.*

There has been strong debate about the role of red meat in colon cancer. Early studies in animals and correlations in humans linked western diets, i.e., those high in red meat and lower in plant foods, with higher risks of colon cancer. Many factors have been explored to find a mechanism by which meat could influence colon cancer risk. Some evidence suggests that the fat content of meat leads to increased amounts of bile acids in the colon, some of which could be converted into carcinogens. Similarly, the types of fat in meat may create a more inflammatory environment, which may be causative of cancer. The microorganisms inhabiting the large intestine may be influenced by dietary intake, with higher numbers of microbes that utilize sulfates present in the colon of people with high meat intakes. These types of microorganisms may alter the mucosal environment of the colon making it more susceptible to cancer development. Other factors, such as less fiber and fewer protective compounds found in fruits and vegetables present in diets containing more meat (and assumedly less plant foods) may be involved.

In 2015, the World Health Organization's International Agency for Research on Cancer (WHO-IARC) went so far as to state that processed meats were carcinogenic to humans and that red meat was probably carcinogenic. Processed meats are defined as foods that have been salted, cured, smoked, fermented, or otherwise processed to modify flavor or for preservation. The WHO-IARC report was not based on any new evidence, and generated much controversy. The scientific community is not in agreement about the connection between red meat and colon cancer and several statistical analyses found the association between red meat and colon cancer risk to be weak and dependent on many variables. The reasons for this lack of consensus lie in the complex nature of assessing the role of a dietary component in a complex and chronic disease such as cancer. There are likely many factors present in processed meats (nitrates that generate N-nitroso-compounds, polycyclic aromatic hydrocarbons (PAHs), and even salt) that have the potential to be carcinogenic. But linking them directly to human colon cancer risk is complicated by the many other both protective and promotive components of the diet, as well as individual genetic and metabolic characteristics.

Cooking methods influence the healthfulness of meat. Frying or cooking at very high temperatures (such as when grilling) or cooking over wood or charcoal (which generate smoke) produces potentially harmful chemicals in the meat (PAHs and heterocyclic amines) whereas slower cooking methods such as roasting or baking do not. It has been difficult to include quantitative assessments of how meat has been cooked, especially over a lifetime of meat consumption, when attempting to correlate meat intake with cancer risk in populations. In the United States, colon cancer remains a very common form of cancer, but according to the CDC the overall incidence rate has declined over the past decade. Higher body fat, a sedentary lifestyle, and other risk factors (smoking, high alcohol intake) are other important contributors to colon cancer risk in addition to diet.

Strong debate continues regarding the role red meat should play in a healthy diet. Groups promoting vegetarianism have started the "Meatless Mondays" movement to encourage reduced consumption of meat. The reasoning for this movement is that eating less meat is more healthful for people and also reduces impacts on the environment. An exploration of the ethical perspectives associated with Meatless Mondays is presented in Expansion Box 4.2.

EXPANSION BOX 4.2

MEATLESS MONDAY CAMPAIGNS

During WWI, there were campaigns for Meatless Tuesday and Wheatless Wednesday to remind US citizens to reduce their consumption of foods in limited supply and to conserve food for the war effort. Meatless days were also encouraged during WWII when meat, sugar, and other foods were rationed. These campaigns were effective in bringing US citizens together and sharing sacrifices for the war effort. During the 1960s when new nutrition research linked certain foods with diseases, such as red meat and dietary fat with heart disease, public campaigns to reduce the intake of these foods were common. A new approach in public campaigning to influence food consumption was launched in 2003. Sid Lerner, an advertising agent, in collaboration with faculty at Johns Hopkins Bloomberg School of Public Health's Center for a Livable Future, created the Meatless Monday campaign. The campaign was initially part of a Healthy Monday initiative to encourage people to give up bad habits from the weekend and start healthier habits at the beginning of the week. The Meatless Monday component of that initiative grabbed a great deal of attention. The platform of the Meatless Monday campaign is that Americans consume too much meat and not eating meat one day a week will improve health. Reducing the impact of meat production on the environment also became part of the platform.

The Meatless Monday campaign gained substantial support from celebrities and is now a global movement. The Meatless Monday website (www.meatlessmonday.com) includes articles and promotional material to encourage groups to create Meatless Monday movements in their communities, schools, and workplaces. Recipes, diet ideas, and suggestions for meatless meals are presented in blogs, magazines, websites, and newspapers by chefs, journalists, nutritionists, and celebrities. The journalist Michael Pollan stated on *The Oprah Winfrey Show* in 2009 that if everybody in America participated in a Meatless Monday, it would have the equivalent effect on the environment of taking 20 million midsize sedans off the road. This statistic is difficult to verify but is easy to remember and repeat. Paul McCartney, a vegetarian, and his daughters started a Meat Free Monday nonprofit organization with the aim of "...raising awareness of the detrimental environmental impact of eating meat, and to encourage people to help slow climate change, preserve precious natural resources and improve their health by having at least one meat free day each week" (www.meatfreemondays.com).

The meat industry and some nutrition professionals question the strength of the scientific evidence that participating in Meatless Mondays has any impact on improving health. The Meatless Monday campaign combines all types of meat together and does not distinguish red meat from poultry. The majority of medical research relating meat intake to health has considered red meats (mainly beef and pork) to be most associated with chronic disease (discussed in Chapter 7: Nutrition and Food Access). Red meat production is most associated with negative impacts on the environment. Perhaps in response to the recommendations from the Dietary Guidelines, and promotion of reduced saturated fat and cholesterol intakes, Americans have been decreasing their intake of red meat since the 1960s (Fig. 4.2). In contrast, poultry intake increased significantly in that timeframe. The meat industry contends that Americans consume meat in the proper quantity relative to

EXPANSION BOX 4.2 (*cont'd*)

their nutritional needs and that meat contains beneficial nutrients, such as iron and zinc that are difficult to obtain from other food sources. In comparison, Americans underconsume vegetables and overconsume added fats and sugars. The North American Meat Institute opposes Meatless Monday campaigns in public schools because they claim meat is a good source of nutrition and children should have a choice of the types of food to consume every day. They feel that meat is portrayed negatively by the campaign, and promotion of vegetables and grains would be a better strategy for healthful eating.

The production and consumption of meat raises a wide range of ethical, cultural, social, and environmental issues, some of which are discussed in this chapter. From an ethical perspective, utilitarian thinking would evaluate Meatless Monday based on the behaviors that produce the most benefits or limit the most harm. The personal benefits of reducing meat consumption, and increasing vegetable consumption, may be a lower calorie diet and increased bioactive compounds from eating more vegetables. A person might save money by substituting beans for beef, pork, or poultry. Some farmers may have higher income from increases in vegetable and bean sales. On the detrimental side of the equation some people may not obtain sufficient protein, iron, or calories if they do not consume meat and people involved in meat production, processing, and sales would have a reduced income.

Virtue ethicists state that individuals will know what food choices are ethical because of the norms, practices, traditions, and institutions valued by a community. Vices such as overeating, unhealthful diets, and environmental damage are avoided if the culture does not support them. Some city governments have adopted Meatless Mondays in an attempt to improve the health of the citizens (for their own good) and the environment (for everyone's good) but this action may be considered paternalistic and in violation of autonomy and rights. The rights of people to have access to food and be well-nourished are not the same as the right to choose the specific foods they eat. This is an important consideration for debate as US consumers become more forceful in demanding changes to the food system. Implementation of any type of group or community food plan should offer some choice (including vegetarian, meat, gluten-free, dairy-free), or at the very least provide the option to participate or not. School lunch options that are not voluntary should involve discussions with parents and students about meal choices, costs, and nutrition. The Meatless Monday campaign provides the opportunity to debate and discuss many issues of the food system.

Suggested websites: http://www.meatlessmonday.com/ and http://www.meatmythcrushers.com/myths/going-meatless-one-day-a-week.html

4.8 ANIMAL DISEASE MANAGEMENT

Over the past several decades a few major animal diseases have caused substantial loss of productivity and profits for farmers that raise livestock. Bovine spongiform encephalitis (BSE), or mad cow disease, is a progressive neurological disorder caused by the unusual transmissible material called a prion. Prions invade the brain and spinal cord of animals, causing damage. The origin of BSE has been tentatively linked to cattle that were fed meat and bone-meal by-products from sheep that were suffering from scrapie, which is a prion-related disease. The disease was then spread by feeding infected bovine meat and bone-meal to young calves. BSE is not contagious, but rather spread by ingesting brain or spinal cord tissue containing prions. Between 1993 and 2010 over 184,500 cases of BSE in cattle occurred in the United Kingdom. BSE caused great concern because a rare human prion-dependent disease, called Creutzfeldt–Jakob disease, arose about 10 years after BSE was found in cattle. It is not clear if the cases of Creutzfeldt–Jakob disease resulted from eating contaminated meat but the similarities in illness between animals and humans made this seem possible. To minimize the potential of human ingestion of tissue from potentially infected animals, strict rules were implemented by the FDA in 1997 to prohibit inclusion of mammalian protein in feed for cattle, and further requirements were implemented in 2009 to prohibit high-risk tissue (brain, spinal cords) from use in all animal feeds. Animal products that enter the human food supply must also be kept separated from brain or spinal tissue during processing. BSE has occurred in only a few animals in the United States and strict monitoring and reporting requirements are in place to avoid an outbreak.

During the spring of 2015 a major outbreak of highly pathogenic avian influenza (HPAI) occurred among US poultry and egg producers. Nationally, more than 50 million chickens, turkeys, and other poultry had to be destroyed due to the disease. The economic impacts included the loss of product revenue, costs of disposing of infected animals, decontaminating facilities and equipment and then restocking the flocks. And thousands of workers were laid off when facilities were shut down. The greatest economic losses were to operations in Iowa ($1.2 billion) and Minnesota ($309.9 million). This situation created environmental concerns associated with disposal of the large number of infected animals without further spreading the disease or contaminating ground or water resources. Consumers were impacted as well, with lower production the price of eggs and poultry increased, costing consumers $3.3 billion. And many countries, including China, Russia, and South Korea, which were the top three importers of US poultry, imposed trade bans that further cut profits. HPAI outbreaks have occurred previously in the United States but were less widespread. A major problem in controlling this disease is that it is thought to be spread by wild birds, especially migratory birds that cross the United States in the spring and fall. Finding solutions to prevent or reduce infections such as HPAI are challenging. APHIS has responded to this crisis by developing standards for facility biosecurity, and working to find an effective vaccine.

The economic impact of such large outbreaks brings to the forefront the significant threats to the US food supply that animal diseases can have, and also illustrates the ability of the system to recover and respond. Diseases will continue to arise and animal producers will need to be vigilant in monitoring and protecting their operations.

Researchers in the USDA as well as in academic institutions are engaged in ongoing studies to find and manage disease outbreaks. Since 1954 the Office of National Laboratory Plum Island Animal Disease Center located off the coast of New York has carried out research to study animal diseases and to develop new vaccines and diagnostic tests to prevent and monitor outbreaks. The Plum Island facility is the only location in the United States where research on some of the most highly contagious animal diseases may be conducted. The Office of Homeland Security oversees the work at the Plum Island facility, which demonstrates the importance of animal disease research to the United States. By 2022, the work being done at Plum Island will be relocated to a new facility being built in Manhattan, Kansas, the National Bio and Agro Defense Facility. The USDA-ARS also operates the National Animal Disease Center (NADC) located in Ames, Iowa (Fig. 4.14). The work carried out at NADC is directed at finding ways to reduce and prevent infectious, genetic, and metabolic diseases in economically important livestock and poultry. For example, work done at the NADC was effective in eradicating a common disease in cattle and hogs, brucellosis, in the 1950s. But concern about the disease persisting in wild animal populations recently has led researchers to develop a vaccine to prevent the disease from spreading to domestic livestock. The benefits to both producers and consumers of long-term research programs to address disease risk in the natural world is clearly evident.

4.9 HORMONES AND GROWTH PROMOTANTS

Administration of growth-promoting hormones to cattle and sheep increases growth rate, feed efficiency, and leanness. Growth promotants are steroid hormones related to androgens (testosterone and trenbolone acetate), estrogens (estradiol and zeranol), and progestins (progesterone and melengestrol acetate). Animals, and humans, naturally produce androgens, estrogens, and progestins throughout their life. Since the 1950s there have been

FIGURE 4.14 The National Animal Disease Center, located in Ames, Iowa, has been at the forefront of research to understand and prevent infectious diseases of livestock. *Source: Photo from the USDA, www.usda.gov.*

over 30 growth-promoting products approved by the FDA for use in beef cattle in the United States. No steroid hormone implants are approved as promotants of growth for dairy cows, veal calves, pigs, or poultry. These treatments are administered as pellets or implants placed under the skin of the ear that release the hormone slowly over time. The ears of treated animals do not enter the food supply. The USDA requires that these treatments are given well before the animals are slaughtered so that there is sufficient time for them to be metabolized and cleared. The FMIA requires the FSIS to routinely test meat for residues of growth-promoting products at the time of harvest. Hence, consuming meat from animals that have been properly treated with growth promotants does not pose a risk to human health. Yet concerns have been raised that the use of these hormones in animal husbandry contributes to problems such as obesity and early puberty in girls. The scientific evidence has not shown any connection between the consumption of meat and animal foods with human growth or reproductive health.

A hormonal treatment developed to increase milk production by dairy cows was approved by the FDA in 1993. Growth hormone, or somatotropin, is naturally produced by lactating dairy cows to regulate the animal's metabolism to support milk production. Somatotropin is a protein and the genes that encode its structure were identified by the Genentech company. Using the tools of molecular biology, Monsanto and several pharmaceutical companies developed a process to insert the genes into *E. coli*, which then produced recombinant bovine somatotropin (rbST). Monsanto was the first company to obtain FDA approval and marketed the rbST under the name Posilac. rbST is identical to the native somatotropin produced by cows and has the same biological effects. With FDA approval, rbST was marketed to dairy farmers as a means to prolong milk production in their herds. Milk produced by cows treated with rbST was shown to be nutritionally identical to milk from untreated cows, and levels of rbST in milk were undetectable compared to the naturally occurring levels of somatotropin produced by the animals. For these reasons, the FDA deemed that milk from rbST-treated cows was safe and wholesome for consumers. The USDA reported in 2000 about 17% of dairy farmers were using rbST and for several years rbST milk was marketed and consumed in the United States. Around that time, concerns about the use of rbST began to arise in the public forum. In animals, somatotropin works in synergy with another hormone, insulin-like growth factor I (IGF-1). IGF-1 activates cellular events associated with growth and maturation. Research being done in the human cancer field was finding that higher levels of IGF-1 were associated with some types of cancer. The assumption was made that rbST-treated cows would generate more IGF-1 and both hormones would end up in the milk. These pieces of information became intertwined leading to a connection between consumption of rbST-treated milk with precocious puberty, obesity, and cancer. Several scientific facts were ignored by these correlations including that the levels of rbST and IGF-1 in milk from treated cows are not significantly higher than from untreated cows, humans produce both somatotropin and IGF-1 at levels higher than those present in treated milk, and both rbST and IGF-1 are proteins that are digested when they are consumed orally (so do not become active hormones). Most medical organizations concluded that rbST did not pose a risk to health, but some consumers remained concerned. In addition to the human health risks, reports were published that suggested rbST-treated cows had higher rates of mastitis (infection of the mammary glands) and had to be treated with more antibiotics, therefore milk from these cows was suspected to contain high levels of antibiotics. As noted previously, the

FDA requires all milk sold to be free of antibiotics, so there was no validity to these concerns. These fears combined with the negative image of genetic modification, made rbST very unpopular. In response to consumer demand, many food processors and major retailers decided to not market milk from rbST-treated cows. Consequently, the use of rbST in the dairy industry today is very low. Some companies saw a market opportunity and began labeling their products as containing "no artificial hormones" or "hormone-free." The latter is, of course, inaccurate because milk from all cows will contain hormones from natural production. Some consumers remain confused by these terms, and may be unaware of the background for why hormones in milk became part of the marketing label.

4.10 THERAPEUTIC AND NONTHERAPEUTIC USE OF ANTIBIOTICS

Antibiotics kill bacteria by blocking aspects of cell growth, metabolism or reproduction. Classes of antibiotics have been developed over the years that target specific groups of bacteria. The first antibiotic discovered was penicillin, by the British bacteriologist Alexander Fleming who isolated it from a mold. Penicillin was recognized for its ability to kill infectious bacteria just after the start of WWII. The value of penicillin to reduce infections in soldiers was soon recognized but British scientists were not able to produce enough quantity of the drug to run clinical trials. They sought help from scientists at the USDA-ARS in Peoria, Illinois, who very quickly devised a method to produce large quantities of penicillin and an even more effective strain. By 1943 human trials proved that penicillin was an effective antibiotic drug and shipments were sent to the front lines during D-Day battles to treat and save many wounded soldiers. Soon other antibiotics were discovered through intense research by pharmaceutical companies. Some were initially isolated from soil bacteria or molds, but others were chemically synthesized.

The application of low-dose antibiotics in food animal production began soon after their discovery. Adding low levels of antibiotics, including pencillins, tetracyclines, and sulfas to the diets of most animals, especially young piglets and chickens, improves growth and feed efficiency. The effects of this nontherapeutic application of antibiotics seems to be associated with reduction of pathogenic bacteria in the intestine, but changes in the ability of nutrients to be absorbed also occurs. Many of the antibiotics used in animal production are not absorbed into the blood, but act only with the animal's intestine. Concerns about this practice were raised soon after it began, mainly associated with potential for developing resistant strains of bacteria.

Oversight of antibiotics used in agriculture is the responsibility of the FDA, USDA, and CDC. The FDA Center for Veterinary Medicine (CVM) approves antimicrobial drugs for use in livestock and determines the minimum withdrawal period required between the use of the drug and the time the animal or products enter the food system. The USDA FSIS continually samples meat and the FDA tests milk for antibiotic residue. Products that test positive for antibiotics are not allowed to enter the food supply. Some consumers may be concerned that meat, milk, or eggs from animals that have received antibiotics will contain those same antibiotics. Antibiotics, either for nontherapeutic or therapeutic use, have always been monitored by the USDA and FDA to ensure that there is adequate time after administration for the drugs to clear the animals' system. Animal foods in the grocery store do not contain antibiotics. Marketing tools, such as antibiotic-free labels, confuse consumers who might not understand that all foods are antibiotic-free by regulatory standard.

4.10.1 Antibiotic-Resistant Bacteria

Antibiotics are used in livestock to treat diseased animals, treat animals that are not yet diseased but may come in contact with a disease (disease prevention), and to enhance growth (growth promotion). The latter two types of use, in which noninfected animals are given antibiotics, have been strongly criticized for the potential to increase the risk of antibiotic resistance. Antibiotic resistance occurs when microbes develop mechanisms to survive a drug's attack. These changes in the microbes arise through mutations or transfer of genetic material with other microbes. A selection process occurs in which susceptible microbes are eliminated allowing resistant microbes to multiply. The US Institute of Medicine first made recommendations to reduce or eliminate antimicrobials in animal feed in 1980 because of concerns about antibiotic resistance. However, they did not provide sufficient data to prove the practice was associated with creating antibiotic-resistant microbes and so no changes were implemented. Recent increases in the number of antibiotic-resistant bacteria have reinitiated the concern that the widespread nontherapeutic use of antibiotics in livestock may contribute to this problem.

Antibiotic-resistant bacteria that arise in livestock have the potential to infect humans. Human bacterial infections resistant to drug treatment were reported over 65 years ago. In the past, there were a sufficient number of different antibiotic drugs with the ability to kill several types of microbes, so if resistance developed to one drug another drug could be used. This situation has now changed as several strains of microbes have developed resistance to all of the available antibiotic drugs. The spread of resistant bacteria from animals can occur if they shed the microbes into the environment allowing them to come in contact with other food (e.g., irrigation water and manure fertilizers used on vegetable crops), animals, and people (e.g., on-farm animal workers as well as off-farm exposure through water and soil; Fig. 4.15).

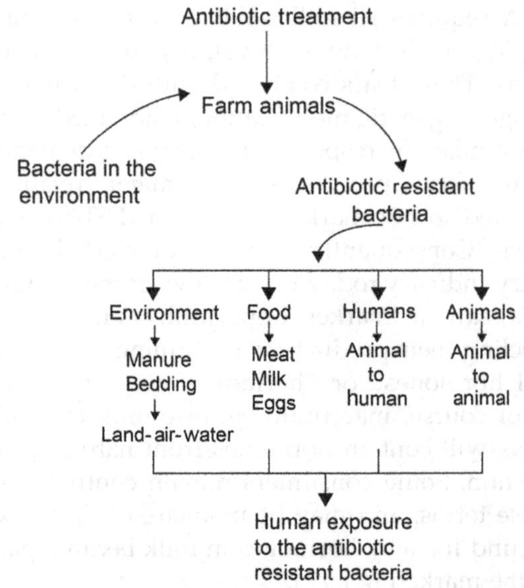

FIGURE 4.15 The process by which antibiotic-resistant bacteria develop and then spread from livestock to people includes distribution through direct contact, water, soil, and food products. *Source: Illustration by Reannon Overbey.*

Some antibiotics used in animals are not effective for or used for treating human illnesses, and so antibiotic resistance to these drugs would be of little concern to human health. Other drugs are used both in animals and humans and are thereby considered medically important (Table 4.3). Should these generate resistant bacteria strains, the risk to human health would be high.

The CDC reported 2 million cases of antibiotic-resistant infections in 2015. An example is methicillin-resistant *Staphylococcus aureus* (MRSA), which causes severe skin infections that can lead to death. There are no effective antibiotics to treat this disease. MRSA first occurred in hospitalized patients and steps were taken to reduce infection through improved hygiene and cleaning protocols. The

TABLE 4.3 Drugs for Animal and Human Therapeutic Use and Animal Growth Promotion

Drug class	Example drug	Livestock therapeutic and nontherapeutic use	Human therapeutic use	Animal growth promotant
Aminopenicillins	Amoxicillin	X	X	
Phosphoglycolipids	Bambermycin	X		X
Polypeptides	Bacitracin	X	X	X
Quinoxalines	Carbadox	X		
Macrolides	Erythromycin	X	X	X
Amphenicols	Florfenicol	X		
Aminoglycosides	Neomycin	X	X	X
Ionophores	Lasalocid			X
Lincosamides	Lincomycin	X		X
Beta-lactams	Penicillin	X	X	X
Arsenicals	Roxarsone	X	X	
Tetracyclines	Tetracycline	X	X	X
Streptogramins	Virginiamycin		X	X

From Marshall, B. M., & Levy, S. B. (2011). Food animals and antimicrobials: Impacts on human health. Clinical Microbiology Reviews, 24(4), *718–733 (Marshall & Levy, 2011).*

spread of MRSA outside of hospitals has become more common. Outbreaks of MRSA have occurred in populations that have close interpersonal contact, such as wrestling and football teams, or through nail salons with insufficient sanitation practices. MRSA has been found in agricultural animals, including swine, and has been shown to be prevalent in farm workers and meat processing employees. The potential of MRSA entering the food supply via meat products was confirmed when a small percentage of retail pork products were found to contain MRSA. This leads to further concern about spread of antibiotic-resistant bacteria through the food supply, which could have much more far-reaching distribution than would human–human or human–animal contact. Note that thoroughly cooking meat does kill the bacteria, so proper food preparation is a way to control spread of MRSA, as it is for most food-borne pathogens.

Three main types of food-related antibiotic-resistant bacteria that have been linked to food include *Salmonella*, *Campylobacter*, and *Staphylococcus*. Since 1996, the FDA's National Antimicrobial Resistance Monitoring System for Enteric Bacteria (NARMS) program collects and evaluates the risk of antibiotic-resistant bacteria in humans, retail meat, and food animals. Data from the NARMS has found that the percent of microbial samples collected from chickens, turkeys, cattle, and swine that have no resistance to 17 different antibiotics decreased slightly (meaning that more samples did show resistance) between 1999 and 2010.

The USDA has also collected data about the use of antibiotics in livestock for many years. From a survey conducted in 2009, they found that the majority of nursery pigs did not receive nontherapeutic (growth-promoting) antibiotics although most were treated with antibiotics to prevent disease (Fig. 4.16). For

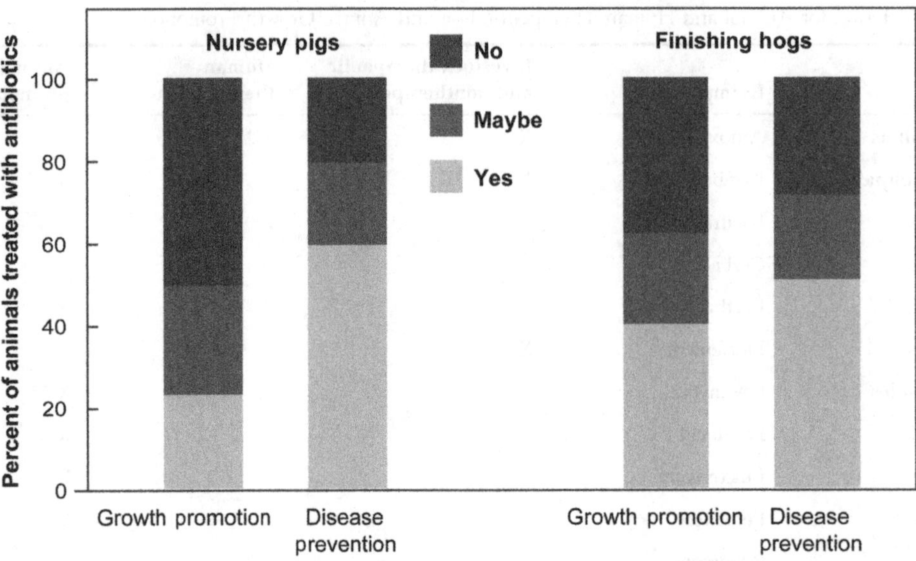

FIGURE 4.16 A USDA survey in 2009 found that the nontherapeutic use of antibiotics was common in both nursery pigs and finishing hogs. Producers responded that over 50% of nursery pigs and 60% of finishing hogs had or may have received antibiotics for growth promotion. Therapeutic use of antibiotics was higher for both nursery pigs and finishing hogs. *Source: USDA Economic Research Service, www.usda.ers.gov, Agricultural Resource Management Survey (ARMS), 2009.*

hogs, antibiotics were used in about 40% of operations for growth promotion, and about 51% used antibiotics to prevent disease. From other studies, the USDA found that about half of broiler producers add antibiotics to feed to promote growth. In beef production, less than 16% used antibiotics to promote growth in young animals, but the majority of animals in feedlots received antibiotics, especially those in large production facilities, to prevent disease. About half of dairy cattle producers used antibiotics in feed to prevent disease and promote growth. The collection of this data, which included the types and routes of administration, was used to inform policies and guidelines to reduce the threat of antimicrobial resistance from the use of antibiotics in livestock.

There is a lack of scientific evidence to directly correlate the development of antibiotic-resistant bacteria with agricultural use of antibiotics. Overuse and improper administration of antibiotics in human medicine is also a route through which antibiotic resistance can occur. It may not be possible to target where a resistance might originate. The large amounts of antibiotics used in agriculture would suggest that reduction and more judicious use would be beneficial to reduce the potential to develop antibiotic resistance. In 2013 the FDA CVM issued a set of guidelines, *The Judicious Use of Medically Important Antimicrobial Drugs in Food-Producing Animals.* The document provided two main principles: the use of medically important antimicrobial drugs in food-producing animals should be limited to those uses that are considered necessary for assuring animal health, and the use of medically important antimicrobial drugs in food-producing animals should be limited to those uses that include veterinary oversight or consultation. The FDA recommended that animal pharmaceutical companies voluntarily remove growth enhancement and feed efficiency uses from their antibiotic products, eliminate over-the-counter distribution, and require veterinary oversight. Simultaneously, the FDA focused on ensuring that physicians and patients appropriately use antibiotics to treat human illnesses. The need to treat animals therapeutically with antibiotics will continue, just as humans will need these drugs to treat illness. For example, mastitis in dairy cows or an infection from a wound must be treated to prevent pain and suffering of the animal. The USDA requires animal producers to treat animals humanely, which includes treating them for illnesses using approved drugs and therapies. But under the new FDA guidance, these treatments will be supervised by veterinarians.

The motivations for companies and producers to comply with these voluntary guidelines are economic, ethical, and political. As more consumers demand to know how animals are raised and express concerns about antimicrobial resistance, producers become pressured to participate in solutions in a transparent manner. Ethically, companies that produce drugs and the farmers that use them realize that antibiotic resistance is a significant problem that has wide-ranging implications for the health of all humans. If the voluntary approach should fail to make sufficient impact, federal regulations would need to be implemented.

4.11 ENVIRONMENTAL ISSUES OF FOOD ANIMALS

Currently the types of housing in animal food production in the United States are under increased scrutiny from the public. There are over 450,000 AFOs in the United States housing all types of livestock. AFOs with more than 1000 animal units (an animal unit is equivalent to 1000 pounds of animal weight which correlates to 1000 head of cattle, 700 dairy cows, 2500 hogs, 125 thousand broilers or 82 thousand hens) are called CAFOs. AFOs and CAFOs have been criticized for having negative effects on the environment and the health and well-being of animals. The majority of beef cattle in production are finished in feedyards or CAFOs (Fig. 4.17). Confinement facilities for hogs and laying hens are also primarily in the CAFO category. CAFOs can provide a low-cost and efficient way to raise animals if well managed, and the low cost of animal foods in the United States is in part due to the use of these systems. When animals are housed in large groups, there is a substantial generation of manure that must be processed or hauled away. It is estimated that on an annual basis livestock waste amounts to between 1.2 and 1.37 billion tons. And, unlike human waste, there is no sanitary sewage treatment facility requirement for animal waste. Manure management on CAFOs is of great importance to both public health and the environment. How the manure is stored on the facility affects gas emissions; for example, closed tanks retain emissions while open ponds or lagoons allow gases to escape. Manure is high in nitrogen and other minerals and these can leach from the facility into ground and surface water if not stored properly. The majority of livestock manure is applied to cropland or pastures as fertilizer. The EPA regulates manure distribution by large producers. Being high in organic matter, nitrogen, and minerals is a positive benefit of manure that contributes to soil fertility and crop production. But ground application has limitations. If the manure is spread on the surface (Fig. 4.18) it will produce more emissions than when injected below the surface. And if the manure is administered when the ground is frozen or right before a heavy rain, runoff into waterways occurs. Finding adequate and suitable land for spreading, and not overapplying, manure can be challenging. When there are many CAFOs within a community, more manure is produced than can be applied to the surrounding land. Shipping manure to other locations is costly and time consuming.

In the United States the beef, dairy, and poultry industries must meet federal guidelines

FIGURE 4.17 Concentrated animal feeding operations (CAFOs) include feedyards where beef cattle are fed high-nutrient diets to reach market weight. A CAFO is defined by the USDA as a site where more than 1000 animal units are held for more than 6 months with no grass or vegetation. *Source: Photo provided by the National Beef Council.*

relative to the environmental impacts of AFOs managed by the EPA. The Clean Water Act's National Pollutant Discharge Elimination System Program for Concentrated Animal Feeding Operations mandates how CAFOs manage their wastes to protect water systems. The Clean Air Act (CAA) enforces regulations on the amount of ammonia released from animal production. In 2002 that amounted to 2.4 billion tons of ammonia.

A variety of air contaminants, including ammonia, nitrous oxide, methane, volatile organic compounds (which produce odors), hydrogen sulfide, and particulate matter, produced from animal production can have negative environmental and human health effects. CAFOs that are near houses or towns may be criticized for producing foul odors, and reducing the quality of life for citizens. Acute and chronic health effects caused by air particulates and gases produced in these operations are also of concern. Although it is rare, farm workers have become asphyxiated while cleaning manure pits due to these gases, and there is some evidence that children living near CAFOs are at higher risk for developing asthma. There are three laws that address CAFO air emissions: the Comprehensive Environmental Response, Compensation, and Liability Act (CERCLA, also known as the Superfund Act), the Emergency Planning and Community Right to Know Act (EPCRA), and the CAA. There has been criticism of the EPA for not strongly enforcing the emission standards in these acts and allowing exemptions for CAFOs. As populations expand and people begin to move into areas were CAFOs

FIGURE 4.18 Farmers spread manure from animal operations on their fields to provide fertilizer for their crops. Manure spreaders distribute dry manure onto the fields, as shown. The amount and timing of manure spreading must be controlled to prevent contamination of water systems. *Source: Image Provided As Courtesy of John Deere.*

have been operating, conflicts arise between the producers and the homeowners about air quality. Researchers are working on strategies that can mitigate both the production of these compounds by the animals (using modified feed systems) and their release into the environment (air filtration and strategic planting of trees and bushes).

A review of lifecycle analyses of animal production in developing countries (Table 4.4) found that beef production was more land-intensive than pork or chicken, and generated more CO_2 equivalents that contribute to climate change. However, these authors noted that this type of assessment does not take into consideration important nuances of animal production. For example, ruminant animals consume feed ingredients (hay and grasses) that do not compete with human food, whereas pork and chicken rations include grains that are suitable for humans. Hence, cattle production using marginal land that is not suitable for producing human foods alters the lifecycle assessment.

There is debate about the amount of methane gas that is produced from ruminant animals (dairy cows and beef cattle, mainly) and the contribution of livestock-generated methane to greenhouse gas (GHG) accumulation and resultant global warming. GHGs are discussed in more detail in Chapter 8, Sustainability of the Food System. An estimate by the Food and Agriculture Organization is that livestock may contribute as much as 18% of GHG emission worldwide. That estimation takes into consideration the emission of carbon dioxide from fossil fuel use in the production

TABLE 4.4 Land Use and Carbon Dioxide Generation by Animal Food Production

	Land use (m^2/kg)	Land use (m^2/kg Protein)	CO_2 equivalents/kg	CO_2 equivalents/kg protein
Pork	8.9–12.1	47–64	3.9–10	21–53
Chicken	8.1–9.9	42–52	3.7–6.9	18–36
Beef	27–49	144–258	14–32	75–170
Milk	1.1–2.0	33–59	0.84–1.3	24–38
Eggs	4.5–6.2	35–48	3.9–4.9	30–38

From deVries, M., & de Boer, I. J. M. (2010). Comparing environmental impacts for livestock products: A review of life cycle assessments. Livestock Science, 128, *1–11 (de Vries & de Boer, 2010).*

system, deforestation for grazing, methane generation from manure and rumination, and nitrous oxide released from fertilizers used in grain production. Livestock are not the only source of methane production in the food system. Rice production is a major source of methane in China and global landfill waste including food materials, generates substantial amounts of methane. The overall impact of ruminant generation of methane is less than these sources, but is a contributing source.

Another way animals impact the environment is that animals consume a large amount of grain, which requires land, water, and chemical inputs to produce. There have been estimates that it requires 6–20 pounds of grain to make 1 pound of beef. However, there are several caveats to this estimate to be considered. For cattle, both beef and dairy, the majority of their diet is forage material that has no human nutritional value. Actually, most of the cattle raised in the United States, and even more globally, consume only small amounts of grain as part of their diet over their lifespan. Grain is included mainly during the finishing period in feedyards. Increasingly, the use of by-products from the grain industry such as corn gluten or distiller's grains, as part of the finishing diet, are fed to cattle rather than ending up in landfills. Another aspect of the animal industry is that cattle generate many products in addition to beef for human consumption. For a 1250-pound steer, about 790 pounds represents the carcass, of which about 600 pounds is edible meat. The remaining parts of the body are used for hundreds of materials including dyes, adhesives, plastics, medicines, insulation, cosmetics, glass, lubricants, and leather to name a few. These secondary markets benefit from the cattle industry with a renewable source of materials that would be difficult replace from other sources.

The nutritional value of animal products for human health is not readily replaced by plant foods. Iron and zinc from beef, calcium from milk, and choline from eggs are some examples of key nutrients that are best provided by animal foods. The protein quality of animal foods is also matched to the amino acid requirements of humans making them high-quality protein foods. Determining the optimum diet for humans that balances nutritional value, economics, and the environment will be critically important as the global population increases.

Animal scientists are actively researching new ways to raise animals to reduce the environmental impact. Studies have found that careful selection of feed ingredients may reduce the amount of nutrients (nitrogen, phosphates, and other metals) excreted by animals in manure. This would increase production efficiency and reduce environmental damage. Also, specific feeding techniques may decrease methane production and eliminate the odors associated with animal facilities.

Better animal housing facilities and processes for utilization of waste streams are continually being developed with the goal of providing safe and healthy food without negative effects on the environment.

4.12 ETHICAL ISSUES IN ANIMAL HOUSING SYSTEMS

The Humane Society of the United States (HSUS), the People for the Ethical Treatment of Animals (PETA), the Animal Liberation Front (ALF), and the Animal Legal Defense Fund (ALDF) have been active in opposing farm animal caging systems using a wide range of media and celebrity spokespersons. These groups engage the public with images of animals in confinement and mistreatment that draw strong emotional responses and raise large amounts of money. While the majority of farm animal producers maintain high standards of humane animal care, incidents of abuse or mishandling do happen. When these are reported in the press and posted on organizations' websites, consumers who may have no first-hand experience with farm animals may think this is the way all animals are treated. Animal producers have begun to publicly share their perspective of humane animal care in which cages and confinement are used to improve animal management and efficiency. There is strong scientific evidence to support improved animal health and production when animal caging and confinement are used appropriately. Public opinion tends to run against this practice, which has led to change in the industry.

In 2013 California passed the California Shell Egg Food Safety regulation, which requires eggs sold in the state to follow testing and vaccination protocols to reduce *S. enteritidis* contamination and defines an increased amount of floor space per hen. This law is the first of its kind in which a state defined animal care and housing regulations independently of federal standards. The implications of this law are significant because even eggs produced in other states and sold in California must meet these housing standards, therefore egg producers nationwide must comply with the California standard in order to sell their eggs to the state. The changes in housing standard, requiring more floor space per hen, means that producers must invest in new caging systems and infrastructures, or reduce the number of animals they house. Either way, production costs are increased. The result of this law has been an increased price of eggs for California consumers, but higher value to egg producers that meet the standards. Hence, a market shift is likely to occur with some egg producers seeing value in changing their housing standards to gain a higher price to sell in California.

Pressure from consumers who have raised concerns about animal housing standards to major restaurant chains has led some to announce their goal to transition to these systems over the next decade or so. Restaurants that have made that pledge include McDonald's, Starbucks, Taco Bell, and Panera and some food manufacturers such as Kellogg's and General Mills have as well. The change in housing standards will put pressure on farmers to transition their production systems or lose these markets. Similarly, another focus has been advocating for the elimination of gestational crates for pigs. Wendy's, Burger King, Safeway, and Kroger among others have stated their intention to require gradual elimination of gestational crates by their suppliers. This will also require farmers to restructure their operations to meet the market demand.

4.12.1 Ethical Views About Animal Rights

The consumption of animal products as food provides an ethical dilemma for some people. In 1975, the philosopher Peter Singer published his views on animal rights in a popular book entitled *Animal Liberation*. His primary theme was that while there are differences between human and nonhuman animals, all have the capacity to suffer and therefore should be treated the same. He argued that although most humans are intellectually superior to animals, some humans such as infants or those with disabilities may not be, yet there is no suggestion that these people should be placed in cages or used for food. In his view, when humans use animals for food or experimentation this is "speciesism," a prejudice of the dominant group over the others. He contended that "...to avoid speciesism we must allow that beings which are similar in all relevant respects have a similar right to life—and mere membership in our own biological species cannot be a morally relevant criterion for this right" (Singer, 1976, p. 21). Singer's ideas are credited with starting the animal rights movement in the United States and led people to think about animal treatment in all aspects of society.

As a result of growing concern for animal rights, there were calls for important changes. The Animal Welfare Act (AWA) of 1966 provided acceptable standards for animal treatment and care, but the law did not cover birds, rats, mice, farm animals, or cold-blooded animals. In 1970, the AWA was amended to include all warm-blooded animals used for experimental research or testing and required humane handling, housing, sanitation, and veterinary care. In 1985, the AWA was further amended to define minimum standards for handling, housing, and feeding animals and minimizing pain. Another requirement was that all research using animal models was required to be supervised by a board comprised of both scientists and nonscientists (Institutional Animal Care and Use Committee) to ensure that experimental animals are cared for and treated humanely, according to the regulations. The APHIS within the USDA is responsible for overseeing the AWA. Farm animals used for food or fiber are not covered by the AWA, however.

An organization that promotes the complete discontinuation of the use of animals for food, research, clothing, or entertainment is PETA. This organization began to recruit members soon after Singer's publication and organized the first World Day for Laboratory Animals protest in 1980. This has grown to become a very large movement and carries out protests against companies that use animals. PETA has been reported to embed undercover agents in animal facilities to capture pictures and videotapes of animal abuse. PETA uses disturbing images on its website and in publications to promote its message against the use of animals. Celebrity spokespersons for PETA such as Paul McCartney and Emmylou Harris bring people to this organization and encourage fundraising. The ALF is an organization with the mission "[t]o effectively allocate resources (time and money) to end the 'property' status of nonhuman animals" with the objective of the mission "[t]o abolish institutionalized animal exploitation because it assumes that animals are property" (www.animalliberationfront.com). The ALF has been responsible for breaking into research laboratories on university campuses to destroy equipment and set animals free. They recruit and promote activism and civil disobedience to achieve their goals. PETA and ALF define animal rights as equal to those of people and disallow any dominance of human needs over those of animals. The HSUS also promotes animal rights and sponsors campaigns to protect animals. HSUS gathers local support and visibility through animal shelters, mainly for cats and dogs, and rescue programs for abused animals. But they are also very active in monitoring, and speaking out against, the housing and treatment of farm animals.

More moderate perspectives on animal rights contend that caring for and treating animals humanely and with compassion is essential, and humans have an obligation to reduce or prevent suffering. When this is done, using animals for human benefit is ethically acceptable to them. For example, animals in research provide valuable tools to develop cures and treatments for human disease, and growing animals for human food provide healthful benefits to humans, therefore justifying the use of animals in these ways.

4.12.2 Assessing Animal Welfare

When most people think about animals, they relate their emotions to their house pets—mainly cats and dogs. Most people have very close emotional connections to their pets and treat them with great care. Pets often live inside their human's house and may even sleep on their human's bed. Pets provide emotional support and enjoyment to their human owners. When people think about how farm animals should be treated, they may consider how they treat their own pets. They envision that farm animals would want to be living in the open, not in cages, and to be able to interact with other animals and humans. Keeping chickens or hogs in cages or preventing dairy cattle from roaming free in the pasture does not fit with that image.

There is currently substantial debate about how to define and measure animal suffering and emotion, especially of farm animals. Because animals cannot verbally tell humans if they are hungry, hurt, sad, or happy, we typically look for changes in behavior that might give us a clue to their needs or state of mind. Dogs may wag their tails when they see their humans or are playing, and cats purr when they seem to be content. Contrarily, when in pain, most animals will stop eating, lie down and curl up, or growl when touched. Scientists have studied farm animal behavior and there is a wide range of literature devoted to understanding their signs and symptoms. Perhaps the most influential animal scientist in terms of understanding animal behavior is Temple Grandin. Dr. Grandin holds a PhD in Animal Science from the University of Illinois and is on the faculty at Colorado State University, and credits her ability to relate to animals to the fact that she is autistic. She contends that autistic people are closer to animals than normal people are, not because of intelligence but because of perception and emotion. Through her research and advocacy work she has been instrumental in changing the design of animal housing and restraining facilities to reduce animal stress and to help animal workers understand the causes of behaviors in their animals. The animal industry has embraced her ideas by modifying their corrals, chutes, and ramps with her designs, which allow animals to calmly enter and move through them without human prodding (Fig. 4.19).

Tools to measure or quantify animal pain or suffering are fairly limited. It is also not clear if animals perceive pain or suffer in the same ways as humans. Can a cow be depressed? Is a chicken aware that its life means nothing more than to be someone's dinner? Do piglets miss their mothers when they are weaned? Do sheep long to run free in the pasture? If so, how would we know? There are clear changes in behavior when an animal is in pain or sick, but assessing subtle pain or stress is difficult. For example, some may argue that keeping hogs on concrete flooring is stressful for them and they would be more comfortable outside in the grass and dirt. But when given a choice, hogs stay on the cool concrete on a hot day and prefer to be inside on a cold night. From a management perspective it is easier to keep concrete clean, which reduces the risk of infection and illness, and hogs inside a building will not be threatened or attacked by fox or coyotes. Animals that are comfortable and feel protected will eat and grow normally, which is observed in well-managed confinement operations.

Chickens are territorial and hierarchal. When allowed to interact with other chickens, they create a "pecking order" of dominance and submission and some win and some lose.

FIGURE 4.19 Dr. Temple Grandin is an animal behavior scientist at Colorado State University and a leading expert on designing corrals and housing facilities to minimize animal stress. She encourages the use of curved fencing and gates to keep animals calm during transitions. *Source: Photo from the American Society of Animal Science image gallery.*

If they are housed in cages, this behavior is reduced so all have equal access to food and freedom from being pecked. Confinement of animals allows careful monitoring of their food intake, growth rates, and overall health. If done with care and humane treatment, some would argue that this is the best way to treat and manage farm animals. Other people believe animals should not be housed together in large groups but rather should be outside in fields to enjoy the sunshine and green grass and given the freedom to behave naturally.

Animals in the food system will continue to generate controversy and ethical debate. The majority of farmers that raise food animals are committed to providing a safe and nutritious product in a humane and environmentally sound manner. Through national organizations, such as the National Pork Producers Association, National Cattlemen's Beef Association, US Poultry and Egg Association, National Dairy Council and others, producers and processors are beginning to engage with consumers about their industries to inform and educate about production practices. Increasingly, consumers ask for information about animal husbandry approaches and demand transparency.

There is currently no FDA-approved standard for labeling foods regarding animal housing practices. For example, "cage-free" or "humanely raised" labels may be found on some meat products in stores. There are no government or industry standards to define these labels, nor guarantees that they are accurate. The USDA Organic Standard does define the types of feed and medical treatments that animals may receive and certain housing criteria and this is described in Chapter 8, Sustainability of the Food System.

4.13 CONSUMER INFLUENCE ON ANIMAL FOOD PRODUCTION

Consumers tend to fall into one of several groups when issues about food are discussed. The Center for Food Integrity (CFI) has conducted surveys of consumers for the past several years (www.foodintegrity.org) to understand how consumers view the food system and the factors that motivate their food choices. CFI has found that a group of consumers they identified as "early adopters" seem to be most engaged with food-related issues. These early adopters tend to be well-educated and financially stable, they are information seekers, and have large interpersonal networks. They use social media and other communication outlets well, which allows them to gather and disseminate ideas quickly. Some of these early adopters utilize food-related blogs, YouTube videos, and other social media to engage with other consumers. In some cases, having a web presence can become a full-time job by gathering financial backing and sponsors. Very effective social media celebrities have arisen who have huge followings to whom they can share their views and opinions.

CFI found that of greatest importance to consumers is that food producers share their values and demonstrate that they are trustworthy. When trust is broken, for example if one egg producer was found to have mismanaged their operation and caused a food-borne illness outbreak, consumers extend their mistrust to the entire industry. When the HPAI outbreak occurred and a massive egg recall was needed, consumers responded with mistrust and were captured on news reports saying they were only going to buy eggs from small, local farmers because they could not trust large egg producing facilities. Even though millions of safe and uninfected eggs had been and were still being produced by large egg producers, that segment of the industry lost consumer confidence. But even if the food industry has done nothing wrong, consumer confidence can be shattered by public input. A prime example of this is the "pink slime" situation, which is described in Expansion Box 4.3.

In a 2015 CFI survey consumers were asked if they agreed with the statement "US meat is derived from humanely treated animals" and about 82% responded with moderate or strong agreement. About 95% agreed with the statement "If farm animals are treated decently and humanely, I have no problem consuming meat, milk, and eggs." This suggests that the majority of consumers have confidence that livestock farmers are treating their animals humanely and that this is an important consideration for them. Consumers are ready and willing to take away their trust, and their purchases, from a food producer if they believe they are not being truthful about their practices.

Farmers who raise livestock are finding themselves in a unique position of having to defend and justify their approach for all aspects of their operations. The burden of communication and trust is placed on the farmer to demonstrate that they are operating in ways that are acceptable to consumers. Breeding and care practices, types and ways of feeding, antibiotics, growth promoting hormones, housing systems, waste removal and management, animal health monitoring, air and water quality, transportation systems, and marketing have been questioned by consumers. Some argue that as Americans have become more removed from the farm there is a lack of understanding of farm practices, especially related to livestock. To mitigate this lack of understanding, communication about food production has now become a major focus for farm producers and processors. Agriculture organizations and associations such as the Farm Bureau, Beef Industry Council, and National Dairy Council engage in public communications about their industries to increase transparency. Farmers have opened their operations to visitors and promote farm tours. The number of people who participate in such events may be small, especially considering the population distribution of the United States. Most farming operations are located in rural areas away from large population centers so reaching urban consumers is challenging. To extend their reach, the US Farmers and Ranchers Alliance provided funding for a movie documentary about farming developed by James Moll. *Farmland* (2014) follows the lives and operations of several livestock producers in different parts of the United States to illustrate the challenges and rewards they encounter. Moll was given full independence to tell the story of these farmers with the intent of demonstrating their commitment to animal welfare and caring for the environment. In contrast, other documentaries have been produced such as *Forks Over Knives* (2011), which encourages consumers to avoid animal foods for health, animal rights, and environmental reasons, and *Meat the Truth* (2008), which links food animal production to climate change and global warming. Negative images of livestock production are abundant on the Internet and on animal rights organizations' websites. Animal food production is arguably the most challenging aspect of our food system because of the social, political, ethical, cultural, economic, and environmental complexities. Balancing the goals of providing safe, healthful, and sustainable animal foods with the rights of animals, protection of the environment, and utilizing natural resources wisely is a challenge for today and into the future.

EXPANSION BOX 4.3

PINK SLIME

Late Night talk show host David Letterman used to have his audience play a game called "Know Your Cuts of Meat." Audience participants were asked to identify cuts of meat from pictures, and most were unable to do so. Butchers working in beef processing facilities trim meat from the carcass according to defined "cuts" or recognizable pieces. Prime rib, sirloin, tenderloin, and rump roast are cuts of beef, for example. In the process of making meat cuts, trimmings remain that are too small to sell individually. Some of these trimmings are used to make ground beef but a significant amount of meat remains associated with the remnants. Over 30 years ago, processes using heat and spinning to collect meat from these trimmings were developed to produce a lean beef product. The trimmings are cut into small pieces and warmed to about 100°F to allow fat to be separated from the meat by a centrifugation process. The fat, or beef tallow, is used in food products. The lean meat is treated with small amounts of either ammonium hydroxide gas (process patented by Beef Products Inc. (BPI); called lean finely textured beef or LFTB) (Fig. 4.20) or citric acid (process patented by Cargill; called finely textured beef or FTB) to reduce risk of pathogenic bacteria. Ammonium hydroxide and citric acid are both approved by the FDA and are used in many other food products. They raise the pH slightly to make an environment unsuitable for microorganisms to grow. The beef product is then flash frozen, chipped into smaller pieces, and packaged for distribution. LFTB and FTB are mixed with ground beef to increase the lean component, or used in prepared and packaged meals containing beef. LFTB and FTB allowed recovery of substantial amounts of consumable meat that would otherwise be wasted, up to 25 pounds/cow, and produced a product

EXPANSION BOX 4.3 (*cont'd*)

that was high in protein and low in fat. Adding LFTB reduced costs and therefore was adopted by many large food chains and products, including those prepared for school lunch programs.

BPI has been producing LFTB since 1993 and by 2010 had four facilities in the Midwest. As it does for all meat processing, the USDA inspected and approved the BPI facilities where LFTB was produced. LFTB was approved by the FDA to be added at up to 10% by weight to ground beef products. Because LFTB is made with meat, it is not considered to be an additive, and therefore not required to be included in a product ingredient label.

In 2002 an FSIS microbiologist Gerald Zirnstein, questioned the appropriateness of LFTB as a beef product. Zirnstein used the term "pink slime" to refer to LFTB in an internal email conversation about the product. Zirnstein and another FSIS colleague, Carl Custer, wrote an internal FSIS report questioning whether the product met the standards to be called beef because they considered it to be mostly connective tissue and not muscle. Their criticism of the product was based on its composition of the product, and not its safety. The USDA, based on their own research and that of other scientists, found that LFTB was a safe and nutritious product and overruled the Zirnstein–Custer report.

In 2009 investigative journalist Michael Moss obtained documents from the FDA, via the Freedom of Information Act, and wrote an article for the *New York Times* questioning the use of LFTB in which he quoted Zirnstein's use of the term "pink slime." The article described the process for making LFTB and raised concerns about its use in fast food and school lunches, but noted that there had been no health risks or food-borne illnesses associated with the product. On April 12, 2011 celebrity chef Jamie Oliver and host of *Jamie Oliver's Food Revolution*, demonized the process and the product on his television show. He put pieces of scrap meat and fat into a washing machine, poured in bottles of liquid ammonia and then showed the audience the sticky product. This was a far

FIGURE 4.20 Lean finely textured beef (LFTB) is produced by Beef Products Inc. using a USDA and FDA approved process. LFTB is used to increase the nutritional value of products containing beef. *Source: Photo provided by Beef Products Incorporated.*

EXPANSION BOX 4.3 (*cont'd*)

stretch from the way LFTB is actually made. Liquid ammonia is not used; rather a small amount of ammonium hydroxide gas is bubbled into the product. Oliver also inferred that the meat trimmings used to make LFTB were unfit for human consumption, which was also inaccurate. The media followed this with a series of news stories aired by ABC News in March 2012 in which LFTB was called a cheap meat filler made from the most contaminated parts of the cow once used only for dog food and cooking oil. This quickly led to public outcry against "pink slime," the manufacturer BPI, the beef industry, and the government.

The rate at which the public response to LFTB occurred was unprecedented in the food industry. Through social media and YouTube videos the public demand to get LFTB out of the food system was overwhelming. There was never any recall of ground beef with LFTB and no illnesses from the product were ever reported. Yet, in response to the fierce public objection, it was renounced by fast-food restaurants, particularly McDonald's, several food companies, and countless supermarkets. Parents were outraged and public schools stopped serving ground beef with LFTB in school cafeterias. As a result of the negative publicity, sales of LFTB plummeted. Production of LFTB decreased from 5 million pounds a week to less than 1 million pounds per week. BPI closed three of its four beef processing plants, lost over $400 million in sales, and was forced to lay off almost 1400 employees.

Dr. Elisabeth Hagen (USDA Under Secretary for Food Safety), The American Meat Institute, BPI, several university meat scientists and the governors of Iowa, Nebraska, Texas, and South Dakota responded with evidence for the safety of LFTB and support for the jobs in the meat processing industry, but to no avail. "Pink slime" had been smeared. The misrepresentation of LFTB as an unhealthy food led BPI to file a lawsuit against ABC News, and their parent company Disney, for defamation, claiming $1.2 billion in damages. Despite the intense negative coverage, LFTB has survived. Today, the term "contains lean finely textured beef" can be seen on ground beef packages, ensuring consumers know that it is being used. BPI re-opened one of its previously closed plants to process LFTB in 2014 and LFTB is slowly regaining markets.

The "pink slime" story illustrates the power of the media to influence consumer perceptions of foods and processes by which foods are made. It shows the willingness of consumers to follow celebrity recommendations for fashion, health, exercise, and food. The contrast between a process that reduces waste, lowers food costs, and increases nutritional value and public perception of collecting meat from leftover trimmings is part of this discussion. Scientific evidence of safety and quality may not be as important as the ideal of how people perceive food should be made. This requires scientific thinking. The comeback of LFTB shows that media frenzies are usually short-lived and that economics and sound science generally overcomes hype and misinformation. But in the process, companies suffered financial disaster, people lost their jobs, food was wasted, and consumers were misled. Consumer behavior in the marketplace has great influence and marketers are wise to follow the trends, but consumers have responsibility to seek factual information before making decisions about food.

Suggested websites: www.beefisbeef.com, http://www.abcnews.go.com/WNT/video/pink-slime-15873068, and *http://www.nytimes.com/2009/12/31/us/31meat.html?_r=2&pagewanted=all*

References

deVries, M., & de Boer, I. J. M. (2010). Comparing environmental impacts for livestock products: A review of life cycle assessments. *Livestock Science, 128*, 1–11.

Farmland. (2014). *Documentary film.* Available from <www.farmland.com>.

Forks Over Knives. (2011). *Documentary film.* Available from <www.forksoverknives.com>.

Hartley, R. M. (1842). *Historical, scientific and practical essay on milk as an article of human sustenance. With a consideration of the effects consequent upon the present unnatural methods of producing it for the supply of larger cities.* New York, NY: Jonathan Leavitt.

Key, N., & McBride, W. (2007). The changing economics of U.S. hog production. *Economic Research Service, USDA Report Number 52.* Washington, DC: U.S. Department of Agriculture.

Marshall, B. M., & Levy, S. B. (2011). Food animals and antimicrobials: Impacts on human health. *Clinical Microbiology Reviews, 24*(4), 718–733.

Meat the Truth. (2008). *Documentary film.* Alalena Media Productions.

Sinclair, U. (1906). *The jungle.* Toronto, ON: McLeod and Allen, 413 p.

Singer, P. (1976). *Animal liberation.* London, UK: Jonathan Cape, 301 p.

Further Reading

Alexander, D. D., Weed, D. L., Cushing, C. A., & Lowe, K. A. (2011). Meta-analysis of prospective studies of red meat consumption and colorectal cancer. *European Journal of Cancer Prevention, 20*(4), 293–307.

Aykan, N. F. (2015). Red meat and colorectal cancer. *Oncology Review, 9*(288), 38–44.

Barkan, I. D. (1985). Industry invites regulation: The passage of the Pure Food and Drug Act of 1906. *American Journal of Public Health, 75*, 18–26.

Bouvard, V., Loomis, D., Guyton, K. Z., Grosse, Y., El Ghissassi, F., Benbrahim-Tallaa, L., & Straif, K. (2015). Carcinogenicity of consumption of red and processed meat. *Lancet Oncology, 16*, 1599–1600.

Bouwman, L., Goldewijk, K. K., Van Der Hoek, K. W., Beusen, A. H. W., Van Vuuren, D. P., Willems, J., & Stehfest, E. (2011). Exploring global changes in nitrogen and phosphorus cycles in agriculture induced by livestock production over 1900-2050 period. *Proceedings of the National Academy of Sciences, 110*(52), 20882–20887.

Boyd, W. (2001). Making meat: Science, technology and American poultry production. *Technology and Culture, 42*(4), 631–664.

Capper, J. L., & Hayes, D. J. (2012). The environmental and economic impact of removing growth-enhancing technologies from U.S. beef production. *Journal of Animal Science, 90*, 3527–3537.

Center for Food Integrity. (2015). *A clear view of transparency.* Available from <www.centerforfoodintegrity.org>.

Clauer, P. (2016). *Modern egg industry. PennStateExtension.* University Park, PA: Pennsylvania State College of Agricultural Sciences. Available from <http://extension.psu.edu/animals/poultry/topics/general-educational-material/the-chicken/modern-egg-industry>.

Dibner, J. J., & Richards, J. D. (2005). Antibiotic growth promoters in agriculture: History and mode of action. *Poultry Science, 84*, 634–643.

Dohner, J. V. (2001). *The encyclopedia of historic and endangered livestock and poultry breeds.* New Haven, CT and London, UK: Yale University Press, 514 p.

Dougherty, T. J., & Pucci, M. J. (Eds.), (2012). *Antibiotic discovery and development* New York, NY: Springer, 1107 p.

Evershed, R. P., Payne, S., Sherratt, A. G., Copley, M. S., Coolidge, J., Urem-Kotsu, D., & Burton, M. M. (2008). Earliest date for milk use in the Near East and southeastern Europe linked to cattle herding. *Nature, 422*, 528–531.

Ferrier, P., & Lamb, R. (2007). Government regulation and quality in the U.S. beef market. *Food Policy, 32*, 84–97.

Gerber, P. J., Vellinga, T. V., & Steinfeld, H. (2010). Issues and options in addressing the environmental consequences of livestock sector's growth. *Meat Science, 84*, 244–247.

He, Y., & Sebranek, J. G. (1996). Functional protein components in lean finely textured tissue from beef and pork. *Journal of Food Science, 61*(6), 1155–1159.

Kaplan, H., Hill, K., Lancaster, J., & Hurtado, A. M. (2000). A theory of human life history evolution: Diet, intelligence, and longevity. *Evolutionary Anthropology Issues News and Reviews, 9*(4), 156–185.

Kritchevsky, D. (1998). History of recommendation to the public about dietary fat. *Journal of Nutrition, 128*, 449S–452S.

Landers, T. F., Cohen, B., Wittum, E., & Larson, E. L. (2012). A review of antibiotic use in food animals: Perspective, policy and potential. *Public Health Report, 127*, 4–22.

Luca, F., Perry, G. H., & Di Rienzo, A. (2010). Evolutionary adaptations to dietary changes. *Annual Reviews in Nutrition, 30*, 291–314.

Magolski, J. D., Shappell, N. W., Vonnnahme, K. A., Anderson, G. M., Newman, D. J., & Berg, E. P. (2014). Consumption of ground beef obtained from cattle that had received steroidal growth promotants does not trigger early onset of estrus in prepubertal pigs. *Journal of Nutrition, 144*, 1718–1724.

Pescatore, T., & Jacob, J. (2013). Kentucky 4-H poultry: Grading eggs. *Cooperative Extension Service.* Lexington,

KY: Department of Animal and Food Sciences, College of Agriculture, Food and Environment, University of Kentucky. Available from <http://www2.ca.uky.edu/agcomm/pubs/4AJ/4AJ05PO/4AJ05PO.pdf>.

Salque, M., Bogucki, P. I., Pyzel, J., Sobkowiak-Tabaka, I., Grygiel, R., Szmyt, M., & Evershed, R. P. (2013). Earliest evidence for cheese making in the sixth millennium BC in northern Europe. *Nature, 493*, 522–525.

Sawyer, G. (1971). *The agribusiness poultry industry: A history of its development*. New York, NY: Exposition Press, 231 p.

Schlesinger, S., Lieb, W., Koch, M., Fedirko, V., Kahm, C. C., Pischon, T., & Aleksandrova, K. (2015). Body weight gain and risk of colorectal cancer: A systematic review and meta-analysis of observational studies. *Obesity Review, 16*, 607–619.

Selitzer, R. (1976). *The dairy industry in America*. New York, NY: Dairy and Ice Cream Field and Books for Industry, 502 p.

Skaggs, J. M. (1986). *Prime cut, livestock raising and meatpacking in the United States 1607–1983*. College Station, TX: Texas A&M Press, 263 p.

Sneeringer, S., MacDonald, J., Key, N., McBride, W., & Mathews, K. (2015). Economics of antibiotic use in U.S. livestock production. *Economic Research Service Report Number 200*. Washington, DC: U.S. Department of Agriculture.

Stewart, H., Dong, D., & Carlson, A. (2013). Why are Americans consuming less fluid milk? A look at generational difference in intake frequency. *Economic Research Service Report Number 149*. Washington, DC: U.S. Department of Agriculture.

Tunick, M. H. (2009). Dairy innovations over the past 100 years. *Journal of Agricultural Food Chemistry, 57*, 8093–8097.

U.S. Department of Agriculture (2012). Controls over shell egg inspections. *Audit Report 50601-0001-23*. Washington, DC: U.S. Department of Agriculture.

U.S. Department of Agriculture (2014). *Antimicrobial resistance action plan*. Washington, DC: U.S. Department of Agriculture. Available from <http://www.usda.gov/documents/usda-antimicrobial-resistance-action-plan.pdf>.

U.S. Department of Agriculture (2015). *Overview of the United States hog industry*. Washington, DC: National Agricultural Statistics Service, Agriculture Statistics Board, U.S. Department of Agriculture.

U.S. Department of Health and Human Services (2009). *Grade "A" pasteurized milk ordinance, 2009 revision*. Washington, DC: Public Health Service and Food and Drug Administration.

U.S. Environmental Protection Agency (2001). *Emissions from animal feeding operations*. Washington, DC: U.S. Environmental Protection Agency.

Von Keyserlingk, M. A. G., Martin, N. P., Kebreab, E., Knowlton, K. F., Grant, R. J., Stephenson, M., & Smith, S. I. (2013). Invited review: Sustainability of the U.S. dairy industry. *Journal of Dairy Science, 96*, 5405–5425.

White, S. (2011). From globalized pig breeds to capitalist pigs: A study in animal cultures and evolutionary history. *Environmental History, 16*(1), 94–120.

Winsten, J. R., Kerchner, C. D., Richardson, A., Lichau, A., & Hyman, M. J. (2010). Trends in the Northeast dairy industry: Large-scale modern confinement feeding and management-intensive grazing. *Journal of Dairy Science, 93*, 1759–1769.

Zeder, M. A., Emshwiller, E., Smith, B. D., & Bradley, D. G. (2006). Documenting domestication: The intersection of genetics and archeology. *Trends in Genetics, 22* (3), 139–155.

CHAPTER

5

Human Resources in the Food System

5.1 FARM LABOR

The United States is a nation of immigrants and all aspects of our country, including agriculture, have been molded by the diversity of the people who have settled here. The colonists and early settlers came to the United States looking for religious and political freedom, as well as financial gain. Some came with the intent to farm and establish a homestead. The history of farming in the United States includes independent, driven people who worked the land to make a living. The complex history of American agriculture includes the many immigrants who came voluntarily to the United States but also indentured servants, slaves, Native Americans, and undocumented migrant workers. In addition, a network of policies and laws shaped US agricultural practices.

In the centuries before Europeans arrived in North America, Native Americans developed agricultural systems based on corn, beans, and squash which were grown mainly by women. Men fished and hunted to provide meat for the diet. In the Great Lakes regions, wild rice was cultivated. When European colonists first arrived to North America, Native Americans assisted them with agriculture and shared their knowledge of crops and farming. This assistance likely allowed the colonists to survive during the early years of settlement. As the population of the colonies expanded, land was taken away from the Native Americans by force and relationships deteriorated.

It was common practice in the colonial period to use captive Native Americans and European indentured servants, who worked in exchange for their passage to America, land, or other goods for agricultural labor. Laws and regulations concerning humane treatment of these workers were few and largely unheeded. As early as 1619 it is thought that Africans were brought to the colonies to work in agriculture and to do other manual labor. The slave trade between Africa and the colonies became very profitable for the slave traders and slave owners. The number of Africans brought to the colonies as slaves quickly outnumbered European immigrants. West and Central Africans had experience with farming and were therefore considered highly valuable workers. It has been estimated that by 1775, there were 500,000 African slaves in the colonies. As the slave population grew, so did the impact of Africans on agriculture. Africans were familiar with growing rice, tobacco, and indigo, which were not common in Europe, and their skills helped make these crops profitable. Many of the slaves brought knowledge of crop rotations, irrigation, and fertilization techniques that were implemented by their owners. But it was the sheer work effort of slaves that

Understanding Food Systems.
DOI: http://dx.doi.org/10.1016/B978-0-12-804445-2.00005-3

created the profitable agricultural system of the Southern plantations and farms along the Atlantic seaboard. In Florida where large cotton and tobacco farms were established, the total population in 1860 was 140,000 and nearly 62,000 were African slaves. The estimates of total slave populations are poorly documented, but possibly as many as 4 million Africans were working in the United States at this time.

Not all areas of the United States engaged slaves in agriculture labor. Settlers in Northern colonies and those moving into the Midwest preferred small farms that could be managed by a single family unit. In these families, which were often multigenerational, everyone, including children, was expected to work and contribute to the farm operations. The concept of the landowner who himself toiled on the land was very different from the Southern plantation owner who never touched a plow. For the most part, the Northern states denounced slavery and abolitionists provided refuge for escaped slaves. During the 1830s the Underground Railroad, which was a network of safe houses and routes leading from the Southern to Northern states, assisted slaves in escaping from their owners. This stark contrast between Southern and Northern philosophies about the acceptance of slavery and the approach to agriculture created a deep philosophical divide in the country that eventually led to the Civil War.

During and after the Civil War, the building of the railroad system brought immigrants from China into the Midwest and West, many of whom settled in California. From about 1860 to 1900, Chinese farm workers contributed significantly to converting the swamplands of the San Francisco Bay region and the Sacramento–San Joaquin River delta regions into fertile farmland. Chinese farm laborers adapted to the needs of fruit and vegetable farming by arriving when crops needed to be harvested. The Chinese labor force was largely not welcomed by the white working class who viewed them as taking their jobs. This tension increased in the latter part of the 1880s as people flooded into California from other areas of the United States seeking work. Under pressure to make jobs available to American settlers, the Geary Act of 1892, which blocked Chinese immigration and imposed deportation of all Chinese illegally present in the United States, was passed. This led to widespread skirmishes across California and thousands of Chinese workers were threatened, killed, or forced to leave the state. In 1902, Congress extended the exclusion of Chinese immigration indefinitely and denied naturalization to Chinese immigrants.

Following the Civil War and passage of the Thirteenth Amendment in 1865, which abolished slavery, working conditions for African Americans (terms used to define African Americans have included Negros and blacks) did not improve. Southern states implemented laws that imposed segregation of blacks from whites in all aspects of society. The so-called "Black Codes" and "Jim Crow" laws prevented blacks from owning property, participating in local governance, and attending public schools. Day-laborers, sharecropping, and lease tenant arrangements were used to keep black workers on the farm, but without allowing them access to ownership. Sharecropping was the process by which plantation owners continued to own the land and provide the inputs of seeds and tools, while the sharecroppers provided the labor to produce the crop. Part of the profit was shared with the sharecroppers. This system allowed more freedom for the sharecroppers, but in reality most sharecroppers remained in debt to the landowners and thereby tied to them for years. Tenant relationships were often similar to sharecropping because the options for rental payments could be defined by the landowner. Not all sharecroppers and tenant farmers were former slaves; many white farmers who had lost their farms and income during the war also were forced to rely on these arrangements to make a living.

Freed blacks who were not engaged in agriculture were typically relegated to low-paying jobs and segregation kept them from educational opportunities. Blacks could be arrested for being unemployed, or for just about any reason, and sent to prisons where they were forced to work on farms or road crews in "chain gangs," which essentially was a return to slave labor. Black children could be assigned to "apprenticeships," which were also a form of indentured servitude. The policies of segregation of blacks in Southern states created racial divides that kept blacks from fully participating in society and eventually erupted in the civil rights turmoil of the 1960s.

As agricultural technology advanced, less manual labor was needed on farms. But a new problem arose: finding enough labor at peak times during planting and harvesting, without the costs of hired hands year-round. In some areas, farm families formed cooperatives to help each other with these tasks. In other areas where the demand was greater, seasonal workers that moved with the crops provided the required labor. Migrant labor camps with minimal housing and services were established, where families would live while working the fields. With no oversight or regulations to ensure worker's rights, migrant laborers were frequently not paid for their work or given less than what they had been promised.

Government programs to assist black farmers were attempted after the Civil War. The Second Morrill Act in 1890 provided funding to establish agricultural colleges for black students (1890 schools). Booker T. Washington and George Washington Carver were leaders in promoting agricultural education among black farmers through their work at Tuskegee University (Chapter 2: History of US Agriculture and Food Production). Washington was well respected by then-President Theodore Roosevelt and the wealthy philanthropist Andrew Carnegie. Through these connections and his work at the local level, Washington developed several programs to support black farmers in purchasing land, improving farm production and diversification of crops. Despite these efforts, the economic conditions from 1920 through the Depression forced many black farmers off the land. The New Deal programs were generally not accessible to black farmers and the number who were able to continue farming decreased. In the 1900–30s about 13%–14% of all farmer operators were black (Fig. 5.1). The distribution of black farmers was predominantly in the Southern states where cotton and tobacco were the primary crops. Black farm operations were generally smaller than those of white farmers, which limited the income potential and restricted the type of farming that was possible.

For many reasons, black farm operators began to decline significantly after 1930, such that by 1960 they represented only about 7% of farmers in the United States. This trend was associated with expansion in the size of farms and increased use of technology nationwide. Black farmers tended to be tenant farmers, and were not financially able to adapt to these changes. Access to financial resources including loans and insurance was a challenge for black farmers in part due to discriminatory practices (discussed in Section 5.4)

5.2 FARM LABORER UNIONS

During the industrialization period in the United States employers were generally not held accountable for worker rights, including wage minimums, working hours, and conditions, and expectations of job security. In factories and manual labor jobs, employees were often placed in dangerous environments where accidents and even deaths would occur. Safeguards for workers were not ensured by federal laws. Attempts to improve the plight of the general working class population in the 1900s included the founding of the Industrial

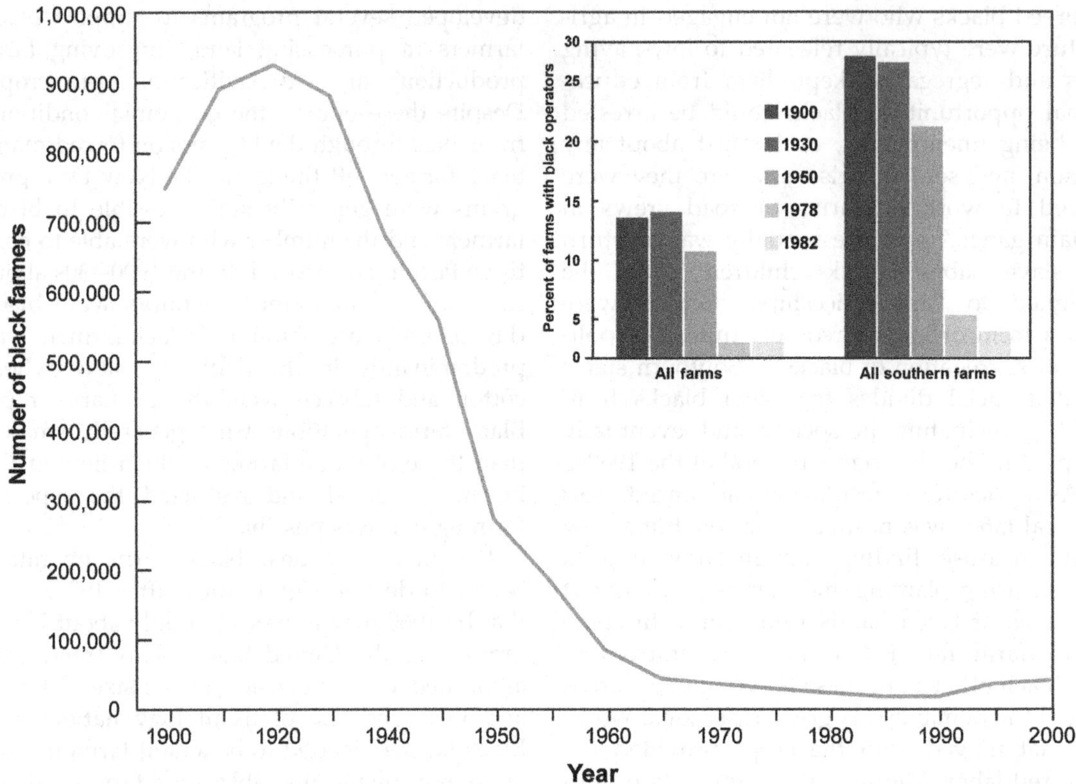

FIGURE 5.1 The number of black farmers in the United States was high in the 1900s but has decreased to less than 34,000 today. As shown in the inset, about 1 in 4 farmers in southern states through the 1950s were black. *Source: Banks, V. J. (1986).* Black farmers and their farms. ERS Agriculture and Rural Economics Division Research Report No. 59. *Washington, DC: Economic Research Service, U.S. Department of Agriculture (Banks, 1986); Cohen, R. L., & Horton, C. (Eds.). (2012).* Black farmers in America: Historical perspective, cooperatives and the Pigford cases. *New York, NY: Nova Science Publishers, Inc. (Cohen & Horton, 2012).*

Workers of the World (IWW, or Wobblies) union. At that time, unions were not recognized by the government as legal entities and employers could fire employees that started or joined a union. The IWW engaged workers from several manufacturing sectors, including agriculture. In California, the IWW sponsored many strikes with agricultural workers to protest labor conditions and wages. But that came to an abrupt end when the United States entered WWI. The federal government arrested thousands of IWW members and accused them of obstructing the war activities of the government. As the United States prepared for the needs of the war, not enough workers were available for agriculture, mining, railroads, and construction projects. Prior to 1917, Mexican workers had entered the United States freely, but a bill passed that year implemented a literacy requirement and $8 tax for anyone over 16 years of age in order to enter the US. This severely limited the number of Mexicans who were either allowed to or could afford to enter the United States. With the increased demand for workers to support the war efforts, the industrial and agricultural sectors put pressure on the government to change this policy and it responded. In an abrupt turnaround in 1918, all restrictions on Mexican workers were lifted. Even after the war ended the Immigration Act was so lax that farm owners had virtually unhampered access to Mexican labor. This led to a rapid shift in the population of Mexican workers across the Southwest, most of whom continued to live in Mexico but came to the United States during the harvests.

With the influx of migrant labor farmers were overwhelmed with finding and hiring workers. Eventually, a contractor system developed whereby a labor contractor negotiated work with the farmer for a group of workers. The contractor earned a fee from the farmer and supervised the field work, taking a percentage of the profit. The farmers liked this system because they did not have to negotiate with individual workers and were not responsible for their working conditions. This system was not ideal for workers as the contractors had great power and often did not treat them fairly. Contractors would frequently keep a large percent of the wages earned by the workers and would deny work to those who complained.

In an effort to protect workers from abuse by employers, unions were formed. A union is a legal representative of a group of employees who pay a fee to belong to the union. The union acts to negotiate or bring awareness of worker issues to the employer. Filipino workers based in Seattle and working in the fishing industry in Alaska, Washington, Oregon, and California formed the Cannery Workers' and Farm Laborers' Union in 1933. This merged with the United Cannery, Agricultural, Packing and Allied Workers of America (UCAPAWA) union, which had been established in 1937 to address worker status during the Depression. UCAPAWA founder Donald Henderson was an economics instructor at Columbia University and a member of the Communist Party, an affiliation that made the group highly controversial. Members of the UCAPAWA included Mexicans, black sharecroppers, and Chinese, Japanese, and Filipino workers from across the country, with the highest number of members in California. Turmoil within the group eventually led to its dissolution in 1950.

During the Depression, the desperate state of the economy and lack of jobs created an environment where workers' rights were ignored. People needing any type of job would work long hours for little pay, children were put to work in dangerous jobs and some employers cheated their employees on wages. To correct this situation, Congress passed the National Labor Relations Act (NLRA) in 1935 (Table 5.1). The NLRA protected the rights of employees and employers, ensured safe work environments, and allowed for collective bargaining. Under the NLRA, workers could form unions to negotiate wages, benefits, working conditions, and other issues with employers without risk of being fired. The National Labor Relations Board (NLRB) enforces NLRA and the rights of workers to bargain collectively. Collective bargaining is negotiation between workers, usually through a union, and employers to reach agreement about wages, hours, benefits, and working conditions. Negotiations are usually brokered by union leaders. While the NLRA was an important legislation for American workers overall, agricultural workers were excepted from the NLRA protections and were not ensured the right to form unions. Despite the lack of protection, formation and dissolution of agricultural unions occurred and were interconnected with political and social upheavals of the emerging nation. The Associated Farmers of California, a group of large landowners, broke up strikes by workers in the early 1930s and prevented migrant workers from organizing. The Southern Tenant Farmers Union was established to protect the rights of sharecroppers in Arkansas in 1934 in place of the UCAPAWA, which had been ostracized in the state due to the group's association with Communist ideals. Later, in the 1950s, young farmers in the Midwest organized the

TABLE 5.1 History of Agriculture Labor Regulations

1935	National Labor Relations Act (NLRA)	Structure for collective bargaining, prevents employers from firing a worker for joining a union. Farm workers were excluded from this protection
1938	Fair Labor Standards Act (NLSA)	Established minimum wage and overtime pay regulations. Farm workers excluded until 1966
1942	Mexican Farm Labor Agreement (Bracero program)	Negotiated agreement to allow Mexicans to work in US agriculture
1966	Fair Labor Standards Act amendment	Lifted exclusion for farm workers from the NLSA
1975	California Agricultural Labor Relations Act (CALRA)	Provided protection to California farm workers to form or belong to a union. Resulted from efforts by Cesar Chavez, and many violent and passive protests by farm workers
1983	Migrant and Seasonal Agricultural Worker Protection Act (MSAWPA)	Regulations to protect rights of farm workers to earned wages, safe housing, and transportation. Requires farm labor contractors to be registered
1986	Immigration Reform and Control Act (IRCA)	Created the H2-A visa for temporary agricultural workers

National Farmers Organization (NFO) based in Corning, Iowa to negotiate with food marketers. Their approach was to withhold product from the markets until the buyers agreed to the prices they demanded. This was ultimately a flawed and ineffective strategy. Because farmers were not able to withhold their perishable products indefinitely, the buyers simply waited them out. When the farmers could hold out no longer and flooded the market with their products, wholesale prices fell and farmers lost money. The NFO sometimes used tactics such as physically blocking farmers from taking their commodities to markets, killing cattle rather than selling them, and dumping milk and grain in roadside ditches to bring attention to their cause. These actions did not gain them widespread support from either the agricultural community or the public, and the NFO changed their tactics. Today the NFO is a nonprofit organization that serves to enhance communications between agriculture producers and the government. Other union groups representing a wide range of agriculture and industry groups rose and fell during the early 1900s while the wages paid to agricultural workers remained low and their working conditions largely unregulated.

The entry of the United States into WWII created new pressures on agricultural labor. As well-paying jobs in cities became available to address wartime industrial needs, agriculture labor became scarce. In response, the United States government negotiated with the government of Mexico to establish the Bracero agreement of 1942. This was an executive order approving the temporary migration of Mexican agricultural workers to the United States. The agreement, called the Mexican Farm Labor Program, was negotiated by the Mexican government with the hope of improving working conditions for its citizens. According to the agreement workers were to be guaranteed payment of a wage comparable to domestic workers (30¢/hour); adequate, sanitary, and free housing; adequate meals at reasonable prices; occupational insurance paid by employers; and transportation back to Mexico at the end of the contract. In practice, most of these regulations were not enforced and labor wages dropped across the country. Between 1942 and 1950 over 500,000 Mexicans

FIGURE 5.2 The Mexican Farm Labor Agreement of 1942, also known as the Bracero program, allowed agricultural workers from Mexico to work in the United States. Mexican workers are shown going through a customs inspection at a Bracero center in Texas in 1957. *Source: Photo from the Department of Labor.*

entered through Bracero contract centers (Fig. 5.2). Originally intended to be in place only during the war, the Bracero agreement became Public Law 78 in 1951, due to lobbying from large farm owners who benefited from the program, and was reapproved biennially until 1964.

It is estimated that in California, Braceros were the main source of farm labor during this time, essentially displacing domestic workers. In addition, illegal immigration outside of the Bracero program was widespread. Outcry from domestic workers about Braceros remaining in the United States after their contracts ended and the increased illegal immigration from Mexico prompted the Immigration and Naturalization Service to implement Operation Wetback in 1954. Teams of Border Patrol agents rounded up over 1 million Mexicans in the first year of the program and sent them back to Mexico. This process continued through 1962. The derogatory term *wetback* was given to Mexicans who crossed the Rio Grande to enter the United States and the term became linked to any Mexican laborer. Throughout this period, despite the government deportation program, illegal migrant farm workers continued to enter the United States and were hired by employers, resulting in no real change in the population of undocumented workers. Operation Wetback did not end Mexican migrant labor in the United States, but it resulted in increased border control between the United States and Mexico and highlights the ongoing struggle to find a balance between the need for farm labor and access to legal work in the United States. Immigration policy concerning Mexican workers remains a controversial issue in US politics to this day.

In an effort to address the issues of undocumented farm laborers, Cesar Chavez founded the National Farm Workers Association in 1962, which became the United Farm Workers (UFW) union. Using nonviolent protests, including strikes and marches, Chavez worked to gain better working conditions and pay for agricultural workers (Fig. 5.3). To raise

"Across the San Joaquin Valley, across California, across the entire Southwest of the United States, wherever there are Mexican people, wherever there are farm workers, our movement is spreading like flames across a dry plain. Our Pilgrimage is the match that will light our cause for all farm workers to see what is happening here, so that they may do as we have done. The time has come for the liberation of the poor farm worker. History is on our side. MAY THE STRIKE GO ON! VIVA LA CAUSA!"
Cesar Chavez

FIGURE 5.3 Cesar Chavez founded the United Farm Workers of America union and led efforts to improve agricultural workers rights in California. Through protests and strikes, workers sought better working conditions and fair labor standards. This poster was printed around 1970 during a nationwide movement to boycott grapes in an effort to bring attention to farm labor concerns. *Source: Photo from the National Museum of American History.*

awareness of his cause, he underwent periodic, prolonged fasts. Some of these were joined by celebrities and political leaders of the time, including Jesse Jackson, Martin Sheen, Emilio Estevez, Kerry Kennedy (the daughter of Robert Kennedy), Danny Glover, and Whoopi Goldberg. UFW was successful in obtaining the first collective bargaining agreement between farm workers and growers in 1966 that included required rest periods, toilets in the fields, clean drinking water, and handwashing facilities, protective clothing during pesticide applications, and testing of workers for pesticide exposure. Senator Robert Kennedy was supportive of the farm workers' movement and met with Chavez, which gained the UFW national recognition. After Kennedy's death, Chavez and Dolores Huerta, a coleader of the UFW, created the Robert F. Kennedy Farmworkers Medical Plan to provide health benefits for farm workers, which still operates as a nonprofit organization today.

The efforts of Chavez and others to bring a focus on agricultural workers in California laid the groundwork for the passage of the California Agricultural Labor Relations Act (CALRA) of 1975 (Table 5.1). This legislation was the first enacted by a state to allow agricultural workers protection to engage in collective bargaining and form unions. Similar to the NLRA, CALRA is implemented by the Agricultural Labor Relations Board, which reviews and settles petitions related to employer–employee union relationships.

Farm operators were also interested in forming unions to increase their collective influence on the political system. Farm groups such as the Farmers Union (organized in 1902) and the American Farm Bureau Federation (AFBF; established in 1919) organized farmers to address low commodity prices and the impending loss of farms and their livelihoods. The Farmers Union was radical, liberal, and willing to take risky actions (strikes, crop withholding, and destruction) and appealed to marginal and struggling farmers. The Farm Bureau was more conservative than the Farmers Union and supported by stable and wealthy farmers. This division between the philosophies of these farm groups is still evident. Today, the National Farmers Union (NFU) has about 200,000 members and 21 state chapters. The American Farm Bureau Federation is significantly larger with over 6 million members and chapters in all 50 states. The mission of both NFU and AFBF is to be a unified voice for agriculture, and to provide education and services to members. Legislative lobbying at the federal and local levels are part of both organizations' activities.

EXPANSION BOX 5.1

UNION WORKER STRIKE AND LOCKOUT

Employers and unions negotiate contracts for wages, benefits, and working conditions. If an agreement cannot be reached, the conflict is sent to "arbitration" by a neutral third party, according to NLRA regulations. The refusal of employees to work is referred to as a "strike" and a "lockout" is a denial of work for employees by the employer. Both are used during contract negotiations to force the other party to accept contract changes. Recently, lockouts in professional sport leagues involving players and referees have generated much publicity. Lockouts and strikes are not uncommon negotiation strategies in agricultural and food processing industries. The lockout of 1300 Bakery, Confectionery, and Tobacco Workers and Grain

EXPANSION BOX 5.1 *(cont'd)*

Millers International Union (BCTGM) employees from American Crystal Sugar Company plants in August 2011 is one example.

The Red River Valley Sugarbeet Growers Association (RRVSGA) was formed in 1926 to represent the farmers who produced sugar beets for the American Beet Sugar Company. RRVSGA is a farmer cooperative corporation with 2800 grower-shareholders, managed by a CEO and board of directors. The farmers in this corporation produce 15% of the sugar sold in the United States. The American Beet Sugar Company, which later became the American Crystal Sugar Company (ACSC), is a publicly held New Jersey corporation in operation since 1899. ACSC built the first sugar beet processing factory in East Grand Forks, Minnesota in 1926. Four other factories were built by ACS in subsequent years: Moorhead, Minnesota (1948); Crookston, Minnesota (1954); Drayton, North Dakota (1965); and Hillsboro, North Dakota (1974).

In 1973, RRVSGA acquired the business and assets of ACSC. The intent of the merger was to have close collaboration between growers and processors so that production acreage, quantity, and quality of sugar could be optimized to allow farmers to better manage market fluctuations. The production and processing of sugar beets has contributed significantly to the economic viability of local communities in the Red River Valley region of northwestern Minnesota and northeastern North Dakota. In 2011, the industry represented an economic impact of $1.7 billion, employed 2273 full-time workers and 18,830 secondary jobs, generated tax collections of $105 million, and contributed $15.4 million in property taxes.

Employees working in the ACSC belonged to the BCTGM union. In 2011, ACSC offered the union employees a contract that included a modest increase in wages, higher worker contributions to health insurance, fewer seniority rights, and allowed ACSC the right to contract out union jobs. Workers were unhappy with the contract, concerned that it would replace union workers with non-union workers, dismantle seniority, and cut healthcare benefits, so they refused to accept the offer. When the negotiations broke down between ACSC and BCTGM, ACSC locked out the union workers and hired temporary replacement workers (Fig. 5.4).

What had been a cordial and productive relationship for decades between the farmers and processing plant workers became a bitter and contentious battle. Communities that relied on both full-time and seasonal employment in the "beet plant," usually the town's largest employer, were torn apart. Management and employees in supervisory positions worked overtime to keep plant operations going as workers picketed outside. Inexperienced replacement workers were not trained on the processing operations and several significant injuries were reported. Neighbors, relatives, and friends who found themselves on opposite sides of the lockout stopped speaking to each other. The AFL-CIO and other unions boycotted ACSC products in solidarity with the BCTGM union. ACSC production and profits decreased about 30%, from $811 to $555 million, affecting not only the workers but the small towns that relied on workers to spend money in their shops and restaurants.

As the lockout wore on, workers had to live on unemployment insurance, odd jobs, or public assistance. About 660 of the original workers decided to retire or find other work. After 20 months, in April 2013, 55% of the remaining

EXPANSION BOX 5.1 (*cont'd*)

FIGURE 5.4 Members of the Bakery, Confectioners, Tobacco and Grain Millers Union were locked out by the American Crystal Sugar Company during a labor dispute. American Crystal Sugar workers from Moorhead, East Grand Forks, Crookston, and Chaska, Minnesota; Hillsboro and Drayton, North Dakota; and Mason City, Iowa protest management on August 11, 2011. *Source: Photo provided by BCTGM International Union.*

union members ratified the original offer, which was enough to end the lockout. The damage done to the company was significant in terms of both financial and personal relationships. The farmers had formed a cooperative for collective bargaining power and the workers formed a union for the same reason. During the lockout sugar beet growers feared that their livelihoods were at risk because of the union workers' demands, and union workers feared their jobs were being marginalized by the corporation. There were legal guidelines for resolving employer–employee disputes, yet the process was divisive and frustrating, and took almost 2 years to settle. These types of conflicts illustrate the interwoven relationships of the components of the US food system that involve farmers, workers, food processing industries, and the complex interactions between economic, social, and political factors.

Suggested websites: www.nlrb.gov/resources/national-labor-relations-act and *http://www.mprnews.org/story/2011/12/14/american-crystal-sugar-lockout-hillsboro*

5.3 GOVERNMENT REGULATIONS FOR MIGRANT FARM WORKERS

Migrant farm work became an essential component of US agriculture following WWII. During this time more Americans were attending college and finding work in urban areas, while farms were being consolidated into fewer, larger operations. The number of people interested in agricultural work was decreasing. Migrant laborers, defined as temporary workers who move along with the crop rotations, became essential to meet the planting and harvesting needs of farmers. Workers were mainly Mexicans in the Southwest, but in Southeastern states American citizens, both black and white, were engaged in this work. Migrant farm laborers were not protected by a union or government regulations. They were paid low wages, forced to live in temporary and inadequate housing, and had limited access to healthcare or education.

During his campaign for president, John F. Kennedy had become aware of the deplorable living conditions of migrant workers. After he was elected to the presidency, Kennedy signed the Migrant Health Act of 1962, which provided federally funded healthcare services in medically underserved areas. Grants were given to over 120 community-based and state organizations to provide comprehensive medical care for migrant and seasonal farm workers and their families. The act was reauthorized in 1966 with the addition of hospitalization services. Further improvement in healthcare for migrant and seasonal farm workers was implemented in 1972 with the formation of the National Advisory Council on Migrant Health. Members of the council are mainly farm workers who also serve as governing board members of local federally funded migrant healthcare programs. The mission of the council is to inform and make recommendations to the Secretary of the Department of Health and Human Services (DHHS) regarding the healthcare needs of farm workers. To protect migrant and seasonal farm workers from exploitation by employers, the Migrant and Seasonal Agricultural Worker Protection Act was passed in 1983. Within this law, farm labor contractors were required to register, and several protections were implemented to prevent farm workers from being unfairly treated regarding wages, transportation, housing, intimidation, and discrimination.

The Healthcare Safety Net Act, authorized in 2008, provided funding for community health centers to serve medically needy populations and to expand healthcare services in rural areas. Today, all but 10 states have at least one migrant health centers (several states have many) that provide care to farm workers, which is administered by the Bureau of Primary Health Care in the Health Resources and Services Administration of the DHHS.

The Affordable Care Act (ACA), signed by President Barack Obama in 2010, required all people living in the United States to have health insurance. Employers with more than 50 employees had to provide health insurance for full-time workers. Farm operations with 50 or more employees needed to determine how to cover the costs of insurance, either to pay it from their profits or pass some of the costs to the employees. Beginning in 2015, agricultural workers on H-2A work permits and who would be in the United States for more than 3 months, were required to purchase health insurance. There were several options for obtaining health insurance that varied by state, which made complying with the law challenging. The ACA provided financial subsidies to assist workers with gaining access to healthcare plans. Workers on H-2A permits are not eligible for Medicare, and the ACA required employers to cover worker's compensation costs for those injured on the job. The ACA is under reconsideration by President Donald Trump following his election in 2017.

Migrant workers suffer from high rates of chronic disease including hypertension, diabetes, and obesity, as does the general US population. Treatment for these conditions is costly. How the health of farm workers will be impacted by the ACA or a policy that might replace it, and the economic effects of this law on farm operators and the food system, will be determined in the coming years.

5.4 FARM WORKERS TODAY

Farming has always been hard work, whether done with mules and horses or computer-managed machinery. At times the work requires long hours, because when the crop is ready to be planted, tilled, treated with pesticides, or harvested, it must be done without delay or the crop will be reduced in quality or lost. Animals

EXPANSION BOX 5.2

HARVEST OF SHAME

In 1960 Edward R. Murrow, Fred Friendly, and David Lowe produced the CBS documentary *Harvest of Shame*, which aired on national television the day after Thanksgiving. The awful plight of migrant workers was brought to the public's attention dramatically with this CBS News report. Their intent was to focus public attention on the impoverished world of migrant farm workers, juxtaposed with the bountiful food supply enjoyed by Americans. The producers spent 9 months traveling from Florida to New Jersey interviewing migrant workers, elected officials, and leaders of farm organizations. The film showed the desperate situation of both white and black workers and their children as they picked the seasonal crops. Families lived in shacks with inadequate sanitation, and children were often left alone while parents worked the fields. The amount of money they earned, even when they had work, was not enough to provide enough food or to improve their living conditions. When weather damaged the crops, there was no work and no pay and the workers were dependent on handouts from the community to survive. Children attended school infrequently when it was available but few were able to graduate from high school. The government officials that were interviewed suggested something needed to be done, but pointed out the costs, challenges, and political nature of addressing the problem. A representative of the American Farm Bureau Federation, interviewed in the documentary, rejected the idea that government policies were needed, instead implying that farmers should be able to manage their workers as they saw fit.

The impact of *Harvest of Shame* was significant because of the timing, and the connection Murrow had with the public. He was a well-respected journalist and he used his position to advocate for change. His closing words in the documentary were "The people you have seen have the strength to harvest your fruit and vegetables. They do not have the strength to influence legislation. Maybe we do." Clearly this was a call to action. Some groups tried to discredit the documentary, saying it was one-sided and inaccurate. A *Time* magazine editorial considered it an "exaggerated portrait." It is likely that the documentary was influential in moving Congress to pass legislation that was already pending to fund health services for migrant workers and education for migrant children.

In the over 50 years since *Harvest of Shame* aired, migrant workers have been the focus

EXPANSION BOX 5.2 (*cont'd*)

of other news stories and documentaries. Journalists from NBC in 1990, CBS in 1995, and CBS in 2010 interviewed workers in the same locations that were visited in the 1960 documentary and found very similar situations. A change in the demographics of workers had occurred, with fewer blacks and more workers from Mexico and Central America. Workers still lined up in the early morning hours to be hired for work and were loaded onto buses to be taken to the fields. Wages remained below the poverty line, education of children was erratic, and most of the workers had no savings or hope for a better future. The role of migrant workers in US agriculture is complicated and politically charged. Paying a living wage to workers would raise the cost of food, but to not do so may be ethically and morally corrupt.

Suggested video: Harvest of Shame *(1960).*

must be fed, watered and cared for throughout the day, and everyday. The unpredictability of weather and market prices are a source of stress for those relying on crops for an income, regardless of the size of their operation. Farmers, ranchers, and other agricultural managers today must interpret and follow governmental regulations and requirements, manage finances and markets, keep up to date with new scientific developments, participate in interest groups, co-ops and unions, and stay abreast of political changes that will affect their operations. Farmers must enjoy working outside, doing physical work, working with plants and animals, and operating and maintaining machinery. The entrepreneurial nature of farming is attractive to those who thrive on managing their own businesses and are driven to innovate and apply new ideas and technologies. Many farm operations are intergenerational so the cultural and family ties associated with farming are also important.

At the turn of the 20th century, 41% of US workers were engaged in agriculture (Table 5.2). Since the 1900s mechanization and other innovations have decreased the need for some types of farm labor, such as planting and harvesting of cotton, corn, wheat, and soybeans, or milking of dairy cows. As described in Chapter 3, Innovations in US Agriculture, farm sizes have increased and farm operations are more specialized than they were a century ago. A consequence of these changes has been a decrease in the percent of US workers involved in agricultural production to less than 2% of the population. Concurrently with the decrease in the number of farmers, rural populations have decreased from 60% to 20% of the total population in this time period. The percent of the gross domestic product (GDP) generated by agriculture also decreased during this time.

The 2011 USDA Agricultural Resource Management Survey found there were 2.1 million farms operated by 3.2 million principal farmers (Table 5.3). Principal farmers are those persons responsible for the day-to-day operations of the farm. The majority of farmers were white (96%) and male (86%) with an average

TABLE 5.2 US Workforce in Agriculture and Value to GDP

	1900	1930	1945	1970	2000
Percent of US workforce in agriculture	41	21.5	16	4	1.9
Agriculture percent of US GDP	—	7.7	6.8	2.3	0.7

USDA Economic Research Service.

TABLE 5.3 Characteristics of Farm Owners in the United States

	Principal operators	Total farmers	Main type of operation	Main size of operation
White, Male	1,821,039	2,210,402		
Women	288,264	969,672	26% Combined crops	76% < $10,000
Black	33,371	44,629	48% Beef cattle	79% < $10,000
Hispanic	67,000	99,734	36% Beef cattle	68% < $10,000
Asian	13,669	22,140	36% Fruits and nuts	43% < $10,000

USDA-National Agricultural Statistics Service, AgCensus 2012.

age of 58 years. A higher percentage of farmers had completed high school (42%) than the United States general population (29%) and nearly 25% of farmers had completed college.

5.4.1 Women and Minorities in Agriculture

About 288,000 women are principal farm operators and 1 million women are spouses of principal operators (Table 5.3). The average hours of farm work are substantial for women principal operators (1097 hours/person/year) and secondary operator spouses (895 hours/person/year). There are also nearly 1 million nonoperator spouses (spouses of farmers who do not make management decisions), who also contribute significant time (818 hours/person/year) to farm operations.

USDA statistics show that the share of US farms operated by women nearly tripled over the past three decades, from 5% in 1978 to 14% in 2007. The majority of women-operated farms, roughly 75%, have annual sales of less than $10,000 and overall accounted for only 16% of US agricultural sales. At the other end of the spectrum, 5% of women-operated farms had sales of $100,000 or more in 2007. Most of these farms specialized in grains and oilseeds, specialty crops, poultry and eggs, beef cattle, or dairy. The poultry and egg specialization alone accounted for roughly half of women-operated farms with sales of $1 million or more. Nearly half of farms operated by women specialized in grazing livestock, including raising beef cattle (23%), horses and other equines (17%), or sheep and goats (6%). Texas has the most women farmers, but Arizona has the highest percentage of women farmers (45% of all farmers in the state).

The number of black farmers decreased significantly after 1920 such that by 1960 they represented only 7% of farmers in the United States (Fig. 5.1). There were many reasons for this decline, including limited access to financial resources. Racial and economic discrimination compounded the challenges faced by black farmers. In 1997 and 1998 the National Black Farmers Association filed two class-action lawsuits against the USDA claiming there had been systematic racial discrimination against African American farmers in approving farms loans for decades. These two cases, *Pigford v. Glickman* and *Brewington v. Glickman,* charged that Secretary of Agriculture Dan Glickman, as the representative of the USDA, had allowed racial discrimination in the awarding of federal farm credit. The lawsuits were eventually settled with the federal government agreeing to add $100 million to the 2008 Farm Bill plus allocating an additional $1.15 billion in 2010 to compensate farmers who had been wrongfully denied farm loans, loan servicing, or other benefits. These lawsuits brought public attention to the discrimination that was widespread among government agents in approving funding for black farmers. The USDA has made changes in its approach to diversity since these lawsuits were filed. It has put more focus on assisting minorities to enter and be successful in farming and to allocate funding for minority and low-resource farmers. The 2014 Farm Bill included funding for the Outreach and Assistance for Socially Disadvantaged Farmers and Ranchers and Veteran Farmers and Ranchers Program, also known as the 2501 Program. These funds were to be distributed as grants to academic and other institutions that work with farmers from underrepresented groups.

The number of minority farmers remains low, but upward trends are evident. In 2007 about 1.4% of all principal farm operators were black and this had increased to 1.6% by 2012. Black farmers are still more common in Southern states. Texas has the highest number of black farmers, but they make up only 3% of the total farmers in that state. The state with the highest percentage of black farmers is Mississippi, where 12% of farmers are black. Other underrepresented groups engaged in agriculture include Hispanic and Asian farmers (Table 5.3). The number of Hispanic farmers has increased 21% since 2007, and the state with the highest percentage of Hispanic farmers is New Mexico with 36%. Asian farmers represent 38% of all farmers in Hawaii, and the majority of Asian farmers specialize in fruits, nuts, vegetables, and greenhouse crops.

Farming can be a stressful occupation. Analysis of 2012 National Violent Death Reporting System statistics from the CDC found in 17 states workers in the farming, fishing, and forestry occupational group had the highest rate of suicide (84.5/100,000 people), followed by workers in construction and extraction (53.3/100,000), and installation, maintenance, and repair (47.9/100,000).

Occupational groups with higher suicide rates might be at risk for a number of reasons, including job-related isolation and demands, stressful work environments, and work–home imbalance, as well as socioeconomic inequities, including lower income, lower education level, and lack of access to health services. Previous research suggests that farmers' chronic exposure to pesticides might affect the neurological system and contribute to depressive symptoms. Other factors that might contribute to suicide among farmers include social isolation, potential for financial losses, barriers to and unwillingness to seek mental health services (which might be limited in rural areas), and access to lethal means.

5.5 ECONOMICS OF FARMING

The USDA Economic Research Service (ERS) gathers and analyzes an extensive range of agricultural economic data. The information spans from the types and amount of crops planted to export trade markets. The ERS generates reports and summaries of these data that are freely available for public use on their website (www.ers.usda.gov). Farmers and agriculture-related industries utilize this information and the reports generated by the ERS staff to make decisions about their operations. Government agencies rely on the data for assessing trends in the agricultural sector that inform policies and set market conditions.

The economic balance of a farming operation can be tenuous due to the volatility in markets and the unpredictability of weather, among other variables. Farmers must make decisions about what to plant or how many livestock to raise before they know what the markets will bear or the prices they will receive. The amount of capital investment in large farm operations can be significant, including the costs for land, machinery, labor, chemical inputs (fertilizer and pesticides), seed, feed for animals, and energy. Farmers rely on economic data and predictions of markets to make their decisions about their operations. Having the ability to interpret and apply such complicated economic data is essential for farming success. On-farm income is generated from agricultural sales, government payments, and farm-related activities. Agricultural sales include the revenue from the sale of crops or livestock produced on the farm. Some farmers obtain conservation payments, direct payments, loan deficiency payments, disaster payments, and other forms of government support that add to their net income. Farm-related activities that generate income may include rent of land or other property, custom work, forest product sales, recreational services, or crop and livestock insurance payments. A balance sheet for income, expenses, and net profit for all US farms combined in 2012 is shown in Table 5.4. The net profit for that year for all US agriculture was $92.3 billion. Government payments contribute a very small amount to overall income compared to the agricultural sales, in contrast to public perception. The average net farm income for the United States followed a generally increasing trend from 2000 to 2015 but then decreased essentially back to the 2000 level by 2016 (Fig. 5.5). Concurrent with the decrease in income has been an increase in farm debt since 2000. These two economic forces created financial stress on farmers but have not risen to the level seen during the 1980 farm crisis.

TABLE 5.4 Income and Expenses for All US Farms Combined

	$ Billions		
Source	**Income ($)**	**Expenses**	**Net profit ($)**
Agricultural sales	394.6		
Government payments	8.1		
Farm-related income	18.5		
Production expenses		328.9	
			92.3

USDA-National Agricultural Statistics Service, AgCensus 2012.

Another tool to gauge the economic status of agriculture is the assessment of overall productivity. Agricultural productivity is defined as the ratio of outputs to inputs. Outputs include yield of crops and livestock and inputs include fertilizer and pesticides, farm machinery, energy, and labor. Improvements in genetics and farm practices have generated an increase in agricultural outputs. The USDA has analyzed the overall productivity of agriculture annually since 1948 (Fig. 5.6). The types of inputs have shifted, from less human labor to more use of technology, but the overall costs of inputs have stayed constant during this time. As a result, agricultural productivity has improved by over 250% since 1948. This growth in productivity has occurred without an increase in the total amount of land being farmed in the United States, but the amount of resources consumed by agriculture, including energy, soil, water, and chemicals derived from fossil fuels has increased. Determining how to continue to increase productivity to meet the future demand for food, while simultaneously conserving these natural resources, is one of the grand challenges facing the agricultural sector.

5.6 HIRED FARM WORKERS

Mechanization of farm operations has reduced the need for farm labor in some types of agriculture. For certain crops, especially fruits and vegetables, manual labor continues to be essential for planting and harvesting. Caring for animals can be somewhat mechanized, for example milking machines, but other aspects require human skills. Overall, about 27% of total farm operating costs are for hired worker wages, but may be as high at

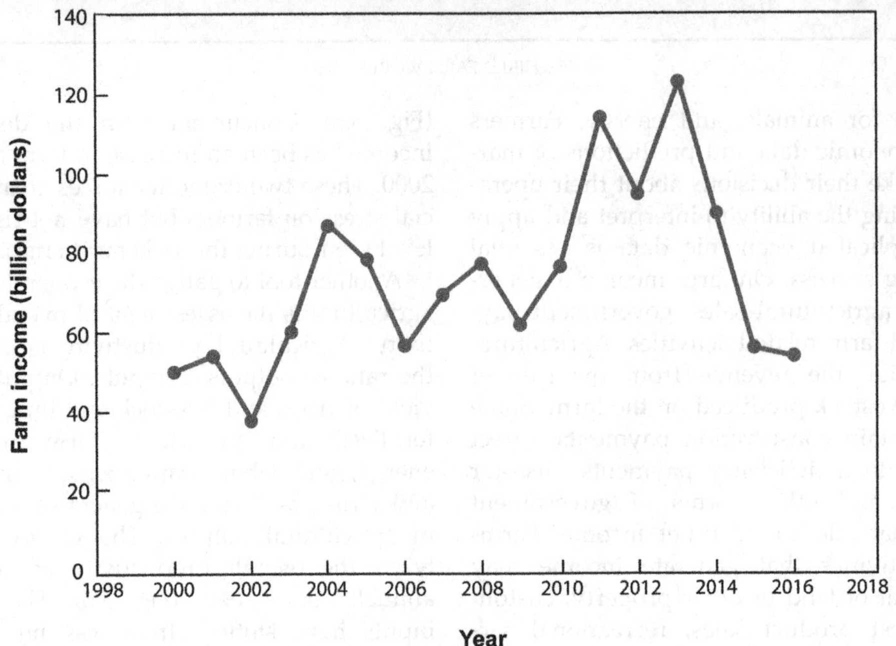

FIGURE 5.5 Net farm income tended to increase from 1999 to 2014 but has decreased since 2014. Farm income is dependent on weather, crop yields, market demands, and input costs, which can vary widely from year to year. *Source: USDA Economic Research Service, www.usda.ers.gov.*

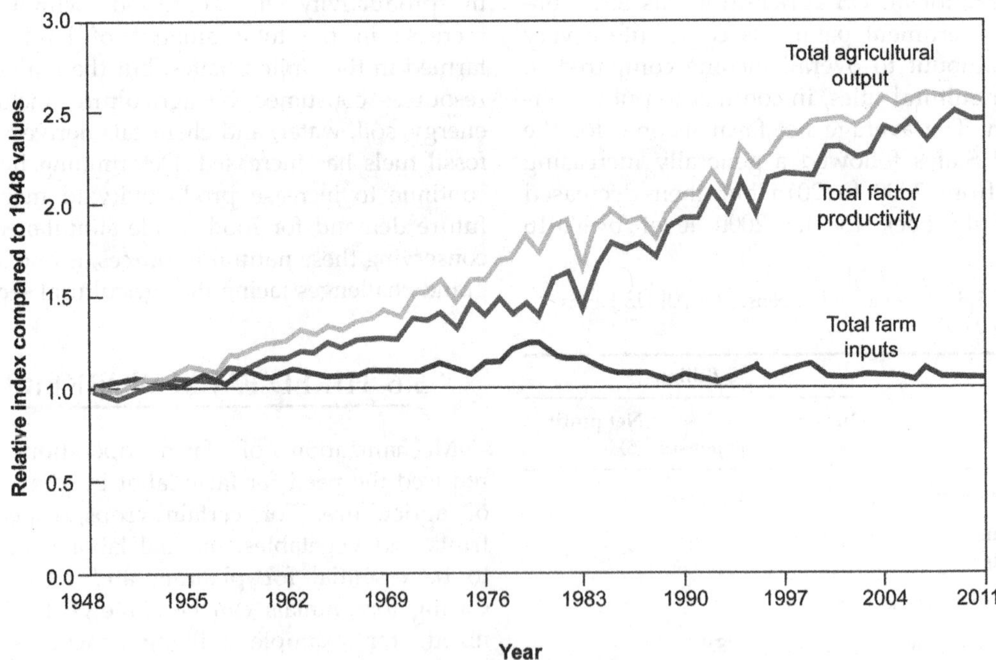

FIGURE 5.6 Total agricultural output in the United States has increased over 2.5-fold since the 1940s while inputs have remained fairly constant. US farmers have benefited from mechanization, high-quality seeds and animal breeds, and scientific advances that have contributed to high productivity. *Source: USDA Economic Research Service, www.usda.ers.gov.*

40% for some crops and animal operations. In 2012, there were over 1 million hired farm workers in the United States. About half of these were employed for the full year, one-fourth worked part time (worked <150 days in a year), and one-fourth were agricultural service workers (brought to the farm by contractors). Hired farm workers were predominantly male (82%), white (92%), US citizens (64%), and married (53%). Only about 25% had attended college and 27% had less than a ninth grade education. About 42% were foreign-born, and 45% defined their ethnicity as Hispanic and spoke Spanish as their native language. Hispanic is defined as persons from Cuba, Mexico, Puerto Rico, South or Central America, or another Spanish culture. The Hispanic population in the United States has risen significantly over the past 30 years from around 15 to 51 million people. In some regions, particularly the Southwestern states of Texas, Arizona, New Mexico, Nevada, and Colorado, Hispanics make up more than 10% of the local population. After 1970, part-time wage workers, usually Hispanic, represented 30%–35% of employees on farms.

Farm work is physically demanding, requires standing and bending to pick crops and carry buckets or trays, and must be done regardless of the weather conditions. Laborers can be exposed to hazardous conditions in the fields from chemical treatments, insects, and machinery. The hours are long and inconsistent. Workers are typically paid by the amount of product picked or hours worked. In 2011, wages paid to farm workers ranged from $8.99 per hour for farm workers and laborers to $20.48 per hour for first-line supervisors. In 2012, the average hourly earnings reported for farm laborers were $10.22 per hour. This may seem like a reasonable hourly pay rate, but farm workers are rarely able to find 40 hours of work per week every week of the year. When the harvest is bad, the workers earn less, which can leave them without sufficient income to meet their financial commitments. In some states, agricultural workers are allowed to be hired at less than the federal minimum wage (currently $7.25/hour). The Migrant and Seasonal Agricultural Worker Protection Act, which was passed in 1983, provides workers with legal protection from being mistreated by employers. The enforcement of these regulations is not uniformly guaranteed and in cases where workers do not have legal documentation abuse can occur.

5.7 US IMMIGRATION POLICIES

A major issue regarding migrant and seasonal farm workers is the high percentage of workers who do not have proper documentation to work in the United States. Immigration laws in the United States are defined in the Immigration and Naturalization Act (INA). This provides for a total of 675,000 new permanent immigrants per year. The main goals of the INA are (1) reunification of families; (2) admitting immigrants with valuable skills of economic importance; (3) protecting refugees; and (4) promoting diversity. Most farm workers would not fall into one of these categories.

Family-based immigration is granted to a maximum of 480,000 people per year, for spouses of US citizens or legal residents, their children, and siblings. Employment-based immigration provides temporary visas (in 20 types) and permanent immigration to those with desirable skills and training (about 150,000/year). Employment immigration has a cap that prevents immigrants from any one country to exceed 7% of the total immigrants in a given year. The president and Congress define the annual limit for refugees who are seeking safety from threats in their home country. In fiscal year 2013 the cap was 70,000, raised by President Obama to 100,000 for fiscal year 2017, but then decreased to 50,000 by executive order of President Trump. Political asylum may be granted by the president without limit. Individuals granted refugee or asylum status may request visas after 1 year. The diversity visa was started in 1990 and allows entry of 55,000 immigrants per year from countries with a low US immigration rate. These are awarded on a lottery basis.

The route to citizenship in the United States begins with securing a US Permanent Resident Card ("green card"), which provides for permanent residency. Green cards may be obtained through family, job, or refugee/asylum status. The process of securing a green card can be challenging, costly, and may take several years. A limited number of green cards are issued each year. Once a green card is held for 5 years, application for US citizenship may be initiated. To be eligible, the individual must be 18 years of age or older, able to demonstrate continuous residency in the United States, and be of good moral character. They must pass an English language and US history and civics exam and pay an application fee. It may take several years to work through the citizenship process.

In part because of the limits and obstacles associated with these immigration laws and the fact that undocumented laborers could find work in agriculture, manufacturing, construction, and service jobs, there has been a steady increase in the number of undocumented persons entering the United States. In 1986, President Ronald Reagan signed the Immigration Reform and Control Act (IRCA; Table 5.1) in order to control and deter illegal immigration to the United States. The IRCA included provision to enact the H2-A visa for temporary workers who perform agricultural labor or services. It also allowed permanent residency status to individuals who had lived in the United States continuously since January 1, 1986; had performed agricultural labor in the United States for at least 90 days in the year ending May 1, 1986; or had performed agricultural labor in each of the 3 years prior to 1986. It is estimated that about 2.7 million people took advantage of this law and became legal residents. The number of H-2A visas issued by the United States has increased from about 56,000 in 2010 to 108,000 in 2015 (Fig. 5.7) although not all of these would be for agricultural workers.

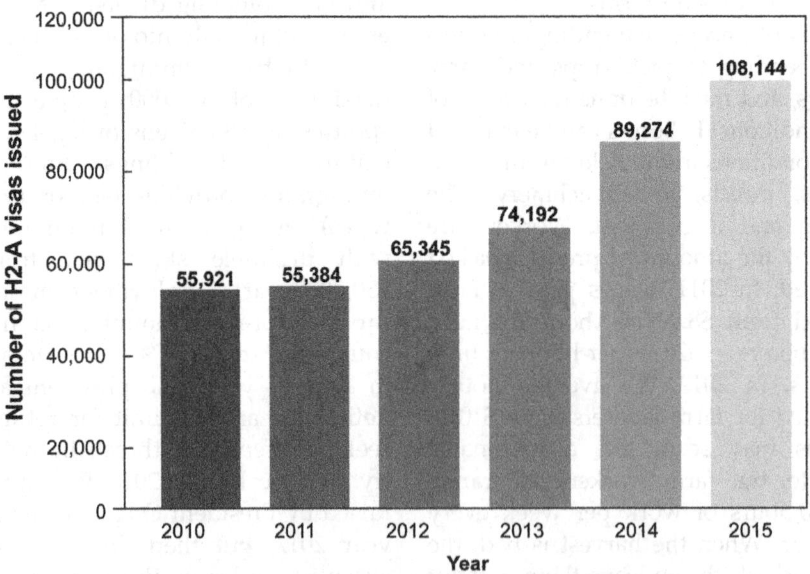

FIGURE 5.7 The number of H2-A visas issued in the United States increased between 2010 and 2015. H2-A visas are issued for temporary workers, including agricultural workers. *Source: U.S. Department of State, www.state.gov.*

IRCA implemented penalties for employers that hired undocumented workers and increased enforcement at US borders. In recent years, programs have been implemented to help employers determine if workers had the proper work authorization before they hired them. One example is the E-Verify system, an Internet-based program where employers can scan records to identify an employee's status. E-Verify has been criticized because employers are charged $150 per scan and the database is not always accurate. It is also cumbersome to use and frequently requires additional follow up to track an individual.

Despite these laws, there has been an increase in the percentage of farm workers who are unauthorized to work in the United States. The USDA estimates that the percentage of hired farm workers who were not legally authorized to work in the United States increased from 15% in 1989 to 55% in 1999, and remained fairly constant thereafter (Fig. 5.8). The highest percent of these workers are from Mexico. Hence, management of undocumented workers in the United States is a major political issue. In recent years, concerns about terrorism have further escalated border security concerns among some Americans and politicians. The southern border of the United States that is shared with Mexico is a primary entry point for undocumented workers because they can move across some parts of the relatively unpopulated border region without being detected. During his campaign for the presidency in 2016, Donald Trump declared that he would "build a wall and make Mexico pay for it" to prevent Mexicans from entering the United States illegally. Efforts to stop undocumented workers from entering the United States have been part of the political debate since the 1960s but with no substantive plan to create a different system.

Over the past decades, many social–political factors have incentivized people from Mexico, and Central and South America, to come to the United States seeking a better future. The economies of some of these countries have become

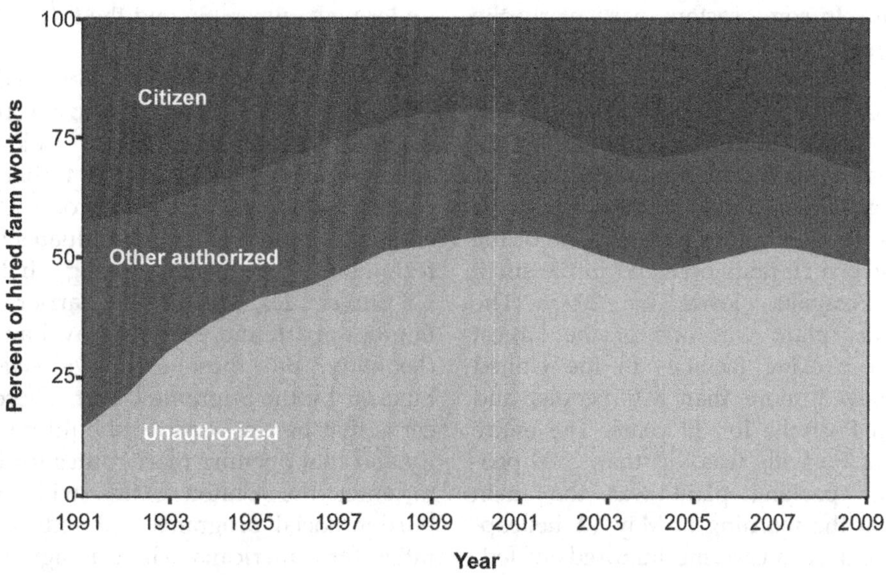

FIGURE 5.8 The number of hired farm workers who are not authorized to work in the United States increased between 1991 and 2009. Overall the number of unauthorized farm workers has been fairly constant since 2000. *Source: USDA Economic Research Service, www.usda.ers.gov; National Agricultural Workers Survey.*

weaker, and crime, drugs, and civil war have increased the flow of immigrants. Working through the United States legal immigration process is difficult, expensive, and takes years. Desperate to find work and better living conditions than in their own country, many thousands attempt to enter the United States illegally across the southern border with Mexico. Those that are not caught by Border Patrol agents stay with relatives or friends, obtain fake identification, and find work doing manual labor. Immigration and Customs Enforcement (ICE) officials estimate that there may be more than 10 million individuals in the United States illegally but are provided with an annual appropriation from Congress that allows only a limited number to be apprehended. This has led ICE to focus on three main priorities: those that have broken criminal laws, have repeatedly violated immigration law or are fugitives from immigration court. In 2011, ICE removed more than 396,000 individuals who fit those criteria.

The federal government has been largely unable to prevent the use of illegal immigrants in agriculture. In some sectors, particularly the meat packing industry, unions have been pushed out to reduce the costs of doing business. Jobs that once paid a reasonable salary are increasingly being done by nonunion, immigrant workers who accept minimum wage or less. Raids on these facilities to round up illegal workers are frequently in the news. One of the most extensive such raids occurred in the small town of Postville, Iowa in 2008. The Agriprocessors plant was one of the largest kosher meat packing facilities in the United States, employed more than 800 people, and had been in Postville for 20 years. The entire population of Postville was less than 2500 people and the packing plant was the main employer. On the morning of May 12, helicopters, buses, and vans carrying hundreds of federal agents descended on the plant and arrested almost 400 employees. Most of the workers were charged with identity theft, given expedited hearings with judges over the next 3 days, sentenced to 5 months in prison and then deported when they were released. The effects on the community were devastating. Families were torn apart, in some cases both parents were arrested and their children were left to be cared for by neighbors with no source of income. Because of the size of the raid, Postville was in the national news and the issues of illegal workers' rights and treatment were publicly debated. Some believed the government was right to enforce the law and arrest those in the country illegally; others thought the workers were being targeted unjustly when the owners should be held accountable. The owners were held accountable and in the following months, criminal charges were filed against many of the administrative staff for harboring illegal persons, use of child labor, nonpayment of overtime, and denial of medical attention for workplace injuries. Agriprocessors folded in bankruptcy soon after the trials resulting in a brief national shortage of kosher meat. Within 2 years the meat packing facility was reopened under new ownership and the town of Postville tried to rebuild.

President Barack Obama attempted to tackle the issue of immigration during the last 2 years of his presidency. He took executive action on immigration in 2014 that included proposals to streamline visa processing for foreign entrepreneurs; retain graduates in science, technology, and engineering fields from US universities; decrease the barriers that keep families apart; and enhance travel and tourism flexibility. But these executive actions were blocked by the Supreme Court. There are concerns that border security should be the priority and that opening more routes for foreigners to enter the United States will excessively burden social programs and reduce opportunities for Americans. There is agreement that immigration reform is needed, but the path to achieve that will be rocky.

As the stories of migrant laborers in the 1960s and illegal meat packing workers in 2008 made clear, agricultural labor practices in the United States face significant challenges. There are workers who are willing to take on low-paying, difficult, and physically taxing jobs in the hopes of building better lives for themselves and their families, and some employers are willing to use that to their advantage. Underpaid and illegal workers have always been part of the US agricultural workforce despite government regulations and enforcement. Consumers today seem to be more focused on the plight of animal rights than human rights in agriculture. Raising the standards for pay and working conditions for agricultural workers will clearly result in higher prices for food. Deciding what values and ethical principles we demand in our food system must include the role of human workers.

5.8 NONFARM AGRICULTURE WORK

Since the 1950s, when the term *agribusiness* was first introduced, the breadth of agriculture-related industries has expanded dramatically. According to the Agriculture Council of America, there are over 22 million people working in agriculture-related fields. The types of work include agribusiness management; agricultural and natural resources communications; building construction management; resource development and management; parks, recreation, and tourism, packaging; horticulture; forestry; food science; and fisheries and wildlife management. Statistics compiled by the USDA-ERS of agriculture-related industries also include the food and beverage sectors, textile and mill products, paper and allied products, leather products, and animal and veterinary

EXPANSION BOX 5.3

FARM WORKER PROGRAM IN ALABAMA

A concern often voiced about undocumented persons living in the United States is that they are utilizing community services and sending their children to school without paying taxes, thereby burdening these systems. Some believe that undocumented people are taking jobs away from US citizens. Others worry that criminals or persons with severe contagious diseases may be coming into the United States. Many political debates have been waged over immigration issues, but finding a working solution has been challenging.

Some states, particularly those near the Mexican border, are frustrated with a lack of federal laws controlling undocumented workers and have decided to take action at the state level. In Alabama, the number of undocumented workers increased from about 25,000 to 120,000 between 2000 and 2010 as jobs in agriculture, meat packing, and construction came available. Concern for the high number of undocumented workers led the state legislature to pass an immigration law in 2011, House Bill 56 (HB56), which allowed law enforcement officers to request anyone at any time to produce their paperwork demonstrating they were legally in the United States. The law also required farmers to use the E-Verify system to track the status of all workers before they were hired. The law made it a felony for a farmer to provide transportation, housing, or meals to an undocumented worker. Soon after the law was passed, the great majority of farm workers, even those with legal work permits, left the state over concerns that they would be arrested and

EXPANSION BOX 5.3 (*cont'd*)

deported. Rural communities were vacated, leaving store owners and businesses without customers.

The Governor of Alabama, Robert Bentley, launched a campaign called Work Alabama, which attempted to connect workers with the jobs left open by the migrant workers. Many of the jobs were in the agricultural sector and required manual labor. Despite a high unemployment rate in the state, few people applied for such work and those that did lasted only a few days before quitting. Unintended consequences, such as having to arrest an executive of the Mercedes-Benz Corporation when he was stopped while driving in Alabama with only a German ID card, hit the national news. Churches and clergy risked criminalization for providing soup kitchens or Spanish-language services to potentially undocumented persons. The effect of this law was reduced crop and livestock productivity leading to economic loss to the state and negative national press. As a result, the law began to unravel. Lawsuits against the state, claiming discrimination and unconstitutionality, were filed that effectively blocked enactment of the law. By 2013 the state had agreed to pay $350,000 to settle various suits and the tenets of HB56 were proven to be unconstitutional, unmanageable, and politically unviable. Within a few years, most of the law was rescinded.

The Alabama example demonstrates the complexity of issues regarding illegal immigrants and migrant farm workers. Producers are increasingly developing mechanization of crop production and harvesting to avoid the regulations and problems associated with seasonal workers. Mechanization can reduce the need for human labor but will also affect the types and quality of food that reach the marketplace. Intensely cultivated horticultural crops, that include the healthy fruits and vegetables consumers are increasingly demanding, require substantial human labor to produce. More fresh fruits and vegetables are being imported from other countries, in part to meet the increased consumer demand, but also because of the labor challenges of production in the United States. As found in Alabama, American workers may not be willing to accept the low-paying, manual labor jobs required to produce these crops. Workers from outside the United States find it difficult to obtain authorization to work in the United States and employers struggle to determine which workers do have legal permits. Finding a way to balance these issues in an economically, politically, and sociologically acceptable manner is challenging.

Suggested website: Alabama HB 56, http://www.ago.state.al.us/Page-Immigration-Act-No-2011-535-Text

services. Additional related industries include finance, insurance, equipment manufacturer and sales, marketing and sales (including cooperatives), transportation, land and real estate management, technical consultants and a wide range of input suppliers. The number of workers associated with agriculture and food production is far greater than the actual number of producers (farmers, ranchers, and growers). In 2014, over 1.4 million people worked in food-related manufacturing based on the US Bureau of Labor Statistics data (Table 5.5). This number is probably an overestimate of the people directly engaged in food production as it includes all aspects of the manufacturing sector, but clearly demonstrates the significant impact

TABLE 5.5 Agriculture-Related Workers and Wages

US Bureau of Labor Statistics category	Number of employees	Annual mean wage ($)
Seafood product preparation and packaging	33,310	31,550
Animal food manufacturing	54,590	40,280
Grain and oilseed milling	59,080	44,700
Sugar and confectionery product manufacturing	69,920	36,830
Dairy product manufacturing	132,820	41,300
Fruit and vegetable preserving and specialty food manufacturing	163,830	36,930
Other food manufacturing	187,470	39,800
Bakeries and tortilla manufacturing	290,820	32,300
Animal slaughtering and processing	485,120	29,590

U.S. Bureau of Labor Statistics.

food manufacturing has on the US economy. The annual mean wage earnings for agriculture-related work ranges from around $30,000 to $45,000 (Table 5.5).

The food service industry is a significant employer within the agriculture-related sector, contributing $1.24 trillion in wages in 2010 according to the USDA. In 2014 over 4.7 million people were employed in the food service industry earning an average of $8.92 per hour. This sector includes restaurants, schools, hospitals, cafeterias, and other dining places. In 1963, about 28% of food expenditures were spent on food consumed away from home, and by 2010 that increased to 48%. More opportunities to purchase food outside the home, especially from fast food outlets, account for a large amount of this spending. The fast food industry leader is McDonald's, which was first opened in 1948. When Ray Kroc took over the franchise in 1954, he implemented assembly line, standardized approaches to food preparation, and marketing strategies that attracted children. These strategies were highly successful. By 1959 there were 100 McDonald's restaurants, by 1983 there were 7778 in 32 countries, and today there are over 36,000 restaurants in 100 countries. It is estimated that McDonald's employs over 1.7 million people worldwide and may be the most well-recognized business in the world. Other fast food chains began in the 1950s, including Taco Bell and Burger King, followed by Wendy's in 1969, showing similar growth and expansion. The fast food section has been widely criticized for providing high-calorie, inexpensive food that contributed to the obesity epidemic, and for not providing a living wage for employees. In contrast, the growth of this sector indicates consumer acceptance of the products, marketing, and employment opportunities of the fast food industry. With their substantial market impact, the types of foods sold by fast food chains directly impact agriculture. For example, when McDonald's decided to offer a salad that contained edamame (green soybeans) they spent several years prior to launching the product working with soybean growers preparing to produce enough edamame to meet their needs. This created a new market niche for those farmers. Similarly, when these large marketers try to meet consumer demands for local and organic foods, cage-free eggs and non-GMO ingredients, they must rely on the agricultural sector to change their production systems to provide these products.

According to the National Restaurant Association, there are over 1 million restaurants in the United States employing over 14 million people, which is roughly 10% of the entire workforce. The median pay for servers, with tips included, was $16.13 per hour. Workers in fast food establishments do not benefit from tips and

TABLE 5.6 The Minimum Wage (MW) as Defined by Each State

			Greater than federal MW		
No MW	**Less than federal MW**	**Federal MW**	**Less than $9.00**	**Less than $10.00**	**$10.00 or more**
AL—No MW	GA—$5.15	IA—$7.25	NM—$7.50	AK—$9.75	CA—$10.00
LA—No MW	WY—$5.15	ID—$7.25	ME—$7.50	CT—$9.60	MA—$10.00
MS—No MW		IN—$7.25	MO—$7.65	MN—$9.50	DC—$11.50
SC—No MW		KS—$7.25	AR—$8.00	NE—$9.00	
TN—No MW		KY—$7.25	AZ—$8.05	NY—$9.00	
		NC—$7.25	FL—$8.05	OR—$9.75	
		ND—$7.25	MT—$8.05	RI—$9.60	
		NH—$7.25	OH—$8.10	VT—$9.60	
		OK—$7.25	IL—$8.25	WA—$9.47	
		PA—$7.25	DE—$8.25		
		TX—$7.25	NV—$8.25		
		UT—$7.25	CO—$8.31		
		VA—$7.25	NJ—$8.38		
		WI—$7.25	HI—$8.50		
			MI—$8.50		
			SD—$8.55		
			MD—$8.75		
			WV—$8.75		

U.S. Department of Labor, www.dol.gov.

there is great variation in the size and distribution of tips within the restaurant sector. As of September 1, 2016, fourteen states followed the federal minimum wage of $7.25 per hour (Table 5.6). The majority of states had rates above the federal level, ranging from $7.50 to $11.50 per hour. Five states have no minimum wage requirement (Alabama, Louisiana, Mississippi, South Carolina, and Tennessee) and two states have rates below the federal minimum (Georgia and Wyoming). The poverty threshold as defined by the US government is $11,770 per year for a single person and $24,250 for a family of four. To achieve this minimum level of income while making minimum wage, the single worker would need to work a total of 203 eight-hour days per year and a family would need to work 418 days per year, assuming no withholdings and taxes. The poverty threshold does not address differences in cost of living within regions of the United States including housing, transportation, and taxes, which can vary dramatically. Controversy around the minimum wage versus a living wage is ongoing. Some argue that raising the minimum wage will increase the costs of goods and services for consumers. Others argue that workers making less than a living wage receive government subsidies in the form of food or housing assistance, which is essentially allowing employers to make more

TABLE 5.7 Top 10 Largest Food and Beverage Companies in 2014

Company	2014 Food sales ($ million)	Example name brands and products
PepsiCo Incorporated	38,224	Pepsi beverages, Frito-Lay snacks, Quaker Oats cereals, Tropicana juices, Lipton tea, Aquafina water
Tyson Foods Incorporated	36,077	Tyson, Jimmy Dean, Ball Park, Hillshire Farms meats and meat products
Nestle	27,978	Nestle and Gerber baby foods, Nescafe coffee, CoffeeMate, Lean Cuisine and Hot Pockets, Carnation milk, Nesquik and Nestea, Dreyer's, Edy's, and Haagen Dazs ice cream
JBS USA	24,000	Beef, pork, chicken, and lamb
Coca-Cola Company	21,462	Coke beverages, Dasani water, Minute Maid, Powerade and Odwalla drinks, Bacardi mixers
Anheuser-Busch InBev	16,093	Budweiser, Michelob, Rolling Rock, Stella Artois, and Beck's beer
ConAgra Foods Incorporated	15,832	Banquet meals, Act II, Orville Redenbacher's and Jiffy Pop popcorn, Blue Bonnet and Fleischmann's margarine, Libby's foods, Marie Callender's products, Peter Pan peanut butter, Slim Jim, Swiss Miss cocoa, SnackPack pudding, Wesson oil
Kraft Heinz Company	14,343	Kraft Macaroni & Cheese, Oscar Mayer meats, Planters snacks, Jell-O, Velveeta, Kool-Aid, Heinz condiments, Cool Whip
Smithfield Foods Incorporated	13,426	Farmland, John Morrell, Nathan's, Cook's meats
General Mills Incorporated	12,502	Cereals (Cheerios, Lucky Charms, Trix, Fiber One), Annie's products, Green Giant frozen vegetables, Gold Medal and Pillsbury flour, Cascadian Farm cereals and granola bars

Food Processing, www.foodprocessing.com.

profits by paying employees less. In response to these concerns, several states have committed recently to a stepwise increase over a period of years to raise their minimum wage levels, but no change in the federal level has been approved.

Another major employer in the food sector is the food industry. Food processing and manufacture is described in more detail in Chapter 6, Food Processing. The types of jobs in the food industry are broad including production and manufacturing, quality control, research and development, packaging, sales and marketing, and regulatory compliance. The food industry includes large multinational corporations (Table 5.7); small, focused operations that produce a single product or ingredient; and everything in between. Food scientists make up a small percent of the total employees in the food industry, but play important roles in product development, safety and quality control, marketing, and new technology development.

The organization that represents food scientists, the Institute of Food Technologists, found from an employment survey of members in 2013 that the level of education was fairly evenly divided across employees with 38% having completed BS degrees, 34% MS, and 24% PhD degrees. The mean salary for employees was $75,000 for those with a BS, $90,000 for those with MS, and $95,000 per year for those with PhD degrees. Most major food companies employ a wide range of personnel, with degrees in business and marketing, human resources, communication, law, engineering, statistics, and basic sciences (biology, chemistry, and math).

For many farmers, the local cooperative, or co-op, provides many services. Co-ops are defined as owned and controlled by their producer members. Rapid growth in co-op size and operations has occurred recently. In 2014 the USDA reported that there were over 135,000 co-op employees, which had increased from 7000 in 2012. Examples of top agricultural cooperatives are listed in Table 5.8. In grain producing areas, co-ops serve as brokers for crops by buying them from farmers, storing them in elevators (silos); processing them into secondary products, animal feed, or ingredients; and marketing the grain and products. Co-ops may also provide farmers with fertilizers, which are often purchased in bulk from international markets. Animal producers utilize co-ops to supply feed, medications, and breeding services, as well as slaughtering and processing of animals. Some farmer co-ops provide transportation, financial, insurance, and consultation services, as well. Increasingly, co-ops are engaging in market analysis and predictions to assist farmers with financial decisions and the use of technology for improving efficiency and productivity. Growth in farmers' co-ops is also occurring in local food systems with small operators in niche markets working together to deliver and market their products.

Expansion of technology in agriculture has created growth in several agribusiness sectors. Large agribusinesses include those engaged in animal breeding and care, inspection and grading, plant and seed development, commodity processing, chemical input manufacturing, and equipment operations and manufacturing. The types of work defined by the US Bureau of Labor Statistics in these areas are shown in Table 5.9.

Corporations associated with agriculture include seed and chemical companies, brokers and processors, farm implement manufacturers, insurance and finance, and farm and data management. Some examples are shown in Table 5.10. Companies such as Monsanto, DuPont Dow, and BASF began their

TABLE 5.8 Top Agricultural Cooperatives in 2014

Cooperative	Location	Products
CHS Inc.	St. Paul, MN	Energy, farm supply, food and grain
Land O'Lakes Inc.	St. Paul, MN	Dairy and food
Dairy Farmers of America	Kansas City, MO	Dairy
GROWMARK Inc.	Bloomington, IL	Farm supply
Ag Processing Inc.	Omaha, NE	Farm supply and grain
California Dairies Inc.	Artesia, CA	Dairy
United Suppliers Inc.	Eldora, IA	Farm supply

Rural Cooperatives Magazine, September/October 2014.

TABLE 5.9 Agriculture-Related Professions

US Bureau of Labor Statistics category	Number of employees	Annual mean wage ($)
Animal scientists	2350	72,590
Pesticide, fertilizer, and other agricultural chemical manufacturing	35,870	54,250
Soil and plant scientists	15,150	64,680
Agriculture and food science technicians	20,640	37,330
Agricultural inspectors	13,800	43,630
Animal breeders	1110	43,470
Graders and sorters, agricultural products	36,100	22,320
Agricultural equipment operators	26,100	28,490

U.S. Bureau of Labor Statistics.

TABLE 5.10 Corporations With Investment in Agriculture

	Headquarters	Products
SEED AND CHEMICAL COMPANIES		
Monsanto	St. Louis, MO	Agriculture and vegetable seeds, plant technology traits, and crop protection chemicals
DuPont Pioneer	Johnston, IA	Animal nutrition and disease prevention products, crop protection and seed
Syngenta	Basel, Switzerland	Crop protection, seeds, seed treatment, farm management
Bayer Crop Science	Cambridge, UK	Crop protection and seeds
Dow AgroSciences	Indianapolis, IN	Crop protection and seeds
BASF Plant Science	Research Triangle Park, NC	Pest control, seeds, and animal health
FERTILIZER COMPANIES		
Koch Ag and Energy Solutions, LLC	Wichita, KS	Fertilizer, energy, and methanol
PotashCorp	Saskatchewan, Canada	Fertilizers
FARM IMPLEMENT COMPANIES		
Deere & Company	Moline, IL	Agriculture, construction, and home equipment
Case-IH (International Harvester)	Racine, WI	Agriculture and construction equipment
New Holland	New Holland, PA	Agriculture and construction equipment
COMMODITY PROCESSORS		
Cargill	Minneapolis, MN	Commodity trading and processing; animal feed and nutrition; meat, food, and beverage ingredients; energy
Archer Daniels Midland	Chicago, IL	Oilseed processing, corn processing, agriculture services, ingredients
Bunge North America	St. Louis, MO	Commodity trading and processing, food ingredients, animal feed and bioenergy
FARM MANAGEMENT COMPANIES		
Hertz Farm Management Inc.	Nevada, IA	Farm management, real estate, and appraisals
Farmers National Company	Omaha, NE	Farm management, real estate, appraisals, insurance, and commodity marketing
Northwestern Farm Management Company	Marshall, MN	Farm management, real estate, and appraisals
AGRICULTURE DATA MANAGEMENT COMPANIES		
AGDATA	Charlotte, NC	Data collection, analysis, and market assessments
OnFarm	Fresno, CA	Data collection systems, analysis, and market assessments
Farm Logs	Ann Arbor, MI	Data collection systems and analysis

businesses producing chemicals or pharmaceuticals, then expanded into agricultural chemicals. As plant biotechnology innovations (hybridization and genetic engineering) were developed these became more profitable and expanded their portfolio of products. People with degrees in plant science, genetics, soil science, agronomy, entomology, and horticulture are employed by these companies. Farm implements, including tractors, harvesters, combines, balers, sprayers, dryers, planters, and irrigation systems are manufactured by companies such as Deere & Company, Case-IH, and New Holland. These companies employ agricultural, electrical, and mechanical engineers, and computer scientists and software designers to develop this equipment. Another segment of the agriculture industry includes large companies engaged in buying and processing commodities, producing ingredients for food and animal feed and, more recently, biofuel production. These companies include Cargill, Archer Daniels Midland, and Bunge North America. Companies such as PotashCorp and Koch Ag & Energy LLC are the main suppliers of fertilizer for the agriculture industry. The scope of these companies is broad and global. Processing plants are located where the crops and animals are produced and products are marketed around the world. Mainly, these companies operate as business-to-business rather than direct-to-consumer operations. Within the past decade, biofuels and biobased chemicals have become important market areas for agriculture-related companies because they have facilities and expertise in commodity processing.

Farm operation management companies, such as Hertz Farm Management Inc., Farmers National Company, and Northwestern Farm Management Company oversee farming operations for owners who do not live on the farm, serve as real estate brokers for agricultural land, and carry out appraisals and assessments of farm operations. The American Society of Farm Managers and Rural Appraisers (ASFMRA) provides accreditation programs for farm managers, rural appraisers, and agricultural consultants, which require work experience, coursework and examinations. Accredited farm managers and appraisers often work as independent contractors.

With recent advances in technology, data analysis and collection have also become major industries in agriculture. Companies such AGDATA, OnFarm, and FarmLogs help farmers collect, manage, and interpret information about their operations. Hardware and software engineers and designers make up the workforce for these companies. The agriculture workforce is diverse and increasingly based on technology. Companies that provide precision agriculture technologies in irrigation, crop management, animal breeding, pest control, and soil assessment and enhancement are becoming part of the industry. These technologies have great potential to improve crop and animal production systems while reducing inputs and environmental impact.

5.9 COMMODITY BOARDS AND CHECKOFFS

Commodity boards are organizations that support the producers of specific agricultural commodities, and a few are listed in Table 5.11. The mission of these organizations typically includes informational, marketing, and promotional information for their products aimed at consumers, educational and support programs for members, and lobbying efforts. In some cases, there are state organizations or chapters of the national organizations as well. Lobbying efforts are supported by membership dues, but promotion and marketing can be funded through checkoff programs.

Research and promotion (R&P) programs, which are approved by Congress and

TABLE 5.11 Examples of National Commodity Organizations

	Year founded	Checkoff program
American Sheep Industry Association	1865	No
National Cattleman's Beef Association	1898	Yes
United Fresh Produce Association	1904	No
American Dairy Association and National Dairy Council	1915	Yes
American Soybean Association	1920	Yes
National Turkey Federation	1941	No
National Association of Wheat Growers	1955	No
National Corn Growers Association	1957	Yes
Cotton Board	1966	Yes
US Potato Board	1971	Yes
American Egg Board	1974	Yes
National Pork Producers Council	1985	Yes
American Sugar Alliance	1987	No
US Rice Producers Association	1997	No
National Peanut Board	2000	Yes
National Alfalfa and Forage Alliance	2006	No
American Goat Federation	2010	No

managed by the Agricultural Marketing Service of the USDA, were established to allow commodity groups to pool resources to enhance marketing, and conduct research within a commodity group. R&P programs, sometimes referred to as checkoff programs, are defined fee assessments on producers. A commodity board comprised of elected or appointed producers must be established to oversee the spending of the funds. The funds must be used for promotional, marketing, education, or research and not on lobbying efforts. An example of an R&P program is the Dairy Research and Promotion Program, or Dairy Checkoff Program. Dairy farmers pay 150¢ per hundred-weight of milk produced and milk importers pay 7.5¢. The Dairy Board, comprised of 36 dairy producers from 12 geographic regions and 2 members that represent dairy importers, operates under Dairy Management Inc. The board is responsible for determining how the checkoff funds are used to promote dairy farmers. Some of the funding supports research at academic institutions and research centers, but the majority is spent in advertising and promotion of dairy products. Examples are the Fuel Up to Play 60 program, which encourages activities and healthy food choices in youth. The Fluid Milk Processors Promotion Program is another checkoff program that assesses large processors that produce more than 3 million pounds of fluid milk per month to pay 20¢ per hundred-weight. The 20 members of the

Fluid Milk Board are appointed by the Secretary of Agriculture and represent 15 geographic regions of the United States. This organization has sponsored familiar ad campaigns to promote milk such as the *got milk?* campaign with the celebrity milk mustaches. The AMS oversees more than 20 checkoff programs across the range of commodities, including not only the main agricultural commodities of eggs, beef, pork, soybeans, cotton, and potatoes, but also smaller crops such as mango, watermelon, avocado, mushroom, popcorn, and honey. Each of these organizations employs staff to develop and implement promotional activities, organize meetings, and oversee research, and to work on legislative issues that are part of the agriculture-related workforce.

The agricultural workforce is diverse and integrated across the US economy. Farmers, growers, and animal producers are the basis of agriculture, but the food system comprises many more types of workers. Technology has become a significant component of agriculture and food production with increasing demand for workers with these skills. Ensuring the continued progress of food production will require advanced educational opportunities and training for future workers.

References

Banks, V. J. (1986). Black farmers and their farms. *ERS Agriculture and Rural Economics Division Research Report No. 59*. Washington, DC: Economic Research Service, U.S. Department of Agriculture.

Cohen, R. L., & Horton, C. (Eds.), (2012). *Black farmers in America: Historical perspective, cooperatives and the Pigford cases* New York, NY: Nova Science Publishers, Inc.

Harvest of Shame. (1960). *Television documentary*. CBS. Directed by Fred W. Friendly. Available from <https://www.youtube.com/watch?v=yJTVF_dya7E>.

Further Reading

AFL-CIO. (2016). *Learn about unions*. Available from <http://www.aflcio.org/>.

American Crystal Sugar Company. (2016). *History*. Available from <https://www.crystalsugar.com/>.

Anonymous. (2014). *Republicans for immigration reform*. Available from <http://www.republicansforimmigrationreform.org/>.

Bakery, Confectionery, Tobacco, and Grain Millers International Union. (2016). *History and purpose*. Available from <http://www.bctgm.org/>.

Bangsund, D. A., Hodur, N. M., & Leistritz, F. L. (2012). Economic contribution of the sugarbeet industry to Minnesota and North Dakota. *AAE Report No. 668*. Fargo, ND: Department of Agribusiness & Applied Economics, North Dakota State University.

Barboza, D. (December 21, 2001). Meatpackers' profits hinge on pool of immigrant labor. *New York Times*. Available from <http://www.nytimes.com/2001/12/21/us/meatpackers-profits-hinge-on-pool-of-immigrant-labor.html>.

Bloom, S. G. (2000). *Postville: A clash of cultures in heartland America*. New York, NY: Harcourt, Inc, 362 p.

Boggess, B., & Bogue, H. O. (2016). The health of U.S. agricultural worker families: A description study of over 790,000 migratory and seasonal agricultural workers and dependents. *Journal of Health Care Poor Underserved, 27*, 778–792.

Bureau of Labor Statistics (2015). *Agricultural and food scientists*. Washington, DC: U.S. Department of Labor. Available from <http://www.bls.gov/ooh/life-physical-and-social-science/agricultural-and-food-scientists.htm>.

Census of Agriculture (2016). *2012 census publications*. Washington, DC: U.S. Department of Agriculture. Available from <https://www.agcensus.usda.gov/Publications/2012/>.

Centers for Disease Control and Prevention. (July 1, 2016). CDC suicide rates by occupational group—17 states, 2012. *Morbidity and Mortality Weekly Report, 65*(25), 641–645. Available from <http://www.cdc.gov/mmwr/volumes/65/wr/mm6525a1.htm>.

Citizenship and Immigration Services (2011). *Green card*. Washington, DC: U.S. Department of Homeland Security. Available from <http://www.uscis.gov/>.

Daniel, C. E. (1981). *Bitter harvest: A history of California farmworkers, 1870-1941*. Ithaca, NY: Cornell University Press, 348 p.

Economic Research Service (2016). *Farm labor*. Washington, DC: U.S. Department of Agriculture. Available from <http://www.ers.usda.gov/topics/farm-economy/farm-labor/background.aspx>.

Employment and Training Administration (2015). *The National Agricultural Workers Survey*. Washington, DC: U.S. Department of Labor. Available from <https://www.doleta.gov/agworker/naws.cfm>.

Eversull, E., Ali, S., & Chesnick, D. (2014). Top 100 Ag Co-ops. Rural Cooperatives. *Rural Cooperatives Magazine, 81*(5), 8–17. Washington, DC: U.S. Department of Agriculture

Rural Development. Available from <http://www.rd.usda.gov/publications/rural-cooperatives-magazine>.

Gunderson, D. (December 14, 2011). In ND town where sugar is king, lockout hits hard. Hillsboro, ND: MPRNews. Available from <http://www.mprnews.org/story/2011/12/14/american-crystal-sugar-lockout-hillsboro>.

HealthCare.gov (2016). *Immigrants-health coverage for immigrants*. Baltimore, MD: U.S. Centers for Medicare and Medicaid Services. Available from <https://www.healthcare.gov/immigrants/>.

Hoppe, R. A. (2014). Structure and finances of U.S. farms: Family farm report. *ERS Economic Information Bulletin Number 132* (2014 ed.). Washington, DC: U.S. Department of Agriculture.

Hoppe, R., & Korb, P. (2013). Characteristics of women farm operators and their farms. *ERS Economic Information Bulletin Number 111*. Washington, DC: U.S. Department of Agriculture. 51 p. Available from <http://www.ers.usda.gov/publications/eib-economic-information-bulletin/eib111/report-summary.aspx>.

Hurt, R. D. (1994). *American agriculture: A brief history*. Ames, IA: Iowa State University Press, 424 p.

Institute of Food Technologists. (2016). *2015 employment and salary survey report*. Available from <http://www.ift.org/CareerCenter/Salary-Survey/2015-Salary-Survey-Report.aspx>.

Jones, M. (July 11, 2012). Postville, Iowa, is up for grabs. *New York Times*. Available from <http://www.nytimes.com/2012/07/15/magazine/postville-iowa-is-up-for-grabs.html?_r=0>.

Majka, L. C., & Majka, T. J. (1982). *Farm workers, agribusiness and the state*. Philadelphia, PA: Temple University Press, 320 p.

National Labor Relations Board. (2016). *National Labor Relations Act*. Available from <https://www.nlrb.gov/resources/national-labor-relations-act>.

Red River Valley Sugar Beet Growers Association. (2016). *History of RRVSBGA*. Available from <http://rrvsga.com/>.

Spencer, J., & Adams, J. (April 15, 2013). Long-running crystal sugar labor dispute ends, but hard feelings may linger. StarTribune. Available from <http://www.startribune.com/crystal-sugar-labor-dispute-s-scars-may-linger/203126391/>.

U.S. District Court for the District of Columbia. (February 18, 2010). *Black farmer discrimination litigation*. Settlement Agreement. 110 p. Available from <https://www.blackfarmercase.com//Documents/SettlementAgreement.pdf> Accessed 26.08.16.

Wang, S. L., Heisey, P., Schimmelpfennig, D., & Ball, E. (2015). Agricultural productivity growth in the United States: Measurement, trends and drivers. *ERS Economic Research Report 189*. Washington, DC: U.S. Department of Agriculture.

White House-President Barack Obama. (2014). *Immigration: Taking action on immigration*. Available from <http://www.whitehouse.gov/issues/immigration>.

CHAPTER

6

Food Processing

6.1 DEFINITION OF FOOD PROCESSING

The role of processed foods in a healthy diet has become increasingly confusing. Some recent headlines suggest "processed foods are responsible for the obesity epidemic" or warn that "eating processed foods lower IQ." Many foods that people consume throughout the day, such as coffee, orange juice, soy milk, cheese, bread, pasta, or pickles, are processed in some manner. The term "processed food," however, has become negatively associated with foods that have low nutritional value, are high in fat, salt, and sugar, that cannot be properly digested, make people addicted to them, and come with excessive packaging. A 2008 International Food Information Council (IFIC) survey revealed that 43% of respondents viewed processed foods unfavorably compared to 18% who viewed processed foods favorably. According to Sloan (2015), most consumers (87% of those surveyed) feel that fresh foods are healthier, 80% believe they are tastier, and 78% are trying to eat more fresh than processed foods. Consumption of fresh foods has grown 20% in the past 10 years and fresh foods accounted for 29% of all US grocery sales for 2014. Consuming healthful and tasty foods should be everyone's goal and processed foods have an important role in ensuring foods are safe, nutritious, and palatable.

Food scientists describe food processing as "one or more of a range of operations, including washing, grinding, mixing, cooling, storing, heating, freezing, filtering, fermenting, extracting, extruding, centrifuging, frying, drying, concentrating, pressurizing, irradiating, microwaving, and packaging" (Floros et al., 2011). Nutritionists and dietitians define processed foods as "any food other than a raw agricultural commodity, including any raw agricultural commodity that has been subject to washing, cleaning, milling, cutting, chopping, heating, pasteurizing, blanching, cooking, canning, freezing, drying, dehydrating, mixing, packaging, or other procedures that alter the food from its natural state" (Fox, 2012). Simply, processed food is any food that has been altered from its state at harvest including washing, cutting, and cooking.

Food processing involves modifications to ingredients or food products in many different ways. A primary role of food processing is to ensure the safety, quality, and availability of perishable foods. Increased shelf-life (the period of time between harvest or preparation and consumption) is achieved through processing. Processing foods at the peak of their ripeness ensures nutrients are retained and food waste is

Understanding Food Systems.
DOI: http://dx.doi.org/10.1016/B978-0-12-804445-2.00006-5

reduced. This results in a more sustainable food production system. Prevention of food-borne pathogens is a goal of processing and involves using heat or other physical measures to reduce microbial contaminants, or adding preservative agents. Foods are processed to enhance their convenience and accessibility, which reduces the amount of time needed to prepare meals. Processing may involve the addition of ingredients or approved additives to the food, such as colors and flavors, or texture enhancers that make the food more palatable, enjoyable, or fun, and allows creation of new food products. Some additives increase the nutritional value of the foods, for example essential nutrients, which contribute to reducing risk of nutritional deficiencies.

Processing of foods, including the addition of ingredients, may reduce, increase, or leave unaffected the physical and nutritional characteristics of raw agricultural commodities. For example, wheat berries (intact seeds with the bran, germ, and endosperm) are a raw agricultural commodity. Whole wheat flour is ground wheat containing the bran, germ, and endosperm. Wheat cereals can be whole grain (such as "Ralston") or only endosperm (mainly starch and protein) that is milled ("Cream of Wheat" and farina), puffed ("Puffed Wheat"), or extruded ("Wheaties" and "Shredded Wheat"). Cakes, bread, and cookies are made with cake, bread, or all-purpose wheat flours of ground endosperm with different protein amounts for the specific characteristics of each type of baked product. Chocolate chip cookies from the bakery and "Twinkies" from a package represent examples of convenience foods made from refined wheat flour (white flour from wheat endosperm).

6.1.1 Degrees of Food Processing

Different degrees of processing are used for different types of foods. "Minimally processed" foods are those that are washed and packaged, such as heads of lettuce and fresh broccoli, washed and precut lettuce, peeled carrots, shredded cheese, roasted and ground coffee, and chopped walnuts. "Ready-to-eat" foods such as breakfast cereals, cookies, and luncheon meats, are processed by extrusion or toasting, emulsification and heating, and baking, respectively, need minimal or no preparation and add to the types of convenient food items enjoyed by US consumers.

Other "convenience foods," such as frozen meats and meals, are fully prepared, then packaged and preserved for use later. The ingredients are combined, baked and frozen as in desserts, pot pies and pizza. These preprepared foods reduce food preparation time and require little skill in cooking. There has been a trend for spending less time in meal preparation over the past decade. The average home-prepared meal for US families is assembled in less than 30 minutes, and many recipes are designed to be prepared in less than 20 minutes. Canned tuna, beans and tomatoes, frozen fruits and vegetables, and jarred pureed baby food are foods processed to help preserve and enhance nutrients and freshness of foods at their peak. Some processed foods combine ingredients and preservatives to improve the safety, flavor, and visual appeal such as cake mixes, instant potatoes, tomato sauce, salad dressings, and sauces. The food processing industry has provided consumers with thousands of new food products for convenient and easy preparation.

6.2 HISTORY OF FOOD PRESERVATION AND PROCESSING

From the beginning of civilization, humans have processed foods to improve digestibility. Cooking was the earliest form of food processing. Raw grains, roots, and tubers are not well digested, and although raw meat can be tender, cooked meat is a wiser choice from a food safety standpoint. Cooked or smoked meat will not spoil as fast as raw meat. People discovered that food could be kept longer if it was dried, smoked, or salted. Wheat, barley, and rice could be made into breads or fermented into beer. Through the process of fermentation, which produced alcohol, bacterial contamination was prevented, and beer could be stored for later use.

Early attempts at effective food preservation included cooking (application of heat), drying (removal of moisture), fermenting (addition of microorganisms), preservation (addition of spices such as mustard seed, sugar, honey), and pickling (addition of acid such as vinegar). These methods have been used for centuries to preserve food at harvest or during times of plenty. Processing allowed food to be saved for seasons in which food was not plentiful or available at all. Caves and cellars were early forms of cool temperature storage. Before refrigeration or rail transport, fruits such as bananas, oranges, and strawberries, and vegetables such as lettuce and spinach were luxuries available only at special holidays or locally during the summer months.

In 1810, Nicolas Appert in France and Peter Durand in England successfully preserved food in bottles and metal "tin" cans, but neither knew the reason for their success. The process of food canning became much safer once the role of microorganisms was understood by Louis Pasteur (in 1864). Food preservation techniques for fish, meat, and fruit developed rapidly in the 20th century. During that time food processing and distribution moved from farms and individual homes to industrial businesses. The first home refrigerator, Frigidaire, was sold in 1925 and Clarence Birdseye developed the industrial plate freezer for frozen foods in 1929. These technological advances dramatically changed the way food was stored and the types of foods that were available.

6.2.1 Development of New Foods

The development of unique foods throughout US history illustrates the ingenuity, business acumen, religious zeal or eccentric health advocacy, and sometimes serendipity, of innovators. Table 6.1 lists the broad scope of innovations in food processing technologies and their impact on the food system. Major innovations in technology include refrigeration and freezing, and microwave ovens, and the introduction of improved packaging techniques and materials, that made food more convenient and easier to prepare.

The stories of food processing innovations are integrated into America's history. Kellogg's and Post cereals arose from a desire (by W.K. Kellogg and C.W. Post) for a more healthful choice than the beef, pork, hominy, and bread that were regularly consumed for breakfast meals in the 1850s. Their precooked shredded wheat, corn flakes, and "Grape Nuts" cereals were sold as health food alternatives. Inventor Gail Borden had several failed attempts to make compact, nutritious foods for explorers before he created "condensed" milk (patented in 1856). Reverend Sylvester Graham, who eschewed gluttony and sensual pleasures such as tasty food, promoted bran as a laxative. Whole wheat bread became known as "Graham bread" and we are still eating graham flour and graham crackers today. Processed cheese (patented in 1917 by J.L. Kraft) began as a way to prolong the shelf-life of cheeses that were delivered by horse-drawn wagons in Chicago. John T. Dorrance (Campbell's soup), Henry J. Heinz (who marketed horseradish and pickles before ketchup and relishes brought the total number of products to "Heinz 57"), Oscar Mayer (who started selling liverwurst, bratwurst, and other German sausages with his brother in a meat market in Chicago in 1883), and J.M. Smucker (who made apple cider and apple butter in Orrville, Ohio in 1897) created products that are still on store shelves today. William Underwood started a condiment company in Boston in 1822 and became famous for Underwood Deviled Ham. His grandson, William Lyman Underwood, and Samuel

TABLE 6.1 Commercial Developments in Types and Marketing of Processed Foods During the 20th Century and Their Impacts on US Food Habits

Change in retail or processed food	Impact
1900 Double-crimped can	Reduced costs for processors and assured an air-tight container for improved food safety
1901 A&P grocery incorporates with 200 stores	Food production begins to move out of individual homes
1913 refrigerators for home use invented; 1923 Frigidaire introduced first self-contained refrigerator	Increased shelf-life of foods and reduced food-borne illness
1916 Piggly Wiggly grocery opens first self-service food store	Increased variety of prepackaged foods available
1935 First canned beer; 1960 first aluminum cans; 1963 first pull-tab cans (beer)	New foods available; lighter containers reduced transportation costs
1940 Home freezers available; 1950 Swanson's frozen chicken pot pie; 1952 Birdseye frozen peas, Mrs. Paul's frozen fish sticks	Frozen foods and convenience foods common
1941 Controlled atmosphere storage	Access to fruits and vegetables year round due to reduced spoilage
1945 Raytheon made first microwave for airlines; 1955 first microwave for home use by Tappan ($1300); 1967 Raytheon (Amana) produced countertop microwave oven for less than $500	Homemakers learned to cook frozen foods; convenience foods more available; new foods developed; by 1997, 90% of US households had microwave oven
1948 McDonald's hamburger restaurants established	Availability of high-quality fast foods, indicative of less home cooking
1954–60s bulk aseptic packaging, storage, and transportation	New types of containers for shipping opened global food trade
1959 Soy protein isolate made by Central Soya Co.	Protein available as functional food ingredient
1960 aluminum can; 1963 foil-laminate fiber can (for frozen OJ); 1966 plastic milk containers	Innovative packaging for more economical processing and transportation
1963 Mechanically deboned chicken; McDonald's introduced Chicken McNuggets in 1980	Fabricated meat product and popular fast food. By 1992, Americans ate more chicken than beef
1990s quick freezing, modified atmosphere packaging, freeze-drying, irradiation	Novel methods of food preservation
1993 SnackWell's cookies and crackers	First low-fat food product

Prescott (who graduated from the Massachusetts Institute of Technology) conducted research to determine the optimum time and temperature exposures for canned meats, vegetables, and seafood in 1895 and are credited with beginning the food technology profession.

Societal changes are reflected in the types of foods available, notably convenience foods, ethnic foods, snacks and microwave products. Various types of processed foods have been marketed over the years and have become favorites in US households (Table 6.2). It may be surprising that many of these products

TABLE 6.2 Commercially Developed Food Products in the US Marketplace Since 1900

Year	Food product
1900	Hershey's chocolate bar
1903	Dole canned pineapple; Kellogg added sugar to corn flakes, boosting popularity; Pepsi Cola introduced
1904	Quaker markets first puffed cereal
1910	Aunt Jemima pancake flour
1912	Oreo cookies; Hellman's mayonnaise
1926	General Mills created Betty Crocker, indicating the importance of advertising
1928	Peter Pan peanut butter; Velveeta cheese
1930	Wonder Bread markets first automatically sliced bread
1932	Fritos corn chips
1937	Kraft Macaroni & Cheese dinner
1942	Dannon yogurt; La Choy canned Chinese food
1946	Maxwell House instant coffee
1951	Swanson produces first frozen meals (pot pies)
1965	Cool Whip and Shake 'n Bake
1970	Hamburger Helper
1981	Stouffer's Lean Cuisine frozen dinners
1982	Bud Light lower calorie beer
1986	Pop Secret microwave popcorn
1989	Garden Salad ready-to-eat salad bags
2003	Wholly Guacamole high-pressure processed avocados

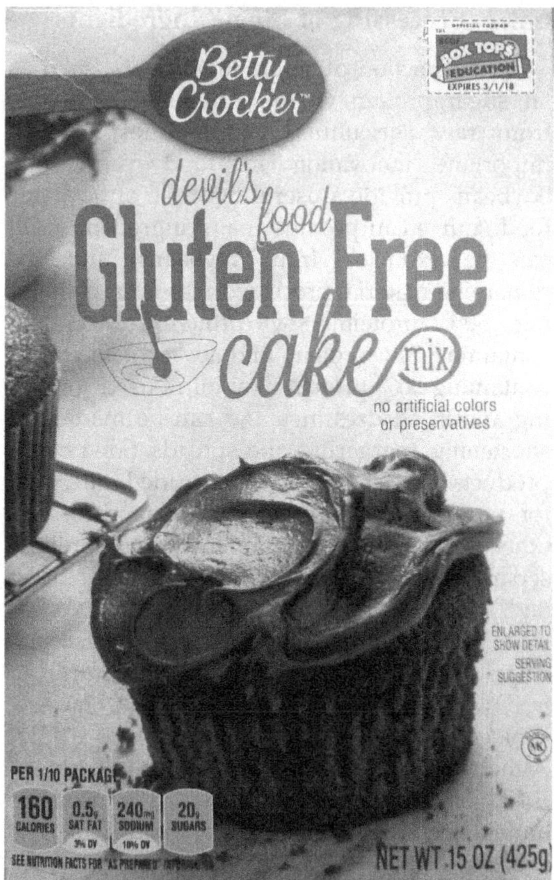

FIGURE 6.1 Food companies offer new products to meet consumers' needs. This gluten-free cake mix was developed to provide an alternative product for consumers with Crohn's disease or those who choose to avoid gluten, which is a protein present in wheat and other grains. *Source: Photo provided by General Mills.*

have been in the marketplace for over 100 years.

Food companies must keep ahead of consumers to retain their market share. Some products lend themselves to what are called line extensions (size or flavor variations of a particular food, such as mini-Oreos, mint-flavored Oreos and "double stuff" Oreos), which draw interest from consumers. They may develop new types of foods in response to consumers' desire for healthful foods (high fiber) and beverages ("Vitamin Water"), special dietary foods (low-fat or gluten-free; Fig. 6.1), international foods (wasabi or curry flavors), and convenience foods (dried fruit and nut snacks, microwave breakfast sandwiches). In today's world of rapid information dissemination, responding to consumers' demands for new foods is a challenge for food companies and requires an environment of speedy product development and marketing.

6.2.2 Processing of Single Ingredients

The extraction of single ingredients, such as fat, sugar, protein, vitamins, minerals, or starch, from raw agricultural commodities was an important innovation in food processing. Soybeans provide a useful example of how one food source can provide many ingredients. Oil can be extracted from soybeans, and the remaining material processed into flour containing 50% protein, soy protein concentrate containing 70% protein, and soy protein isolate containing 90% protein. Soy oil is used for frying and salad dressings and can be made into shortening, margarine, and spreads. Soy protein products can be texturized or extruded into fibers for use as meat analogs, meat extenders, and other vegetarian foods. Similarly, cow's milk is separated into the protein components casein and whey, lactose, milkfat, and minerals. Whey protein was once a waste product discarded into rivers as a by-product of cheese manufacture, but is now used as an additive in breads, crackers, cookies, nutrition bars, and sports drinks. Although the majority of "dent" corn (the type of field corn grown in the Midwest) is used for animal feed and ethanol production, it is also processed for corn oil, corn meal, and flour for tortillas and snack chips, starch used as a thickening agent, corn syrup (100% glucose), and high fructose corn syrup (HFCS). The type of corn we consume as corn-on-the-cob and canned and frozen corn is "sweet" corn, a different variety of corn that is grown on a very much smaller scale than dent corn. "Value-added" processing of crops and animal foods for human food production has opened new markets and expanded the food industry.

Screening, centrifugation, and filtration are physical or mechanical methods used to separate food components in applications such as juice extraction, sucrose crystallization, and recovery of proteins from whey. Distillation of oils and flavors (for beverages), solvent extraction of oils (for refined vegetable oils), evaporation of tomato sauce (for tomato paste), and supercritical fluid extraction of caffeine from coffee beans and tea leaves (for caffeine-free coffee and tea) are additional methods to extract a variety of molecules, resulting in new types of foods. Membrane technology is rapidly being implemented for separation of food components because it requires low capital investment in equipment as well as less water and energy consumption than other methods. Extracted and purified ingredients are more stable biochemically and less susceptible to spoilage microorganisms. "De-constructed" grains, oil seeds, and animal products allow new formulations to be developed, such as low-calorie snacks, cholesterol-lowering spreads, and high-protein drinks.

6.2.3 Current Food Processing and Manufacturing

As described in Chapter 5, Human Resources in the Food System, the food system involves many types of businesses and workers. As a rough estimate, there are 300,000 individual companies in the US food system. The food processing or food manufacturing segment of the food system is comprised of 30,000 individual food and beverage plants owned by 24,500 companies, representing 14% of US manufacturing. The food industry contributes 8%–10% of the gross domestic product and employs about 15% of the US workforce. Thirty-two percent of food industry workers are in meat processing, 17% in bakeries, and 11% in fruit and vegetable processing.

California has the highest number of food processing plants (4510), followed by New York (2186) and Texas (1774). Large facilities (with over 100 workers) are 12% of the total number of plants but they produce 77% of the total value of food shipments. There are many more small plants (<20 workers) but they produce only 4% of the value of shipments. Just as there is increased consolidation in food production, there is increased concentration in food processing, especially of beef, pork, and milk products. From 1997 to 2007, 43% fewer farms produced over twice as much milk per farm and, during the same time, 21% fewer milk facilities processed 27% more milk per plant. Pork and broiler industries are vertically integrated, that is, production, processing, and marketing are controlled by one company, with increased reliance on contract growers (Chapter 3: Innovations in US Agriculture).

Major areas of food processing include grain and oilseed milling, dairy products, fruit and vegetable preservation, animal foods, sugar and confectionery products, seafood products, and bakeries and tortillas. In 2014, the top 10 companies with the most sales were

1. PepsiCo
2. Tyson Foods, Inc.
3. Nestlé
4. JBS USA
5. Coca-Cola Co.
6. Anheuser-Busch InBev
7. ConAgra
8. Kraft Foods
9. Smithfield Foods Inc.
10. General Mills

It is notable that beverages and meats are the primary products of 8 of the top 10 companies, likely due to the profitability of these types of products. Other large food processing companies include:

- Hershey's Co.
- Dean Foods
- Kellogg's
- Land O'Lakes
- McCormick and Co.
- MOM Brands
- Ocean Spray
- H.J. Heinz
- Hormel
- Frontier Natural Products Co-op
- Starbucks
- M&M/Mars
- Campbell Soup Co.
- Dr. Pepper Snapple Group
- J.M. Smucker Co.
- McDonald's
- Ocean Spray and Odwalla

Current trends in consumer dietary habits have resulted in a decline in sales of packaged food products and food companies have adapted to consumers' desires for unprocessed and fresher foods. Growth has been limited in boxed and canned products, particularly in bake mixes and refrigerated doughs, because busy consumers prefer ready-made cookies, pies, and cakes. The entire baking mix business fell 4.6% in 2013 and 6.4% in 2014. General Mills, maker of Betty Crocker cake mixes, has responded with lower prices and simplified products targeted for specific audiences and seasonal cooking experiences. In addition, General Mills and other companies have learned to connect with consumers via digital technology and now offer recipes, cookbooks, video instructions, and baking suggestions on their websites to encourage use of their products.

According to the Bureau of Labor Statistics employment classifications, there were 15,400 college graduate–level food scientists and food technologists in the United States working in food manufacturing facilities (45%), scientific research and development for food companies (13%), management (10%), research in colleges and universities (6%), and testing laboratories (3%).

Food science professionals strive to provide novel, safe, healthful, and nutritious foods to the public. Food companies are businesses that must make a profit in order to continue operating. The United States has a consumer-driven economy and food companies must provide products that consumers will buy. Consumers choose, buy, and eat foods that they like and can afford.

A survey conducted by the IFIC in 2016 found that taste was the number one factor in food-buying decisions for consumers, followed by price, healthfulness, convenience, and sustainability. Convenience was the primary reason consumers purchased packaged foods. Recent food industry initiatives to increase dietary fiber, reduce fat, reduce carbohydrates (sugar), and reduce sodium have created a wide variety of products touting these features based on consumer demand. For example, when nutritionists claimed dietary fiber could reduce cancer risk, a plethora of high-fiber foods appeared on the store shelves. When fat was determined to be a cause of heart disease, low-fat products dominated the markets. Today, there are foods made without HFCS, gluten, GMOs, or artificial colors and flavors because these are issues in the forefront of consumers' minds.

The food industry is constantly searching for ways to attract consumers and provide foods they will buy. There are over 20,000 new food product introductions each year. Candy, gum, and snacks; beverages; condiments; and processed meat are the food categories with the most new product introductions. However, from 2006 to 2010, introductions of candy, gum, and snack products have declined while new fruits and vegetables, dairy products, and cereals have increased. New advertising categories, such as "natural," "sustainable," "single serving," and "fresh," are among the health- and convenience-related attributes that account for one-third of all new product claims.

Processed foods are used in restaurants, schools, hospitals, airlines, military, food service institutions, vending machines, and concessions. Forty-seven percent of the US consumer's food dollar is spent in over 1 million US restaurants, representing expenditures of $1.8 billion every day. The restaurant industry projected sales of $709 billion for 2015, 4% of the US GDP. Many food companies manufacture food products in retail versions for consumers and larger sizes for restaurants and food service operations. Processed foods in the form of precut vegetables, prepared salads, portioned meats, frozen soups and sauces and desserts provide savings in labor, storage, and purchase of special equipment as well as consistent quality and less inedible waste.

6.3 METHODS OF FOOD PROCESSING

The history of food processing in the United States parallels advances in science and technology. Machinery built to produce food and food ingredients was first powered by water, then steam, and eventually electricity. Innovations in home appliances led to novel types of food products. Following WWII, innovations in technology were applied to food processing. Refrigeration and freezing, microwave ovens, irradiation, extrusion, and other approaches were used to create new products to entice consumers. Science and technology in food was viewed positively. During the space race of the 1960s, the powdered orange juice drink Tang became widely popular because it was used by the astronauts. TV dinners, Jiffy Pop popcorn (made on the stove in a pop-up foil wrapper), colorful breakfast cereals with cartoon characters, and many other innovations were market successes. Developments in chemistry created new food additives, flavors, and colors, and products to increase the shelf-life of foods. Convenient packages were developed including the pop-top can and plastic milk bottles. More people owned cars and did their shopping in grocery stores, and drive-in restaurants were everywhere. This environment allowed the food industry to prosper and expand.

6.3.1 Processing Cereal Grains by Milling

In the early years of grain processing, local mills served farming communities throughout the country but, with the building of railroads and improved transportation during the late 1800s, grain processing moved to larger cities. With ready access to major waterways and railroads to deliver grain and distribute products, Minneapolis, Chicago, and Omaha became the nation's largest grain processing centers (Fig. 6.2).

Milling is a process in which grains such as oats, wheat, rice, and corn are dehulled and ground into smaller pieces or flours to improve palatability, reduce cooking time, and create food products. Each type of grain has a unique processing method that yields a wide range of products. The milling process usually includes removal of the outer hull which contains tough fibrous material. The grains may then be toasted, soaked, or cooked to soften and release the starch and other carbohydrates. Flaking, crushing, or grinding the grains is done to generate the desired product. For example, whole wheat flour is made using the entire wheat kernel that includes both the germ (part of the kernel that contains most of the vitamins and minerals) and the endosperm (inner part of the kernel that is mostly starch). White flour is made from just the endosperm. Because unsaturated fatty acids contribute to rancidity and subsequent off-flavors, removal of the germ (which contains about 10% fat in most cereal grains) from the rest of the kernel improves the storage time of white flour compared to wheat flour. The germ, however, contains most of the B vitamins and minerals, so whole wheat flour has more of these nutrients than white flour. A process of bleaching white flour, by exposed it to small amounts of chlorine gas or benzoyl peroxide, was found to improve the elasticity of the dough made from the flour and inhibited mold growth. This made bleached white flour a desired product.

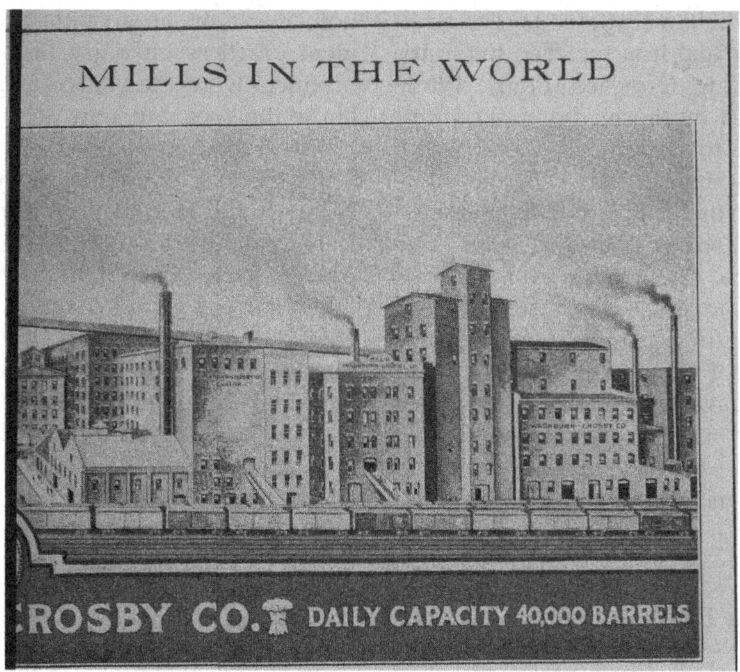

FIGURE 6.2 The city of Minneapolis was a hub for grain milling in the early 1900s because of its location on the Mississippi River and proximity to wheat growing regions. The Washburn-Crosby Milling Complex in Minneapolis later became General Mills, which marketed Gold Medal Flour. *Source: Photo from the Library of Congress image library.*

In addition to producing flours with different compositions, modern milling techniques can separate flour particles of different sizes to produce flours with exact protein contents for cakes, cookies, doughnuts, and breads. Different types of wheat are also processed for pastas, cereals, and additives for a wide variety of foods. Processing of grains can create more convenient food products. For example, whole brown rice requires about 50 minutes of cooking to become softened enough to eat. Polishing the grain to remove the hull allows the white rice to be cooked and ready to eat in about 20 minutes. Fully cooked and dried rice (instant rice) can be cooked (rehydrated) in a minute. Reduced cooking times provide convenience and saves time and energy.

Before industrial breadmaking was widespread, bread was made at home using whole grain flours. At that time, commercially made white bread, made from bleached white flour, was considered a modern food and was more expensive than whole wheat breads. The addition of B vitamins and iron replaced the nutrients lost in milling (enriched flour). White bread was favored by those who could afford it and being able to serve white bread was a bit of a status symbol. In the 1920s the white bread product "Wonder Bread" was marketed as "helping build strong bodies 12 ways" due to the addition of 12 vitamins and minerals and was heavily advertised as a healthy food for children. With further understanding of the nutritional value of whole wheat, white bread has become less favored. Today, whole wheat bread is more expensive than white, refined bread, illustrating the changing dynamic of food technology and nutrition science on food choices.

6.3.2 Processing Food by Pasteurization

Pasteurization (thermal pasteurization) is a mild heat treatment (140–212°F) designed to kill the most heat-resistant vegetative pathogens (disease-causing microorganisms) and undesirable enzymes. Pasteurization improves the safety of milk, cheeses, and fruit juices and extends the shelf-life of the food. For milk, pasteurization is described by the Grade "A" Pasteurized Milk Ordinance as a specific temperature for a defined time such as 161°F for 15 seconds for fluid milk. This scheme is the minimum processing needed to kill *Coxiella burnettii*, the organism that causes Q fever in humans, which is the most heat-resistant pathogen currently recognized in milk. By using the length of time required to kill the most resistant organism it is inferred that all of the other, less heat-tolerant organisms will also be destroyed. At the turn of the 20th century, the milk-borne illnesses of typhoid fever, scarlet fever, tuberculosis, septic sore throat, diphtheria, and diarrheal diseases were common. These illnesses were virtually eliminated with the commercial implementation of pasteurization, in combination with improved management practices on dairy farms. In 1938, milk products were the source of 25% of all food-borne illnesses, but now account for less than 1% of all food-borne illnesses.

Pasteurization is not a sterilization technique (milk sterilization requires 285°F for 15 seconds) so foods must be refrigerated to prevent other bacterial growth. Orange and other juices, liquid eggs, carbonated beverages, and beer are pasteurized at a range of temperatures (140–212°F for <1 minute). Ultrahigh temperature (UHT) pasteurization uses higher temperatures than pasteurization (but not as high as sterilization) resulting in longer shelf-life for the products. UHT can generate off-flavors but is popular in Europe, Canada, and Asia. Products treated with UHT, such as milk, can be held at room temperature until opened.

The current interest in "raw" or unpasteurized milk and milk products presents a food safety problem. The CDC reported that unpasteurized milk is 150 times more likely to cause illness than pasteurized milk. From 1993 to 2006, 1500 people became ill from drinking raw milk or eating cheese made from raw milk. Consuming unpasteurized foods is especially dangerous for immune-compromised people, including the elderly, children, and pregnant women. There have been recent cases of food-borne illness associated with products that were marketed to consumers as unprocessed with claims that this made the product more healthful. In 1996, unpasteurized fruit juices made by Odwalla caused illnesses in 66 people and 1 child died from kidney failure caused by *Escherichia coli* infection. Odwalla now "uses a pasteurization process to eliminate harmful bacteria while preserving great taste and vital nutrients."

6.3.3 Processing Food by Canning or Thermal Sterilization

Canning preserves food by heating food in a jar or can (originally "canisters") and sealing with a vacuum (anaerobic or no oxygen) that occurs when the container cools. Because there is no oxygen in the container, aerobic microorganisms cannot grow but any anaerobic microbes that survive the heating process can grow in this environment. The heating time is designed to destroy any microorganisms that could grow without oxygen (anaerobic). Acidic foods (pH <4.5), such as fruits and tomatoes, naturally contain citric, malic, tartaric, and other organic acids that limit the growth of many pathogens and are less of a concern for food safety. Low-acid foods (pH 4.5–7) such as meat, fish, mixed soups, and vegetables provide a favorable environment for microorganisms and must be canned at higher temperatures. Boiling water reaches a temperature of 212°F (100°C) but water and food in a pressure canner or retort achieve higher temperatures (230–250°F). Time and temperature calculations for canning processes are designed to destroy the most heat-resistant microorganism, specifically the spores of *Clostridium botulinum*, which can produce a deadly toxin if they sprout and grow under the anaerobic conditions of a canned food. Acid foods with pH less than 4.5 can be processed safely at 212°F, the temperature of boiling water, because *C. botulinum* spores cannot grow in acidic conditions.

Botulism is a paralytic illness caused by ingestion of the toxin produced by *C. botulinum*; death results from muscle paralysis and respiratory failure. ("Botox" is the application of this nerve toxin for cosmetic and medical purposes.) *Clostridium botulinum* spores (resistant reproductive bodies) are present in the soil and can contaminant fruits and vegetables. Spores will not germinate in the presence of oxygen or in acidic conditions but can germinate in sealed jars of low-acid foods such as asparagus, green beans, beets, and corn, if not destroyed by temperatures above boiling achieved under pressure during the canning process. The food industry is highly regulated to ensure canned foods are processed correctly to avoid this contamination. Home canning of foods can be a risk if not done properly. From 1996 to 2008, there were 116 outbreaks of food-borne botulism, with 48 of these from home-canned foods. In 2015, botulism poisoning occurred at a church potluck dinner; the cause was identified as a potato salad made with improperly processed home-canned potatoes. Home canners are urged to use sanitary procedures, follow USDA guidelines, and process low-acid vegetables and meats in a pressure canner. The botulinum toxin is destroyed by boiling so it is also recommended that all home-canned foods be boiled for 10 minutes before consumption.

Properly preserved canned food is safe and wholesome for consumption for months and even years. Preservation of food at home by canning was a necessity for early settlers and was promoted during WWI and WWII for economic reasons. Home canning has experienced a resurgence in popularity recently as more people are returning to home gardening and are interested in processing their own fruits and vegetables. It is essential that proper canning techniques and equipment are used to avoid food-borne illness when doing home canning. Methods such as heating jars of food in the oven or even a dishwasher are unsafe and very risky.

The heating used to process canned foods does change the color, texture, and flavor of foods. Also, salt may be added to canned vegetables or sugar to fruits to retain color and texture. Many consumers prefer the firmer texture and more intense flavors of frozen or fresh foods over canned foods, and may want to avoid added salt or sugar. New technologies to reduce the amount of heat used to preserve food, such as aseptic heating, are being developed. Beverages, dairy products, wine, sauces, and soups can be rapidly heated in thin films or as highly agitated fluids at pasteurization or sterilization temperatures, then aseptically (sterilely) packaged in foil laminate packages or pouches. Ohmic heating and microwave-assisted heating also are used to eliminate pathogenic bacteria and reduce the amount of heat damage to retain the quality of the food.

6.3.4 Processing Food by Refrigeration

Fruits and vegetables are living plants and contain enzymes that cause softening of their texture, the conversion of starch to sugar, or sugar to starch as ripening occurs. With time, these enzymes also cause deterioration and spoilage. By chilling food to temperatures of 32–40°F, spoilage processes are slowed or reduced. Enzymes, the catalysts for biochemical reactions found naturally in plants, are less active at low temperatures and less able to break down cell walls and cause spoilage.

Before mechanical refrigeration, blocks of ice cut from rivers and lakes were delivered to food processing facilities and household ice boxes to provide cooling of foods. Problems with ice availability and sanitation, especially in the meat, dairy, and brewing industries, made the development of mechanical ice-making and refrigeration a necessity. Mechanical refrigeration revolutionized food processing, transportation, and storage, and, with the adoption of household refrigerators (by 1950, more than 80% of farm homes and over 90% of urban homes had one), changed the way Americans ate. Instead of relying on daily shopping for fresh foods or relying on preserved foods, consumers could store perishable food in home refrigerators and freezers. New foods such as ice cream and frozen dinners were possible. Today, foods are sourced from around the world, and transported by refrigerated air, train, and truck to be available in supermarkets all year long.

6.3.5 Processing Food by Freezing

Freezing kills some microorganisms, but not all. Some bacteria continue to grow, albeit at a slower rate, during freezing. Enzymes that cause deterioration are also slowed by freezing. In commercial freezing technology, vegetables and some fruits are blanched by a short heat treatment (<212°F for 2–3 minutes) before freezing. Blanching inhibits enzymes that cause the color, texture, and flavor of the food to deteriorate during storage and kills surface bacteria. This is typically done in home freezing as well. The major difference between home and commercial freezing is that commercial processors use blast freezers that generate very low temperatures (−40°F) to freeze the water in the food very quickly. This rapid freezing prevents the breakdown of the cellular structure that occurs with the slower freezing process in a home freezer. Commercial freezers use cold air or liquid refrigerants in cabinets, plates, heat exchangers, and air-blast equipment for indirect freezing, and air-blast, fluidized bed, immersion, and conveyers in direct contact freezing.

The temperature of the freezer, as well as the type of food packaging, are the most important factors that determine the quality of frozen food. Freezers should be less than 0°F and packaging should be moisture/vapor-proof. The quality of frozen foods stored in frost-free freezers (which use a cycle of increased and decreased temperatures) will be negatively affected if packaging is not adequate to prevent freezer-burn (loss of water). Frozen foods are of the highest quality if used within a few months.

Frozen foods have become the preferred type of processed food among US consumers because of the high quality of the products and ease of preparation. Ice cream and other desserts, meats, microwave-ready vegetables, main dish meals, breads, baked products, snacks, fruit mixtures, pizza and other ethnic dishes are mainstays of many American meals. Freezing is the most expensive type of food preservation because of the continual use of electricity to maintain the frozen conditions. Canning, drying, and other food preservation methods are less energy-intensive because the energy use occurs once. Freezing is the easiest type of food preservation (in terms of labor and special equipment needed) and is often used by home gardeners to preserve surplus fruits and vegetables. Frozen foods are not as susceptible to food-borne pathogens as canned foods due to the limited ability of microorganisms to grow at low temperatures. Sanitary processing, including blanching prior to freezing are important to reduce any potential pathogenic or spoilage microorganisms in the food.

6.3.6 Processing Foods by Dehydration

Removal of water from food reduces the potential for spoilage because bacteria are less able to grow without water and degrading enzymes are inactivated. Historically, foods were dried by exposing them to the sun or hanging over a fire. Modern drying techniques involve controlled heat and air flow for more rapid removal of water and better preservation of the quality and texture of the foods. Dried or dehydrated foods are usually cut into small pieces, blanched or treated with antioxidants to preserve the color, then exposed to circulating dry air to remove moisture from the food by evaporation. Food that is dried in a solar dryer is referred to as "dried" while foods dried under controlled conditions with electric heaters and fans are referred to as "dehydrated." Commercial dehydrating processes also employ evaporation of water under vacuum or reduced pressure (freeze-drying or spray-drying).

In spray-drying, small droplets of fluid are forced into drying chambers, resulting in powdered products such as nonfat dry milk, instant coffee, and fruit drinks. Freeze-drying occurs when water is removed from frozen foods by sublimation (directly from frozen to water vapor), and this process occurs under vacuum. The texture and flavor of dried foods may be very different from the fresh product and vitamin loss can be great if the exposure to heat and air during drying is prolonged. Dehydrated foods are very convenient for camping and other outdoor activities and do provide variety and interest to the diet.

Intermediate moisture (IM) foods, such as raisins and other dried fruit, are moist enough to be consumed without rehydration. Water activity (a_w) is the availability of water for microbial, enzymatic, or chemical activity in foods. The water activity of a food can be measured using hygrometers and is often monitored during food processing. Bacteria need at least a_w 0.9 to grow, fungi and enzymes 0.8, and yeasts 0.6. Reducing the a_w to less than 0.6 limits microbial spoilage. IM foods contain 10%–50% water but water activity is decreased (0.65 a_w) by the addition of glycerol or sorbitol. Sugar and salt also decrease water activity and may be added to foods to enhance their shelf-life for this reason.

6.3.7 Processing Foods by Fermentation

Fermented foods are preserved by the production of citric, lactic, or acetic acids by beneficial bacteria such as *Lactobacillus*, *Streptococcus*, *Bacillus*, and *Pseudomonas*, yeasts, and fungi that use the food as a substrate for their growth and metabolism. Most bacteria grow best between pH 5.5 and 7.0 and do not thrive in acidic environments (below pH 4.5). This type of food preservation has been used for centuries and provides unique flavors and textures to foods. Commonly consumed fermented foods include sausages, sauerkraut, pickles, yogurt, cheeses, soy sauce, and balsamic vinegar. The fermentation process generates new flavors and textures to foods and keeps them from spoiling. The addition of salt and heat during canning can prolong the storage time of fermented foods even longer.

Beer, wine, and hard apple cider are examples of beverages in which alcohol and acid are by-products of fermentation. The alcohol is produced by yeasts, such as *Saccharomyces*, from the naturally occurring sugars in the beverages.

6.3.8 Processing Foods by Curing (Salting or Brining) and Smoking

Salt (sodium chloride) preserves meats and fish, dairy foods (butter and cheese), and vegetables (cabbage/sauerkraut, olives, cucumbers) by removing moisture and suppressing undesirable microorganisms. Mixtures of salt, sugar, sodium nitrate, and sodium nitrite are used to pickle and cure meats. Ham, bacon, corned beef, frankfurters, and many sausages are cured using these ingredients. The nitrites preserve the red color of meat (myoglobin) and prevent the growth of *C. botulinum*, which is a significant concern due to the anaerobic environment of these cured meats. Celery powder, which is naturally high in nitrate, acts similarly to nitrite as a curing agent and has been promoted as a 'natural' curing method. But the levels of nitrite in celery are variable. Current regulations require that meats cured with celery powder be labeled "uncured" to distinguish them from conventionally preserved meats due to this lack of control in the amount of nitrite. If the statement "no nitrates or nitrites added" is used, it must also say "other than those which naturally occur in celery powder."

Some meat products, as well as cheeses and fish, are smoked for additional flavor and improved palatability. In modern facilities, the temperature of smoke generation is controlled to reduce the formation of carcinogenic compounds. Old-fashioned smoking procedures with hardwood fires preserved meat because the surface was sterilized by heat, there was a reduction in moisture content, and the salt concentration was high enough to limit microbial growth. Smoked foods today are generally preserved using lower temperatures with added preservatives.

6.3.9 Processing Foods by Irradiation

In 1958, the Food Additives Amendment to the Federal Food Drug, and Cosmetic Act approved food irradiation as a means of preserving foods. Unlike all previous types of food processing, irradiation is considered an additive and foods processed with irradiation must be labeled with the Radura symbol and the words "treated with radiation" or "treated by irradiation" (Fig. 6.3).

The reasons for this response likely reflect the world situation at the time the legislation was passed. The United States had ended WWII by dropping two atomic bombs on Japan and radiation was widely feared as a weapon.

FIGURE 6.3 Irradiated foods must be labeled with the Radura symbol or have the words "treated with irradiation" or "treated by irradiation" on the package according to FDA regulations. *Source: FDA, www.fda.gov.*

It was misunderstood that foods do not become radioactive, and no residual radiation exists in irradiated foods. Irradiation is the process of exposing foods to ionizing radiation to reduce microbial load, destroy pathogens, extend shelf-life of perishable products, and remove infestation of produce. Foods are irradiated with gamma rays from cobalt-60, cesium-137, beta rays from an electron beam, or X-rays. Irradiation damages the DNA of insects and bacteria but is of insufficient strength to cause any structural damage to the food.

Foods for space travel and the military were among the first foods to be treated with irradiation and continue to be so today. The FDA has approved irradiation at 10 kGy for prevention of sprouting in potatoes; destruction of insects in wheat, cocoa, fruits, and spices; reduction of bacterial spoilage of tropical fruits, fresh ground meat, poultry, and seafood; and treatment of packaging materials. Irradiation replaced chemical fumigation (by ethylene oxide and methyl bromide) to kill microorganisms and insects in spices and tropical fruits that are imported from Hawaii and foreign countries. This reduced exposure to these chemicals by handlers and consumers. Higher doses of irradiation, similar in results to heat processing, are approved for preservation of meats, fish, and poultry. Irradiation can destroy *Trichina* parasites in pork and *Salmonella* in poultry as well as delay ripening and spoilage of fresh fruits and vegetables.

Irradiation has been shown to be an effective food processing technology for more than 50 years, with countless research studies to support its safety and efficacy. Over 30 countries have approved irradiation of some foods and the Food and Agriculture Organization, World Health Organization, and International Atomic Energy Agency concluded that food irradiated to any dose appropriate to achieve the intended technological objective is both safe to consume and nutritionally adequate. Perhaps because of the need to label irradiated foods and the connotation with radiation, irradiation has been controversial and resisted by consumers. Fears that irradiation allows unwholesome foods to be processed for sale, or that unusual chemicals are produced in the food, are unscientific and unfounded. Irradiation could be a means of reducing the use of chemical additives and preservatives in food processing, and spoilage that contributes to food waste. Other benefits are also possible; for example, irradiated ground beef has much longer shelf-life than untreated meat, which would allow more stores, including convenience stores, to sell ground beef, making it more available to consumers. Consumer-driven concern has limited this potentially beneficial industry from expanding. Irradiated strawberries and a few other fruits are available and some private brands, for example Schwan's, sell irradiated meat. Irradiation is an example of a technology that has been tested extensively by respected scientists and documented as safe, yet is not accepted by consumers due to confusion and misunderstanding.

EXPANSION BOX 6.1

FOOD IN SPACE

The first space food was "cubes and tubes" and not very palatable. These astronauts spent only a few days in space, and food was intended to just keep them healthy for that short time. When people began to spend months in space, tasty food that was similar to their diets on Earth became more important. Today, astronauts to the International Space Station (ISS) have a wide variety of tasty, nutritious, and safe foods. Foods are similar to foods prepared on Earth and many are available commercially. Astronauts select their own menus, and diets are designed to supply each astronaut with 100% of the daily value of vitamins and minerals they need, as determined by food scientists and dietitians at the Johnson Space Center in Houston, Texas.

Astronauts need the same number of calories and most of the vitamins and minerals in space as they would on Earth. Astronauts have fewer red blood cells while in space so the requirement for iron is less. To retain bone mass, which is reduced while in zero gravity, sodium is limited and vitamin D is supplemented.

On ISS, there is an 8-day menu cycle of three meals per day plus snacks. Half the food system is from the United States and the other half from Russia, with other foods added if the crew includes Japanese and Canadian members. Crew members usually eat breakfast and dinner together. Food is prepared in the Russian *Zvezda* service module. Russian cans and packages are heated in a specially designed warmer and US foods are heated in a food warmer "suitcase."

The types of food processing required of foods for space travel are similar to the types of food processing technologies available on Earth, i.e., fresh, freeze-dried, canned, and irradiated. In NASA terms these are referred to as natural form, rehydratable, thermostabilized, and irradiated. Foods are packaged in single-service, disposable containers to eliminate the need for a dishwasher (Fig. 6.4).

Electrical power for ISS is generated from solar panels so no extra water is generated from fuel cells, as it was on the Shuttle vehicles. Water is recycled from cabin air but there is not enough for use in foods so the amount of rehydratable foods was decreased and thermostabilized food increased compared to the Shuttle foods. All food is precooked or processed, requires no refrigeration and is ready-to-eat or prepared by addition of water or heating. Fresh fruits and vegetables must be eaten within the first few days of arrival or spoilage will occur. Natural form foods such as nuts, granola bars, and cookies are ready-to-eat, packaged in flexible pouches, and require no preparation.

Rehydratable foods include soups, casseroles (such as Macaroni & Cheese), shrimp cocktail appetizer, scrambled eggs, and cereals. Rehydratable food packages are flexible to aid in trash compaction. Beverages are in powdered form and include coffee, tea, apple cider, orange juice, and lemonade. The beverage package is a foil laminate, similar to commercial juice boxes with straws, with maximum barrier properties for longer shelf life. There is a septum that fits with the water dispenser and a straw is inserted for consumption. Only ambient, warm, and hot water is available on the ISS.

Thermostabilized foods are heat-treated for safety and are preserved in cans, plastic cups, or flexible retort pouches. These foods include fruits, tuna, salmon, puddings, and entrées such as beef and mushrooms, tomatoes and eggplant, Chicken à la King, and ham. Pouches

EXPANSION BOX 6.1 (*cont'd*)

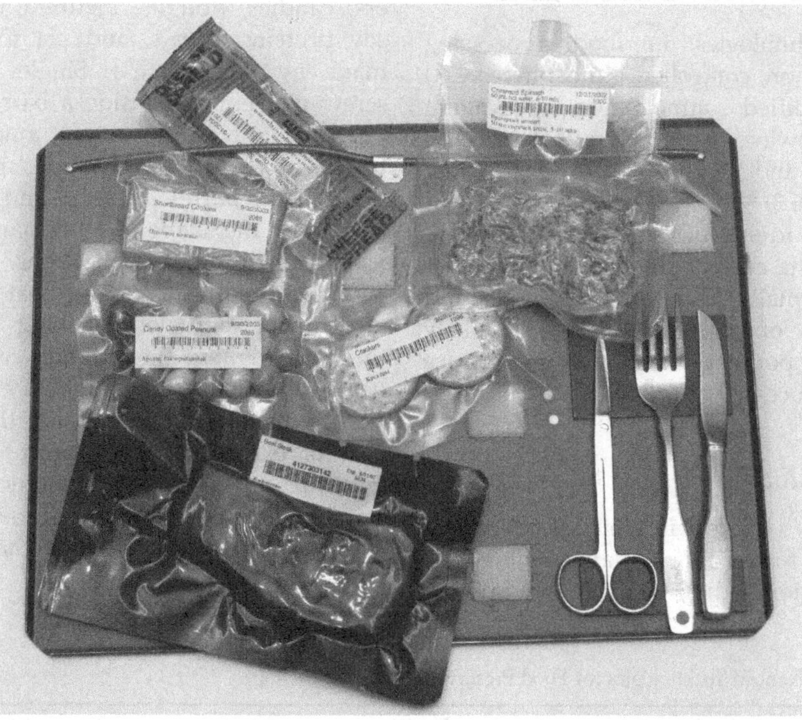

FIGURE 6.4 Food scientists have created a variety of food items and meals for astronauts. The challenges for space food development include the need for a long shelf-life without refrigeration, extreme sanitation, and minimal water or preparation. Food packaging innovations, such as the retort pouch, were developed for NASA but have become part of terrestrial food products (juice boxes for example). Shown here is an example food tray with typical packaging systems used on the International Space Station (ISS). *Source: Photo from the Iowa State University NASA Food Technology Commercial Space Center.*

are heated, cut open, and eaten directly from the container. Irradiated foods in flexible pouches, such as meats and main dishes, are also ready to eat and only require warming before consumption.

Future space travel to the moon or Mars would require food with a shelf life of 3–5 years. In transit, foods similar to those for ISS could be used, but once residence on a planetary surface is established, food could be grown in climate-controlled, hydroponic laboratories (not unlike the 2015 movie *The Martian*). It is proposed that this food system would be similar to a vegetarian diet. Sweet and white potatoes, soybeans, wheat, peanuts, dried beans, lettuce, spinach, tomatoes, herbs, carrots, radishes, cabbage, and rice are crops that could be grown. Produce would be processed into edible ingredients for immediate consumption or stored. Issues related to water and energy conservation, nutrient retention, microbial safety, packaging, and waste reduction would be of utmost importance, similar to the concerns of food processors on Earth.

Suggested reading: Bourland, Kloeris, Rice, and Vodovotz (1999).

6.4 NEWER FOOD PRESERVATION AND PROCESSING TECHNOLOGIES

Recent technologies in food processing include extrusion, controlled atmosphere (CA) storage, modified atmosphere packaging (MAP), microwave heating, pulsed electric field (PEF), and high-pressure processing (HPP) (Table 6.3). The combination of two or more methods to produce safe and more fresh-like foods is the current trend in food processing. Nonthermal preservation methods and reduced use of preservatives will likely become more popular techniques to produce "minimally processed" products.

6.4.1 Extrusion

Extrusion is a process in which a mixture of ingredients is forced through an opening in a perforated plate (or die), then cut to a specific size or shape. Sausages and pasta have been made by extrusion for over 100 years. Many types of breads, cereals, snacks, cookies, crackers, candies, doughs, textured vegetable and soy proteins, meats, and pet foods are now made by this process. Single- and double-screw mechanisms inside a barrel control the pressure, rate of movement, amount of moisture, temperature, and mixing of components through the barrel. The size and length of the barrel and the shape of screws and dies can be varied for different applications. Extrusion can be a noncooking process or heat can be generated during extrusion. Cooking is done inside the barrel where the product creates its own heat and friction generated by the pressure exerted by the screw(s) inside the barrel. The heat of extrusion can denature proteins and gelatinize starch. Many products with high starch content expand or puff as they exit the extruder, creating the unique characteristics of corn curls and puffed cereals.

TABLE 6.3 Advanced Technologies for Food Preservation and Processing

Technology	Mechanism	Function	Types of foods
Extrusion	Temperature, pressure, and shear by screw press	HTST treatment, changes shape and texture	Pasta, cereals, snack foods
Controlled atmosphere storage	Reduced oxygen and increased carbon dioxide in storage	Slow ripening of fruits by inhibiting enzymes	Apples, bananas, tomatoes, melons
Modified atmosphere packaging	Selective packaging to limit, maintain, or remove gases	Reduce deterioration during refrigeration	Meats, salad greens
Microwave sterilization	Sterilization by heat; pasteurization; dehydration	Aseptic packaging in flexible pouches for military and space	Sweet potato puree
Pulsed electric field[a]	Very high voltage for microseconds	Pasteurization treatment, speed drying	Fruit juices and peeled fruits and vegetables
High-pressure processing	Very high pressure without temperature increase	Destruction of microbes by pressure, little alteration of food	Guacamole, raw oysters, meats, juices, baby food
Infrared heating	Electromagnetic vibrations	Destruction of microbes without long heating	Baked goods, roasting, peeling

[a] *Process not approved by FDA.*

6.4.2 Controlled Atmosphere Storage

To prolong the shelf-life of fruits and vegetables, it is possible to reduce the respiration rate in these foods by lowering the levels of oxygen, increasing carbon dioxide and adding nitrogen around them. This controlled atmosphere (CA) storage is effective in delaying ripening of "climacteric" fruits and vegetables. Apples, tomatoes, bananas, melons, kiwi, plums, avocados, peaches, pears, and apricots are climacteric foods that can be harvested before the onset of ripening and stored in CA conditions. They continue to ripen after being harvested. Some of these fruits and vegetables naturally produce ethylene gas during the ripening period, and this can be either removed or added to delay or hasten ripening. Hastening the ripening of tomatoes by placing them in a paper bag works because the ethylene produced by the fruit is concentrated inside the bag. A banana added to the bag contributes more ethylene and speeds ripening. Commercial management of ethylene gas allows fruits and vegetables to stay fresh longer so that food waste is reduced.

CA storage also slows the spread of microbial diseases such as fungal diseases in cabbage and reduces the incidence of some physiological disorders such as browning of cabbage, pitting of oranges, sprouting of potatoes, and development of bitter flavor in carrots. CA storage can be used in conjunction with vacuum conditions during transport of tropical fruits such as bananas, mangoes, papayas, and guavas to keep them from spoiling before reaching the market.

6.4.3 Modified Atmosphere Packaging

Similar to CA, modified atmosphere packaging (MAP) is used to slow down respiration and prolong the storage life of fruits, vegetables, and salad greens. Selective packaging films and materials with specific gas permeability are used to create defined oxygen and nitrogen levels around the food. Elimination of all oxygen would result in pickled products due to anaerobic respiration so the packaging must allow gases to escape and oxygen to enter. Each plant has a different rate of metabolism so finding the correct conditions is necessary. Newer technologies in plastics have also provided materials that change permeability based on temperature, which allows additional levels of control.

MAP is used along with measured levels of carbon dioxide (CO_2), oxygen, nitrogen, and/or carbon monoxide (CO) in packaging fresh meats. Beef and pork will undergo a loss of red color during storage, causing it to look brown and unappealing. Consumers confuse this lack of red color with spoilage, even though it is not. The loss of red color is from the release of oxygen from myoglobin (the muscle form of hemoglobin). CO binds to myoglobin more tightly than oxygen and keeps the red meat color. CO_2 is used to limit microorganism growth and nitrogen is used as a carrier gas to allow other gases to be mixed in the correct percentage. The most common modified atmospheres used for red meat consist of 80% oxygen and 20% carbon dioxide, or 0.4% carbon monoxide, 30% carbon dioxide, and 69.6% nitrogen.

6.4.4 Microwave Sterilization

Microwave ovens use nonionizing radiation in the electromagnetic spectrum (2450 and 915 MHz with heating capacities between 10 and 200 kW) to excite polar molecules, such as water, in food. This results in the generation of heat. Commercial microwave ovens are used in food processing facilities for precooking bacon, tempering frozen meats, and precooking other foods. Microwave processing requires less heating time, uses less energy, and retains more nutrients in foods than other thermal processing methods. Because it is a faster heating process, food retains more fresh characteristics and has similar microbiological reduction properties as commercial heat and pressure treatments. Used in combination with prepackaging, microwave sterilization can increase the shelf-life of foods without refrigeration.

6.4.5 Pulsed Electric Field

Pulsed electric field (PEF) processing is a nonthermal treatment (although the temperature of the food does increase during treatment) of food in a chamber with a high-voltage electrical field (20–400 kW) that causes increased permeability or rupture of biological cell membranes (microorganisms and plants). PEF kills vegetative bacterial cells but does not inactivate spores. Pasteurization of fruit juices, soups, milk, and other liquids can be accomplished by treatment with PEF. PEF is used to extract juices from apples, grapes, and carrots; make beet, broccoli, and kale mashes; enhance peeling of tomatoes and prunes; and accelerate drying of potatoes, onions, and peppers. PEF is capable of replacing some traditional food processing methods with lower energy use and shorter processing times, making it an increasingly attractive new technology.

6.4.6 High-Pressure Processing

The application of high pressure (up to 600 MPa) to juices and beverages destroys bacteria at a similar level to pasteurization. HPP does not generate heat and foods do not undergo flavor or color changes. Pressure with a small amount of heat (<240°F) can be used to sterilize low-acid products such as sweet potatoes and mashed potatoes, which can then be stored at room temperature. Because of the high quality and nutrient retention that results, HPP is used to preserve baby foods and specialty products such as guacamole, deli meats, salsa, and seafood. HPP is an attractive alternative to the use of heat and chemical additives in foods but is an expensive process due to the high cost of the equipment.

6.4.7 Infrared Heating

Infrared (IR) radiation releases energy in electromagnetic wave form and causes molecular vibration of food components (water, organic compounds, biological polymers) in the 2.5–10 μm wavelength range. IR heating was first used in the 1930s for curing rubber in the automotive industry and later in the manufacturing and electronics industries. IR is used for baking breads and other baked goods, roasting nuts and browning meats, and has potential for blanching and peeling fruits and vegetables; drying fruits, herbs, nuts, shrimp, and cereals; and destroying pathogens on foods. Because IR does not require a heating liquid, less water is used and improved energy and processing efficiency is achieved compared to traditional canning. IR is most suitable for heating thin layers of food materials and heats only a few millimeters below the surface.

6.5 FUNCTIONAL ADDITIVES IN PROCESSED FOODS

Food scientists developed methods to not only preserve foods, but also to maintain and even enhance the color, flavor, and texture of food during storage. Legally, the term "food additive" refers to "any substance where the intended use of which results—directly or indirectly—in its becoming a component or otherwise affecting the characteristics of any food" (Food, Drug, and Cosmetic Act). This definition includes any substance used in the production, processing, treatment, packaging, transportation, or storage of food. The primary categories of additives are preservatives, flavors and spices, coloring, fat replacers and stabilizers, nonnutritive sweeteners, thickeners, and texturizers. These are added to improve the quality and shelf-life of foods, for preservation, ease of manufacture, and improvement of appearance, texture, or flavor. Vitamins and minerals added to improve the nutrient content of foods are also additives.

6.5.1 Regulation for Food Additives

The Food and Drug Administration (FDA) has the primary and legal responsibility for approving food additives as safe. Today, food and color additives are more regulated and monitored than at any time in the past. When evaluating the safety of food additives, the following criteria are examined by the FDA: the composition and properties of the substance, the amount that would typically be consumed, immediate and long-term health effects, and various safety factors. The term *generally recognized as safe* (GRAS) is used by the FDA to describe various products that are added to food. Over 700 additives on the GRAS list are accepted as safe, either by a long history of safe use or by extensive testing. Substances can be removed from the GRAS list if new evidence indicates that they may be harmful. If the FDA does not approve a product, it will take action to prevent the distribution of that product. Products must be demonstrated as safe for the intended use before they are allowed to be used in foods.

6.5.2 Preservatives in Processed Foods

Food preservatives are specific additives to prevent deterioration from enzymes, microorganisms, and exposure to oxygen. All chemical preservatives must be nontoxic and readily soluble, not impart off-flavors, exhibit antimicrobial properties over the pH range of the food, and be economical and practical.

Sugar, salt, nitrites, butylated hydroxy anisol (BHA), butylated hydroxyl toluene (BHT), tert-butylhydroquinone (TBHQ), vinegar, citric acid, and calcium propionate are all chemicals that preserve foods. Salt, sodium nitrite, spices, vinegar, and alcohol have been used to preserve foods for centuries. Sodium benzoate, calcium propionate, and potassium sorbate are used to prevent microbial growth that causes spoilage and to slow changes in color, texture, and flavor. Potassium sorbate and sodium benzoate both prevent spoilage by inhibiting mold and yeast. Sodium benzoate may be in foods such as salad dressings, soft drinks, canned tuna, and mixed dried fruit. Potassium sorbate is found in cheese, wine, and dried meats. BHA and BHT are antioxidants that prevent rancidity of fats and are added to shortening, margarine, and fried snacks such as potato chips.

Consumers have raised concerns about the use of preservatives in foods that have complicated chemical names that make them seem more appropriate for a chemistry experiment than a meal. Sodium benzoate, BHA, BHT, and TBHQ have especially been targets of consumer apprehension. These compounds have been approved for their safe use in foods and have not been linked to any human illness or complications for the general public. As is the nature of scientific inquiry, reports of adverse effects of these compounds can be found in the literature. The abundance of evidence suggests that the risks of these compounds, which are used in small amounts, to human health are insignificant. And, in contrast to having a negative impact on health, BHA and BHT have been linked with having a positive effect due to their antioxidant capacity. Weighing the risk/benefits of using these chemicals in foods is an ongoing debate and the FDA, food companies, and consumers must all participate. No food, additive, or ingredient will be 100% safe for 100% of the people. Using scientific thinking to consider these complicated decisions is essential to avoid emotional reactions based on misinformation.

6.5.3 Flavorings and Spices

Natural flavoring, artificial flavor, spices, and monosodium glutamate (MSG) improve the palatability of foods. Natural and artificial flavors are found in a number of products like granola bars and flavored juices and beverages. MSG, a common additive in soup, barbecue sauce, and seasoning mixtures, gives food enhanced flavor, similar to adding soy protein, mushrooms, or other savory ingredients. There has been some concern among consumers that MSG causes allergic-type reactions. There is little scientific evidence to support that concern and the FDA considers MSG to be a safe food additive. Some people may experience a mild reaction to MSG, but these are usually short-term and do not require medical treatment.

6.5.4 Color Additives in Processed Foods

Color additives are used in foods and beverages to enhance and correct colors already present, provide color identity to colorless foods, and account for color loss during storage. Any dye, pigment, or substance made or obtained from a vegetable, animal, mineral, or other source capable of coloring a food is a color additive, according to the FDA. Synthetic or artificial colors are derived from petroleum or coal and identified with FD&C numbers, such as FD&C Blue No. 1. Natural color additives are derived from plants, animals, or minerals. Both types of color additives are regulated by the FDA. Synthetic colors must be certified for identity and purity by the FDA, while natural color additives do not require certification but must meet identity standards and specifications. Annatto extract, beet powder, sodium copper chlorophyllin, grape extract, carrot oil, paprika, titanium dioxide, iron oxide, and many other "natural colors" are example of colors exempt from certification.

Concerns about the potential health effects of synthetic food colors have been in the public arena for several years. In the 1970s, pediatrician Dr. Benjamin Feingold raised the hypothesis that hyperactivity in children was caused by additives in food. Around this time, changes in how attention deficit hyperactivity disorder (ADHD) and autism were diagnosed and managed were being made by mental health professionals, and more children were being diagnosed with these conditions. The Feingold Diet, which required avoidance of all food colors and other food additives, became popular with parents as a means of addressing behaviors in their children. Debate as to whether there was any scientific rigor to the claims that synthetic food colors affected behavior continued for the next decade. In 1982, the National Institutes of Health conducted a broad review of the literature and concluded that for a small percentage of children with ADHD and confirmed food allergy, dietary modification may produce some improvement in behavior. This conclusion was made despite the lack of scientific evidence to predict which children would benefit from a dietary restriction, or strong evidence that food colors were responsible for the changes in behavior. In 1997 another scientific review concluded that there was minimal evidence of effectiveness of dietary restriction of food colors on behavior in children and noted the extreme difficulty of getting children and adolescents to comply with restricted diets.

In 2007, synthetic certified color additives again came under scrutiny following publication of a study commissioned by the UK Food Standards Agency to investigate whether certain color additives cause hyperactivity in children. Both the US Food and Drug Administration and the European Food Safety

Authority independently concluded that the study did not show a substantial link between color additives and behavioral effects. But despite these negative conclusions, the European Union currently requires foods that contain any of the six colors tested in the report, including Yellow 5 and 6, to be labeled as "may have an adverse effect on activity and attention in children."

Yellow 5 (tartrazine) and Yellow 6 are synthetic FD&C yellow dyes used in foods, candies, drugs, and cosmetics. Yellow 5 is Lemon Yellow and Yellow 6 is Sunset Yellow. Some consumers (0.01%–0.1% of the population) cannot tolerate tartrazine. In these consumers, tartrazine may cause symptoms similar to an allergic reaction, including hives and swelling, but the reaction is not considered a true allergy. Tartrazine was also thought to be associated with the onset of asthma attacks, but recent scientific evidence indicated that tartrazine is an unlikely cause of asthma symptoms. To help protect people who may be intolerant to tartrazine, the FDA requires that any food for human use that contains Yellow No. 5 must specifically declare it as an ingredient. The addition of any coloring must be identified on food packaging and labeled as "contains artificial color," "colored with," or "color added." FD&C colors must be identified by name. "Natural color" is not allowed as the FDA considers all color additives to be artificial but a definition for "natural" on food labels is being reviewed. In response to consumer demand, many food companies are replacing synthetic color additives with natural colors. These uncertified color additives may vary in purity, quality, and safety. Because they are derived from plants, animals, or soil, they must be tested for microbial contamination. Also, organic solvents are often used to extract, purify, and concentrate these natural colorants and it is not yet known if these chemicals are fully removed during processing. The trade-off of switching from well-studied and tested synthetic colors to natural colors with limited oversight and safety testing is just beginning to play out in the food system.

EXPANSION BOX 6.2

KRAFT MACARONI & CHEESE

Kraft Foods Group Brands LLC is one of North America's largest consumer packaged food and beverage companies, with annual revenues of more than $18 billion. The company's iconic brands include Kraft, Capri Sun, Jell-O, Kool-Aid, Lunchables, Maxwell House, Oscar Mayer, Philadelphia, Planters, and Velveeta. Oscar Mayer became part of General Foods in 1981 and General Foods was purchased by Kraft Foods in 1989. In 2015, Kraft Foods and H.J. Heinz Co. merged to create the third largest food and beverage company in North America and the fifth largest food and beverage company in the world.

Kraft Foods introduced packaged Macaroni & Cheese in 1937 and the product became immediately popular. Today the "Blue Box" line includes many varieties, including different cheese flavors such as Buffalo Cheddar, Cheesy Southwest Chipotle, Garlic & Herb Alfredo, Three Cheese Jalapeño, Cheddar Explosion, Thick and Creamy, Three Cheese, and White Cheddar, and kid-friendly shapes of pasta including Minions, SpongeBob, Star Wars, and Teenage Mutant Ninja Turtles. Kraft Macaroni & Cheese comes in Whole Grain, Organic White Cheddar, and Organic Cheddar. In addition, there is the Homestyle line with 7 options,

EXPANSION BOX 6.2 *(cont'd)*

Microwavable Macaroni & Cheese with 17 options of sizes and flavors, and the Deluxe line with 9 options.

Kraft Macaroni & Cheese is a popular food with children and families because it is inexpensive, easy and quick to prepare, nutritious, and tasty, and the packaged product has a long shelf life. The two basic ingredients of Kraft Macaroni & Cheese are:

- Enriched macaroni, which contains
  - wheat flour
  - durum flour
  - vitamins and minerals (niacin, ferrous sulfate (iron), thiamin mononitrate (vitamin B1), riboflavin (vitamin B2), and folic acid)
- Cheese sauce mix, which contains
  - whey
  - milkfat
  - milk protein concentrate
  - salt
  - sodium tripolyphosphate
  - less than 2% citric acid, lactic acid, sodium phosphate, and calcium phosphate
  - spices paprika and turmeric
  - annatto added for color
  - enzymes
  - cheese culture

Consumers have criticized Kraft for using artificial colors and preservatives. In March 2016, Kraft responded by changing the Macaroni & Cheese recipe. Calories and fat were decreased 12% and 14%, respectively, sodium was decreased slightly, protein and fiber amounts were increased and artificial colors (Yellow 5 and Yellow 6) were replaced with annatto, paprika, and turmeric. The label lists all ingredients (all additives are GRAS) in Kraft Macaroni & Cheese and the organic versions also offer consumers a choice of ingredients.

What are the functions for the "chemicals" in Macaroni & Cheese? Niacin, iron, vitamins B1 and B2, and folic acid are nutrients added to refined wheat flour for enrichment. The cheese sauce mix is composed of cheese components that, as dry ingredients, can maintain high quality and safety for a long time. Whey and milk protein concentrate are proteins from milk (whey fraction and cheese curd, respectively) and are added to the cheese mixture for nutrition and thickness. Sodium tripolyphosphate is an emulsifier to help bind fat and water together and make the sauce creamy. Citric acid, lactic acid, sodium phosphate, calcium phosphate, enzymes, and cheese culture are byproducts of the cheese-making process and contribute to flavor. Citric and lactic acids also act as preservatives. Chymosin is the primary enzyme in rennet, used to coagulate casein curds from milk. *Lactobacillus* and *Streptococcus* are the bacterial cultures used to make many cheeses.

These ingredients have all been approved for use by the FDA and have a specific function that gives Kraft Macaroni & Cheese its characteristic taste and look. Consumers have become interested in foods that are "clean" meaning they do not contain "unnecessary" ingredients or chemicals or have only a few ingredients. Products such as Kraft Macaroni & Cheese may seem like they have a lot of ingredients with complicated names. An understanding of what these ingredients are and why they are present in the foods can alleviate some of the hesitation consumers may have about consuming processed foods.

6.5.5 Thickeners and Texturizers in Processed Foods

Gelatin, pectin, gums, and protein concentrates and isolates are stabilizers, thickeners, and texturizers that provide body to foods such as soups, sauces, salad dressings, and desserts. These ingredients give foods a smooth, creamy texture that we enjoy in ice cream, yogurt, cheese, and soups. Gums that are extracted from plants, seeds, shrubs, and seaweed, or produced by bacteria, are a natural source of thickeners. Examples of extracted gums include guar gum, locust bean gum, gum acacia, carrageenan gum, and xanthan gum. Gelatin (from animal collagen) is used in "Jell-O," cream cheese, and frozen desserts, and pectin (extracted from apple pomace and citrus rinds) is used in jellies, jams, and candies. Consumers consider gelatin and pectin gums as more "natural" (likely because their names are familiar and easy to pronounce) and manufacturers often use these gums in food because they are more consumer-friendly. But all gums are obtained from natural sources.

6.5.6 Nonnutritive Sweeteners in Processed Foods

Consumers have long sought replacements for sugar (4 kcal/g) in order to reduce the calorie content of desserts, beverages, and baked products. Saccharin and acesulfame-K are nonnutritive sweeteners approved for use in foods and are about 200 times sweeter than sugar, and sucralose is a nonnutritive sweetener that is 500–1000 times sweeter than sucrose. Aspartame (4 kcal/g) and sugar alcohols such as mannitol, xylitol, and sortibol (1.6–2.6 kcal/g) have calories but are several hundred times sweeter than sucrose so a very small amount can be used to sweeten foods. In no-sugar foods, the physical bulk of sugar is usually replaced by a noncaloric fiber such as maltodextrin or polydextrose. Other functional properties of sugar (browning, water binding, crystallization) are not replaced by substitute sweeteners and must be provided by other ingredients.

6.5.7 Fat Replacers and Stabilizers in Processed Foods

The function of fat in foods is to provide creamy mouthfeel, rich flavor, and smooth texture. Foods made to be low in fat, do not have these characteristics and are less enjoyable to eat. Olestra, cellulose gel, carrageenan gum, modified food starch, guar gum, and whey protein concentrate are ingredients used to replace the characteristics of fats in food. Gums are a common fat replacer because they thicken, have a gel-like consistency, and give products stability. Fat replacers such as xanthan gum are used in fat-free salad dressings, sauces, and ice cream.

Chemically modified food starches can be used to thicken many different types of products. Gravies, sauces, and pie fillings are common items that contain modified corn starch as a thickener. Chemically modified starches provide stability during canning and freezing. For example, tapioca starch is used in making home-made cherry pie filling because it is translucent. When frozen, tapioca starch becomes opaque and when canned, becomes clumpy and does not provide an appetizing appearance for cherries. By using a modified starch that is both clear and remains thick and stable during processing, frozen cherry pies and canned cherry pie filling can be produced without these side effects. Modified starches can be made to have unique functional characteristics, to allow new products such as instant pudding to be created. In food processing, modified starches are widely used to ensure food quality and stability.

6.6 BENEFITS OF PROCESSED FOODS

The benefits of food processing to consumers include increased efficiency of food preparation, improved safety, and enhanced palatability (Table 6.4). Processed foods can be high-quality foods that are healthy, safe, convenient and enjoyable. The societal benefits of food processing include keeping food costs low by optimizing food retention, reducing food waste, and adding nutritional value to the food supply. Food processing facilities employ thousands of workers, which aids the local and national economy. The application of newer technologies to food processing is aimed at reducing the use of additives and chemicals in foods, minimizing water use, and conserving energy. Negative aspects of food processing include the high use of water and energy needed in food processing, transportation and fuel use to deliver foods from processors to consumers, and the control of food processing by large food corporations. Consumers have concerns about chemicals and additives in foods; the ready availability of foods high in fat, salt, and sugar; and the ways foods are marketed to children. These latter topics are discussed in more detail in Chapter 7, Nutrition and Food Access.

TABLE 6.4 Comparison of Commercial Brands of Macaroni and Cheese

Brands	Package size (oz)	Servings per container (1 cup)	Cost per package ($)
Kraft	6	3	1.39
Annie's Homegrown	6	2.5	2.29
Hodgson Mill	7.25	3	2.29
Pasta Roni	7.2	2.5	2.19
Back to Nature	6	2.5	2.23
Our Family	5.5	2	0.89

6.6.1 Processing Increases Efficiency

Processing reduces the cost of food by limiting postharvest losses. In developing countries where food processing is not readily available, nearly 50% of foods that are grown are not consumed due to postharvest damage from pests, spoilage, or deterioration. In the United States, processing facilities located near agricultural production sites have reduced this type of loss significantly. Top-quality fruits and vegetables are transported to markets for direct consumer sales and seconds or less desirable products are canned, frozen, or made into juices. Efficient and effective tools to reduce microbial spoilage of foods further reduce food waste. It has been suggested that packaged foods result in 2.5% product loss to waste while fresh foods may have up to 50% product loss to waste.

Processing facilities increase the efficient use of by-products that might otherwise be wasted. For example, citrus oil for flavoring and pectin can be extracted from citrus peels, beta carotene can be extracted from palm kernel oil, and glycosaminoglycan protein can be extracted from egg shells. Agricultural by-products can be collected from processing plants and added to animal feed or used for biofuels or other uses, whereas excess food material (broccoli stalks, meat fat, potato peelings) produced at home is generally discarded and sent to landfills.

6.6.2 Processing Improves Food Safety

The United States has one of the safest food supplies in the world. The food industry follows defined regulations that require sanitary food handling and packaging, as well as testing, to assure safety. Standards are in place to ensure foods are properly processed to reduce the risks of food-borne illness and other hazards. The FDA defines Current Good Manufacturing Practices (CGMPs), food identity and microbiological quality standards, and the processing methods and equipment for food manufacturers. The FDA and Food Safety and Inspection Service of the US Department of Agriculture (FSIS-USDA) inspect and monitor US food production, processing, and marketing. We take for granted that our salads, fruit, and other uncooked foods are safe to eat, and that meat and dairy foods are wholesome and fresh. Because of this safety assurance, when a food safety issue does arise it becomes headline news.

The population of the United States is about 324 million people; if it is assumed each person eats three meals a day, that equals 974 million meals consumed per day. The Centers for Disease Control and Prevention (CDC) estimates that there are about 48 million food-borne illnesses a year. This roughly predicts that the chances of contracting a food-borne illness are less than 0.01% or less than 1 in 10,000 meals.

The number of illnesses, hospitalizations, and deaths caused by food-borne pathogens is difficult to quantify because many people do not see a doctor when sick with gastrointestinal disorders, the doctor is unable to diagnose the problem, or the illness is not reported to the CDC. It is important for healthcare providers and government agencies to know the number and sources of food-borne illnesses in order to prevent contamination problems throughout the food chain. Scientists use a method called food-borne illness source attribution as a way to obtain a more accurate estimate of the number of illnesses associated with specific foods. County health departments across the country report diagnosed cases of food-borne infections to state health departments who, in turn, report to the CDC. These actual numbers of food-borne illnesses are usually lower than "estimates" of food-borne diseases.

In 1995, the USDA, CDC, ten state health departments, and the FDA established the Foodborne Diseases Active Surveillance Network (FoodNet) to provide annual data from designated surveillance sites on actual numbers of laboratory-diagnosed cases of predominantly food-borne bacterial (*Campylobacter*, *Listeria*, *Salmonella*, STEC-O157 or Shiga toxin-producing *E. coli*, *Yersinia*, and non-O157 STEC) and parasitic pathogens (*Cryptosporidium*, *Cyclospora*). FoodNet collaborators conduct active, population-based surveillance at ten US sites (representing 15% of the US population or 46 million people) for confirmed cases of food-borne illnesses. FoodNet scientists monitor trends, identify sources, implement epidemiological studies to determine risk and protective factors, and conduct surveys about behavior regarding food handling, consumption, and prevention of food-borne illness.

Despite these safeguards and monitoring, pathogens do enter the food system. About 89% of all food-borne illnesses that occurred between 2006 and 2010 in the United States were from noroviruses (49%) and bacteria (40%; Fig. 6.5). Norovirus contamination is spread through contact with environmental factors (soil, water) and infected people. The most common food-borne bacterial pathogens are *Salmonella*, *Clostridium*, and *Campylobacter* (Table 6.5). The risk of disease from *E. coli* or *Listeria* contamination is low and the risk is even lower from the toxin of *C. botulinum*. Seafoods, improperly cooked meats and eggs, and cross-contamination of produce are the most likely sources of food-borne illness. Children younger than 5 years of age, pregnant women, adults over 65 years of age, and people with weak immune systems are more likely to become ill from contaminated food than other people, and the effects of illness may be more serious.

Salmonella infections are often associated with poultry and eggs. Eggs can be contaminated on their surface from contact with manure, and the yolk of an egg can become contaminated during development from an infected hen. *Escherichia coli* infections occur from meats when, during the slaughter process, pathogens on the animal's hide or

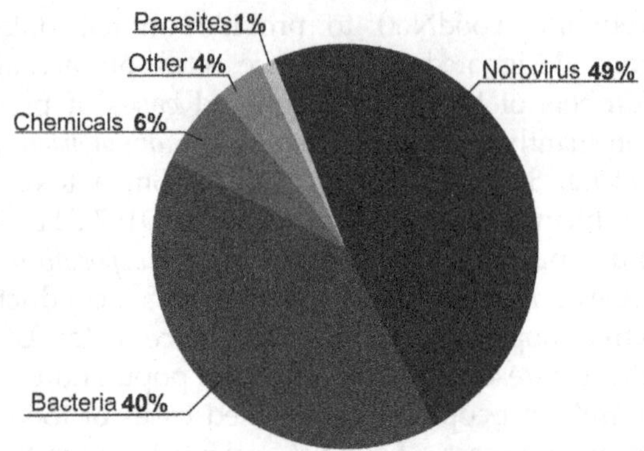

FIGURE 6.5 According to the Centers for Disease Control and Prevention (CDC) norovirus and bacteria are the main causes of food-borne illnesses in the United States. Norovirus is a highly contagious virus that is readily spread from person to person or through contaminated foods. Hand washing, and proper handling and storage of foods at home, are effective ways to avoid food-borne illness. *Source: CDC, www.cdc.gov.*

intestines contaminate the meat product. *Salmonella* and *E. coli* are, as are most bacteria, destroyed by heat, so proper cooking of eggs and meat will reduce the risk of food-borne illness. Fresh produce can be contaminated with *Listeria* or *Shigella* that are present in water or soil, or by *E. coli* from contact with animal or human fecal material.

Spoilage microorganisms, molds, yeasts, and bacteria reduce the quality and palatability of foods but do not cause illness. Molds can cause spoilage in fairly low-moisture foods (about 12% or water activity a_w 0.80) such as jams and jellies, especially when exposed to oxygen. Yeasts require higher moisture conditions to grow (about 30% or 0.88 a_w) and can thrive in both aerobic and anaerobic environments. Bacteria require the most moisture (about 30% or 0.91 a_w), can be aerobic or anaerobic, and can grow in a range of temperatures: thermophiles, above 113°F, mesophiles about 68–113°F, and psychrophiles below 50°F (refrigeration temperature or less). Food preservation techniques are designed to limit or destroy both pathogenic and spoilage organisms that are likely to occur in a particular type of food.

TABLE 6.5 Benefits of Processed Foods

Reason	Example
Efficiency increased	Commercially canned vegetables
Shelf-life increased	Refrigerated vegetables, bananas in CA, frozen meat
Quality improved	Frozen peas, tortillas, and breads
Availability increased	Kiwi, pineapple, strawberries, frozen orange juice
Waste reduced	Packaged meat, bagged salad greens, frozen broccoli
Safety improved	Pasteurized milk, canned vegetables, cured meats
Packaging protection	Refrigerated doughs, fresh meat, snack-size crackers
Variety enhanced	Instant oatmeal, pastas, frozen waffles
Nutrition added	Enriched bread, fortified milk, cereals with folate
Flavor improved	Potato chips, low-fat ice cream, pudding with aspartame
Convenience provided	Breakfast cereals, frozen pizza, reduced calorie frozen meals

The food industry uses strategies such as risk-based prevention controls, monitoring procedures, verification, and record-keeping to reduce the risk of food-borne illness. The Hazard Analysis Critical Control Point (HACCP) system, a method of monitoring food safety by anticipation and prevention of problems, is required for all facilities that "manufacture, process, pack or hold food and that are required to register with the FDA" (Section 415 of the Food, Drug, and Cosmetic Act). Food processors develop a HACCP plan by assessing all possible hazards during the production process, eliminate avoidable hazards, and set limits for hazards that are difficult or impossible to eliminate. Processing facilities are required to have food safety personnel trained in the HACCP process, as well as quality assurance staff who monitor sanitation, personnel, suppliers, transportation, storage, and plans for managing complaints, recalls, and traceability. HACCP has been shown to greatly reduce the prevalence of pathogens in foods.

Challenges to food safety are continually arising. New food production technologies require safety testing and evaluation, imported foods must be monitored and tested, extreme weather events such as floods and droughts change the types of pathogens present in the environment, and antibiotic-resistant bacteria are emerging. Each new food product or ingredient brings new potential sources of contamination. Ensuring that the millions of food preparers, including those who prepare food in their own homes, are well-educated and trained in proper food handling is a major undertaking. Even with all the government and food industry controls, the most important factor in prevention of food-borne illness are the humans who handle and prepare food. The simple act of washing hands effectively remains the primary way to prevent food-borne illness.

6.6.3 Benefits of Food Packaging

Prior to WWI, food was transported and sold out of bulk barrels, tanks, or bags. Consumers brought containers from home or merchants would wrap food in paper for transport. Milk was delivered in open tanks and dispensed on the street corner into containers consumers brought from their homes. Advances in transportation, storage, and food packaging materials, especially plastic, changed how food was offered to the consumer.

Food packaging is designed to prevent contamination, protect food during handling and storage, maintain freshness and reduce spoilage, and provide information about the food. Effective packaging is a primary reason for the safe and high-quality food available in the United States. Marketing is a very important part of packaging, as a walk through a supermarket will show. In addition to the required labeling, packaging provides cooking instructions, recipes, health information, pricing, coupons, packing and "use by" dates, and coding to facilitate identification of the batch. Other convenience features of packaging include ability to reseal, container portability, ease of opening, convenient preparation, and product visibility. Although we may enjoy the experience of scooping rolled oats from a bin or weighing pasta on a scale before packaging it ourselves, oats and pasta packaged at a processing facility are cleaner and fresher than products stored in open containers being handled by many people.

Food packaging is made from glass, metals (primarily steel, with tin and chromium lining to prevent rust, and aluminum), paper and paperboard, plastics (polyolefins and polyesters are the most common), and combinations of these materials (Fig. 6.6). In the United States, over 50% of all packaging, including food, is paper-based. Packaging is regulated by the FDA in Section 409 of the Food, Drug and Cosmetic Act. Food contact substances (FCS), and any material used in manufacture, packing, holding, packaging, and transport that could reasonably be expected to migrate into food under conditions of intended use, must be either GRAS or regulated as food additives.

Food manufacturers must balance the cost of packaging materials with their functional properties (barrier to air and moisture, flexibility, weight, ability to withstand processing, and strength). It is important for food technologists involved in packaging to understand all the physical, chemical, microbiological, and biochemical characteristics of the food and the likely changes that will occur over time in storage, including interactions between the food and the package materials. These interactions include the movement of gases and water that could potentially result in contamination of the

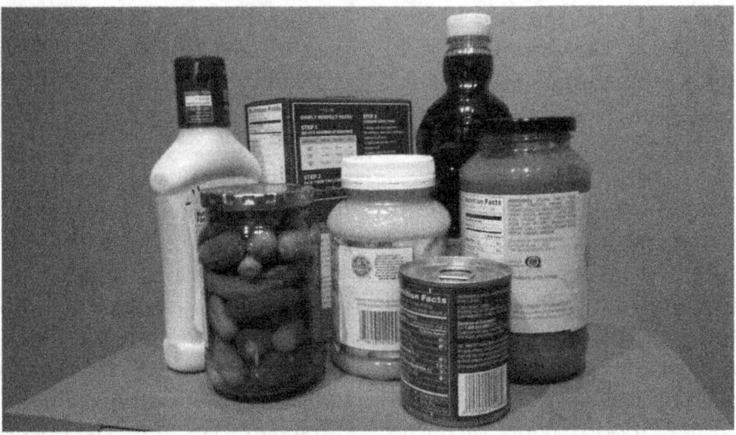

FIGURE 6.6 Foods are packaged in paper, glass, aluminum, plastic, and combinations of these materials for safety and convenience. The type of food will determine how it should be packaged and processed to ensure quality and safety are retained. *Source: Photo by authors.*

food, loss of package integrity, or loss of food quality. Flavors and aromas may move between the food and the packaging materials and some food components, such as fats or acids, may affect the integrity of the package. Low molecular weight substances such as plasticizers, stabilizers, antioxidants, and monomers and oligomers from plastic materials may migrate into foods.

Glass, aluminum, paper, plastic, and metals are reusable and recyclable types of packaging materials. Americans express the desire to use environmentally friendly packaging yet recycle only 40% of all packaging materials. Recycling any material involves costs for collecting, cleaning, and processing the material, all of which require water and energy. Paper and paperboard is the most recycled material, but use in foods is limited because of contaminants remaining in the paper and the weaker structure of recycled paper. Glass and aluminum retain their strength after recycling but contaminants can remain in glass. Plastics are the least recycled packaging materials (only 13%) due to physical changes in the plastic, incompatibility of plastic types, and difficulty of cleaning and removal of contaminants. Plastic is derived from petroleum, so the price of oil directly influences the economic value of recycled plastics. When oil is cheap, it is more cost effective to make new plastic than to recycle used plastic. The increased use of plastics in food packaging, notably water and drink bottles, have generated problems for landfills. Recently, some cities have been forced to ban the sale of beverages in plastic bottles because the capacity of the landfills are being exceeded. As an alternative to petroleum-based plastics, biopolymers from starch, cellulose, proteins, and polylactic acid produced by bacteria can be formulated into packaging that is biodegradable. There are challenges to the economical production of food packaging with similar characteristics to conventional packaging from these materials, but research in this area has progressed. Coca-Cola recently announced that a 100% biosourced plastic bottle would replace their current plastic bottle of polyethylene terephthalate.

Sausage casings and rice paper for candies are examples of edible film packaging that have been in use for many years. Edible films of polysaccharides (cellulose, pectin, starches, gums), proteins (casein, whey, soy, corn zein, gelatin), and lipids (waxes, fatty alcohols, fatty acids) have been produced but applications in food are limited. Some examples of edible films include breath strips with active ingredients and flavorings, sore throat treatments and toothpaste, cake decorations, pouches for vitamin and mineral enrichment, and glaze sheets for ham. There are likely to be more applications of edible packaging as a means of reducing packaging waste and in response to consumer demand for more sustainable food products.

Smart or active packaging uses materials that sense and change with the food and environment. Oxygen scavengers, used to prevent oxidation of oil, flavor, and color ingredients, and vitamins, can be incorporated into the package or included as a sachet. Moisture absorbers, ethylene scavengers, antimicrobial agent releasers, and flavor and odor absorbers are used in specific packaging applications to prolong the shelf-life of a variety of foods. Self-heating packages for coffee, tea, and ready-to-eat meals are commercially available and self-cooling packages have been produced.

Nanotechnology, which is the application of technology at the atomic scale, is being used to develop sensors to detect contaminants in food. The molecular form of plastic bags made of side-chain crystallizable polymer (SCCP) changes in response to temperature to modify the permeation rate of environmental gases into and out of the bags. This could be applied to delay ripening of bananas during transport, for example. Metal organic framework (MOF) sachets can be embedded with materials that scavenge volatiles, regulate water, and absorb oxygen, or with natural antimicrobials such as wasabi to keep beef jerky flexible and salad greens fresh. Food scientists at the US Army Soldier Research, Development, and Engineering Center in Natick, Massachusetts have been researching microwave-assisted thermal sterilization (MATS) with pouch and lidded tray packages to provide 4-year storage life for MRE (Meals Ready to Eat) combat rations. PepsiCo, Inc. has developed specially designed Gatorade bottles for professional soccer players. The Gatorade mixture was custom-engineered for each player in bottles with sensors that monitor the amount of Gatorade consumed during a game. Sensors imbedded in packaging are being developed that will detect spoilage or contamination of foods, and track the temperature foods are exposed to during storage.

6.6.4 Processing Enhances Nutritional Value of Foods

The digestibility of foods is improved by milling cereal grains, curing and aging meats, and cooking legumes. The bioavailability (uptake and utilization by the body) of some nutrients can be improved by food processing. For example, the carotenoid lycopene, which is considered an important antioxidant, is much better absorbed from processed tomato sauce than from raw tomatoes. Through enrichment and fortification, the nutrient value of foods can be enhanced. Enrichment is the term used for adding back nutrients that were removed during processing (Fig. 6.7) and fortification is the process of adding nutrients to foods in which the nutrients are not normally present.

The Dietary Guidelines for Americans recommend that consumers increase their intake of dietary fiber, vitamin D, calcium, folate, potassium, iron, and vitamin B12. Data from the NHANES survey of 2003–08 found that processed foods contribute 40%–50% of the intake of fiber, calcium, potassium, and vitamin B12, and over 60% of iron and folate by consumers. In a study by Fulgoni and others (2011), 88% of participants consumed less than the estimated average requirement (EAR) for folate from natural sources. The consumption of fortified or enriched foods raised folate intake such that only 10.7% of participants fell below the EAR. This is clearly a nutritional benefit to the consumption of processed foods.

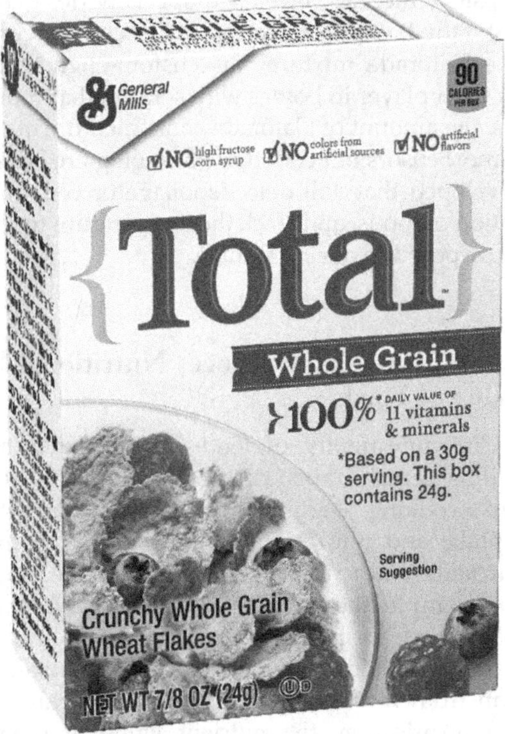

FIGURE 6.7 Breakfast cereals, such as Total, are fortified with vitamins and minerals to enhance their nutritional value. Vitamin and mineral deficiencies became rare in the United States in part because of food fortification and enrichment programs. *Source: Photo provided by General Mills.*

Processing can both positively and negatively impact the nutrient content of some foods. Nutrient retention in fresh foods is maintained by freezing or canning foods within hours of harvest, while nutrients are lost from fresh foods held for long periods of time. Some nutrients are destroyed during processing by heat or exposure to acids, alkali, light, oxygen, or air, while others are more stable to these treatments (Table 6.6). Most foods are processed in neutral conditions, neither acidic nor alkaline, but there are exceptions. Alkaline treatment of corn, called nixtamalization, actually increases the bioavailability of niacin, and has been used in making masa for tortillas by Central American cultures for thousands of years.

The nutritional value of fats, protein, carbohydrates, and minerals are rarely affected by processing (Table 6.7). Fat-soluble vitamins such as vitamins A, D, E, and K are quite stable to heat but may be oxidized under some conditions (exposure to light and prolonged heating). The water-soluble vitamins can be lost from food by migrating into the liquid during blanching, canning or cooking. Ascorbic acid (vitamin C) is the nutrient most sensitive to loss during food processing, especially heat. Vitamin C loss is higher in canned foods than in frozen foods.

6.7 NEGATIVE ISSUES IN PROCESSED FOODS

The Dietary Guidelines for Americans recommend that consumers reduce their intake of salt, refined or added sugars, fat (especially saturated fat), and calories. Processed foods are the primary dietary source for these nutrients. The addition of sodium, sugars, and fats enhances the flavor of processed foods.

6.7.1 Salt

Salt refers to sodium chloride, but sodium is the specific health-related target. Processed and prepared foods are the main sources of sodium in the diet. Breads and rolls, cured meats, pizza, soups, cheese, snack foods, and prepared meat and pasta dishes provide more than 40% of the sodium in the average American's diet. Sodium is added to foods in the form of sodium chloride, sodium nitrate, sodium bicarbonate, and MSG. Sodium chloride and MSG are flavor enhancers, sodium nitrate is a preservative for meats, and sodium bicarbonate, or baking soda, is a leavening

TABLE 6.6 Occurrence and Control of Common Food-Borne Pathogens

Pathogen	Likely source of contamination	Control	CDC statistics (each year in US)
Norovirus	Contaminated person, food, water, or surfaces	Hand-washing and general cleanliness	19–21 million illnesses, 56,000–71,000 hospitalizations, 570–800 deaths
Salmonella	Foods of animal origin; reptiles and birds	Cook poultry, meat, eggs; do not eat raw eggs or drink unpasteurized milk	1.2 million illnesses, 19,000 hospitalizations, and 450 deaths
Listeria monocytogenes	Soil, water, animal foods	Wash fruits and vegetables; do not eat unpasteurized milk products	1600 illnesses, 260 deaths
Shiga toxin-producing *Escherichia coli* (STEC)	Intestines of people and humans	Wash hands and cooking areas; cook meats	265,000 infections
Shigella	Human feces, contaminated water	Wash hands	500,000 cases
Campylobacter	Poultry, unpasteurized dairy products, contaminated produce and water	Cook poultry; wash hands; prevent cross-contamination	1.3 million cases (est.), 76 deaths (est.)
Clostridium botulinum	Low-acid home-canned food	Preserve vegetables with pressure canner (240°F)	145 cases (15% are food-borne types)

agent used in baked goods. High sodium intake is associated with increased blood pressure (hypertension), which can lead to a wide range of cardiovascular disorders including heart attack, stroke, and kidney failure. It has been estimated that 1 in 3 people in the United States have hypertension (Chapter 7: Nutrition and Food Access). The FDA has recently established voluntary sodium reduction goals for the food industry, which include recommendations to reformulate food products over time to include less sodium. Removing sodium from processed foods is a challenge for the food industry, not only because of the flavor aspects of sodium-containing compounds, but more importantly because of the functional and safety aspects these compounds contribute. As a curing agent, sodium nitrate is uniquely able to inhibit the growth of pathogenic bacteria, especially *C. botulinum*, which is not easily achieved by other compounds. Sodium chloride adds not only salty taste to food, but also enhances sweetness, and masks bitter or off-flavors. Sodium chloride enhances and stabilizes the texture and color and retains moisture in foods.

Food companies, recognizing that consumers are aware of the high sodium content of processed foods, are introducing more reduced-salt products. Frito Lay, for example, is working to reduce artificial ingredients, use 30% less salt, and keep the ingredient list short. Low-sodium, reduced sodium, and no-salt meats, breads, snack foods, desserts, and frozen entrees have been available for many years, but consumers have not responded favorably to low-sodium products because it is difficult to replace the flavor associated with salt. Ready-to-eat meals, sauces and seasonings, cereal products, snacks, and soup with reduced salt and sodium have

TABLE 6.7 Comparison of Methods of Food Preservation

Method of food preservation	Quality changes	Nutritional changes	Energy consumption
Canning	Flavors decreased Texture softened	Vitamin C reduced by heat Fat-soluble vitamins stable Minerals unchanged	High energy requirement for heating but no energy for storage of canned food
Freezing	Minimal flavor and texture changes during processing; depends on packaging and storage temperature Food most similar to fresh	Slight vitamin C reduction by blanching Nutrient content most similar to raw	Low energy requirement for processing but high energy use during storage
Dehydration	Flavor and texture dramatically affected by removal of water during processing	Vitamin C greatly reduced Some reduction of niacin and riboflavin due to prolonged heating and exposure to oxygen	Energy required for heating and removal of moisture but no energy required during storage
Irradiation	Very little change in flavor at low doses; slight softening of fruits	Slight nutritional change, similar to thermal processing Protein, lipid, carbohydrate unaffected Amino acids cysteine and tryptophan reduced; B vitamins increased due to release of bound vitamin; thiamine and vitamin C reduced at high doses	Refrigeration required to slow enzyme activity

found limited success in the marketplace due to technical difficulties as well as consumer resistance. Food scientists will need to utilize new approaches to formulate foods that taste, look, and feel the same without these forms of sodium.

The goal of reducing sodium intake has been politicized. Recently, the New York City Department of Health and Mental Hygiene implemented a Sodium Warning Label Rule that would require food service operations in New York City with more than 15 locations to provide a warning label on menu items that contain more than 2300 mg of sodium. The National Restaurant Association has filed suit against the labeling requirement stating that the City of New York does not have the authority to define food labeling.

Taking a more balanced approach, the voluntary National Salt Reduction Initiative (NSRI) has been coordinated by the New York City Health Department. Several large companies have joined the initiative including Kraft-Heinz, Mars Food US, Starbucks, Subway, Unilever, McCain Foods, Hain Celestial, Boar's Head, FreshDirect, Goya, Au Bon Pain, LiDestri, Red Gold, Uno Chicago Grill, and White Rose. The initiative seeks to reduce salt in packaged and restaurant food by 25% over 5 years, which could reduce Americans' salt consumption by 20%.

6.7.2 Sugar

Sugar exists naturally in food, usually in the form of sucrose, but also as glucose, fructose, lactose, or maltose, and is added as an ingredient from cane and beet sugar, maple syrup, corn syrup, or honey. Refined sugar is added to breads and baked products, soft drinks, candies, canned fruit, frozen desserts, and some milk products. The amount of added sugar can be considerable and contributes to excess calories and dental caries. Sugar is an important component of food processing for both the flavor and functional characteristics it provides. Sugar enhances the aeration of batters and provides nutrients for the growth and fermentation of yeasts in breads and lactic fermentation of dairy products, modifies the structural characteristics of gluten in baked products, and stabilizes egg white foams. The type of sugar in a cookie influences the texture of the product. Cookies made with table sugar (sucrose) are crispy while cookies made with brown sugar are soft and chewy. Sugar acts as a preservative by binding water and limiting microbial growth, for example in jams, jellies, candies, cakes, cookies, and frozen desserts. Removal of sugar affects more than sweetness and calorie content of a food product and reformulation for a successful no-sugar item requires knowledge of the functions of sugar.

There are several types of sugar used in food processing. White table sugar is comprised of the disaccharide sucrose, which contains two monosaccharides (glucose and fructose) in equal proportion. Sugar beets contain about 20% sucrose and are grown in northern temperate climates. At harvest, the roots are washed, sliced, and mixed with water to extract the sugar juice. The juice is mixed with calcium hydroxide and carbon dioxide to separate the nonsugar components, filtered, and boiled under vacuum to evaporate most of the water. The resulting syrup is filtered and boiled in a vacuum pan until sugar crystals begin to form. Molasses syrup and crystals are separated by centrifugation, and the sugar crystals are washed, dried, and passed through screens to separate the sizes of crystals to make granulated and powdered sugar. The product, beet sugar, is about 99% sucrose. Brown sugar is a minimum of 88% sucrose with 4.5%–6.5% retained molasses syrup. Sugar cane is a grass that grows in tropical and warm temperate climates and contains 12%–16% sucrose. Processing of the syrup to crystalline sucrose is similar to beet processing, but juice is extracted from the cane by pressing and rolling.

High fructose corn syrup (HFCS) is produced by hydrolyzing corn starch to glucose syrup (corn syrup), which is treated with a naturally occurring isomerase enzyme that changes about half of the glucose to fructose. The name "high fructose" corn syrup is an accurate term based on the fact that corn syrup is mainly glucose and during the process a higher amount of fructose is produced. HFCS is made to have the about the same ratio of fructose and glucose as table sugar. Fructose is a sweeter monosaccharide than glucose, which is why sugar is sweeter than starch. Honey (48% fructose and 52% glucose) and agave nectar (75% fructose and 20% glucose) are other types of sugar syrups that are used in foods. As HFCS became available, food manufacturers have replaced sucrose with HFCS in soft drinks and many other foods because it is lower in cost; available in liquid form, making it easier to measure and transport; and provides desirable functional benefits to foods. HFCS browns more readily, prevents crystallization in candies more effectively, and absorbs more moisture in baked products than sucrose. Since 1966, sucrose consumption has decreased while consumption of HFCS has increased (Fig. 6.8). The total sweetener consumption by Americans has increased 20% over that timeframe to 130 pounds per capita per year.

HFCS, beet sugar and cane sugar are similar in chemical composition, provide the same caloric value (4 kcal/g), and are metabolized in the body similarly. During the past few years, the higher use of HFCS in foods raised concerns that HFCS was specifically associated with increased rates of diabetes, obesity, and other chronic illnesses. Consumers began to demand that food processors remove HFCS from their products. There is no evidence that HFCS is any different from sugar when consumed as part of the diet. Both HFCS and sugar contribute

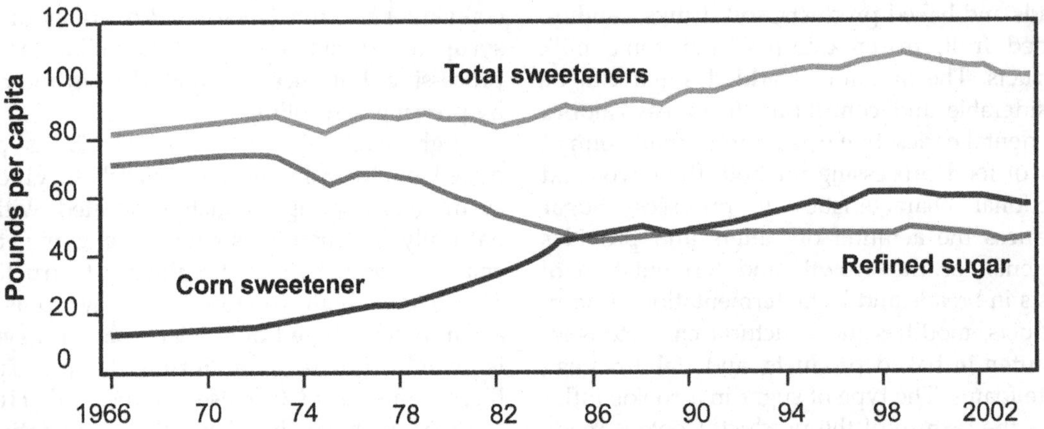

FIGURE 6.8 In the United States, consumption of caloric sweeteners was 130 pounds per person per year in 2012, which is about 50 pounds more than in 1966. The use of high fructose corn syrup (corn sweetener or HFCS) increased after 1970 while the use of refined sugar decreased. Nutritionists recommend reducing the consumption of added sugars from all sources to prevent weight gain and risk of diabetes. *Source: USDA Economic Research Service, www.usda.ers.gov.*

calories to food and therefore should be consumed in moderation. Replacing HFCS in food products with sugar, honey, or agave nectar does not make the food more healthful or nutritious.

In May 2016 the FDA announced that the Nutrition Facts panel on food packages will need to include the amount of added sugars by July 2018 (Chapter 7: Nutrition and Food Access). This change reflects the Dietary Guidelines for Americans recommendations to reduce the amount of added sugars consumed. With this information available on food packages consumers will be more aware of the amounts and types of sugars in foods and manufacturers will likely develop new formulations that reduce the amounts of added sugars in their products.

6.7.3 Fat

Fat occurs naturally in meats, milk, and cheese and is added to foods as vegetable oils, lard, butter, margarine, or shortening. Salad dressings, mayonnaise, ice cream, cheese, chocolate, and sour cream are food products that contain a high amount of fat. Fat provides a wide range of features to food. One of the most important functions of fat is to tenderize baked products and limit the amount of structure developed by gluten in pastry, cookies, cakes, and bread. Fat aids aeration of batters and doughs; contributes to the emulsion structure of mayonnaise, salad dressings, and baked products; limits crystallization of candies and frozen desserts; and contributes the foam structure to whipped cream. Fat provides its own flavor (butter, bacon, olive oil) and also absorbs and delivers other flavors in foods. Fat-soluble vitamins bind to fats and are absorbed more efficiently from foods when consumed with a source of fat. For this reason, the nutrients in salads are better absorbed when consumed with a fat-containing dressing compared to a fat-free dressing. Fats and oils are excellent media for frying and provide the brown color, crispy texture, and desirable flavors that make French-fried potatoes so popular.

Fat is the most calorie-dense of the macronutrients, providing 9 kcal/g, so that foods that contain more fat are naturally higher in calories. As discussed in more detail in Chapter 7,

Nutrition and Food Access, fats exist in saturated or unsaturated forms. Saturated fats are solid at room temperature and unsaturated fats are liquid at room temperature. Fats from animal sources tend to be higher in saturated fats while fats from vegetable sources tend to be higher in unsaturated fats. Consuming a diet containing a high amount of saturated fat has been associated with greater risk of cardiovascular disease. The Dietary Guidelines have recommended Americans reduce their total intake of fat as a means of reducing calories, but specifically reduce intake of saturated fats. Replacing fat in processed foods can and has been done fairly effectively. A wide range of low-fat or fat-free foods are readily available in stores. Consumers tend to view low-fat foods as less desirable than full-fat products because it is a challenge to exactly replicate the mouthfeel and creamy taste of fats.

6.8 GOVERNMENT OVERSIGHT OF PROCESSED FOODS

Since passage of the Pure Food and Drug Act in 1906, ensuring food safety and sanitation have been important roles of the government. A history of regulations for additives and labeling of food products, and for the prevention of contamination by pathogens and toxins, is summarized in Table 6.8. Food safety is a major concern of several government agencies, specifically the FDA, CDC, and USDA. These agencies work together to prevent and reduce the number of illnesses and deaths from food-borne causes. The FDA is the primary food safety agency in the United States and oversees the safety of both foods produced in the United States and imported from other countries by monitoring pathogens, toxins, pesticides, and other contaminants. The FDA regulates ingredients, additives, packaging, allergens, dietary ingredients, dietary supplements, and labeling of foods; and provides guidance manuals and regulatory information for facilities, CGMPs, and HAACP protocols for food processors, as well as for all retail food and food service industries. FDA officials are responsible for recalls and alerts when a food product is mislabeled, contaminated, adulterated, or has caused an outbreak of a food-borne illness. FDA scientists conduct studies to assess potential exposure and risk of pathogens and contaminants and laboratory research to determine standards, methods of analysis, and good manufacturing safety. Public education for safe storage, use, and disposal of food during public emergencies and other food safety information is provided by the FDA in collaboration with the USDA (Table 6.9).

The Food Safety and Inspection Service (FSIS) of the USDA ensures that the US supply of meat, poultry, and egg products is safe, wholesome, and correctly labeled and packaged, as required by the Federal Meat Inspection Act, the Poultry Products Inspection Act, and the Egg Products Inspection Act (also discussed in Chapter 4: Animals in the Food System). The USDA, in collaboration with the FDA, collects information about food consumption patterns and the availability and distribution of foods. The CDC tracks the occurrence of food-borne illness and provides educational material to improve food safety within their role as a public health agency. CDC scientists collaborate with state and local health departments, and federal and international agencies, to investigate food-related illnesses, provide testing and detection of the source, and provide guidance for mitigation and control.

The Food Safety Modernization Act (FSMA) of 2011 was a significant legislative policy intended to improve the security and safety of the US food supply by focusing on prevention of food-borne pathogens throughout the food system, including agricultural producers, food and animal transporters, and food importers.

TABLE 6.8 Impact of Processing on Selected Nutrients in Foods

Nutrient	Chemical name and features	Food sources	Processing conditions	Nutritional impact of heat
Carbohydrate	Sugars, starches, fibers	Cereals, beans, fruits, vegetables	Starch gelatinized, softens	Little change
Lipid (fat)	Fatty acids	Meat, dairy, nuts, seeds, fish	Oxidation	Little impact
Protein	Amino acids	Meat, dairy, fish, beans, cereals, nuts	Denatured by heat; some amino acids affected by heat, acid, alkali	Little change
Vitamin A[a]	Carotene, retinol Fat soluble	Milkfat, liver, orange fruits, green leafy vegetables	Unstable in heat, acid, oxygen, UV light	Moderate loss
Vitamin D[a]	7-Dehydro-cholesterol Fat soluble	Milk, egg yolk	Stable in acid; destroyed in heat, light, oxygen, alkali	Moderate loss
Vitamin E	Tocopherols Fat soluble	Vegetable oils, nuts	Stable to acid; light-sensitive	Moderate loss
Vitamin K[a]	Fat soluble	Green leafy vegetables, eggs	Stable to heat and oxygen; labile to acid, alkali, light	Minimal loss
Thiamin	Vitamin B-1 Water soluble	Whole and enriched cereals, beans	Stable to light and acid; degraded by heat, alkali, and oxygen	Significant loss
Riboflavin	Vitamin B-2 Water soluble	Whole and enriched cereals, milk, eggs, meat, green leafy vegetables	Stable to acid, oxygen; degraded by light, alkali, heat	Significant loss
Riboflavin	Vitamin B-2 Water soluble	Whole and enriched cereals, milk, eggs, meat, green leafy vegetables	Stable to acid, oxygen; degraded by light, alkali, heat	Significant loss
Niacin	Water soluble	Meat, fish, eggs, whole grains	Stable	Little loss
Vitamin B-6	Pyridoxine Water soluble	Meat, fish, eggs	Stable to alkali, acid; degraded by light and heat	Little loss
Folacin[a]	Folic acid Water soluble	Green leafy vegetables, nuts, beans	Sensitive to acid, alkali with light, oxygen, heat	Susceptible to high loss
Vitamin B-12	Cobalamin Water soluble	Meat, fish, dairy	Stable to heat, acid, and alkali; unstable in light and oxygen	Little loss
Vitamin C	Ascorbic acid Water soluble	Citrus fruits, some fruits and vegetables	Stable in acid; readily destroyed by light, heat, alkali, oxygen	Susceptible to high loss
Minerals	Calcium,[a] sodium, iron,[a] zinc, iodine[a]	Dairy, cereals, meat, fish	Stable	Little loss

[a]*Nutrient likely to be limited to US diets.*

TABLE 6.9 Government Regulation of Processed Foods

Legislation	Significance
1906 Pure Food and Drug Act	Prohibited food adulteration and misbranding
1906 Meat Inspection Act	Required federal inspection of slaughterhouses
1907 First Certified Color Regulations	Approved seven colors for use in food
1930 McNary-Mapes Amendment	Authorized FDA to establish standards of quality and fill-of-container for canned foods
1938 Food Drug and Cosmetic Act	Established standards of quality and procedures for inspections
1939 Food Standards	First food standards for canned tomatoes, pureed and paste tomatoes
1949 Procedures for the Appraisal of the Toxicity of Chemicals in Food	Provided handbook for guidance of food industry
1954 Miller Pesticide Amendment	Required safety limits for pesticides on raw agricultural commodities
1958 Food Additives Amendment with Delaney proviso	Established safety regulations for new food additives
	Banned food additives shown to induce cancer in laboratory animals or humans
1960 Color Additive Amendment	Required manufacturers to establish safety of color additives
1969 White House Conference on Food, Nutrition, and Health	Review of all generally recognized as safe (GRAS) substances
1990 Nutrition Labeling and Education Act (NLEA)	Required all packaged foods to bear nutrition labeling (Nutrition Facts panel) and allowed some health claims
1992 Nutrition Facts panel	Ruling to list most important nutrients on Nutrition Facts panel
1994 Dietary Supplement Health and Education Act	Dietary supplements and dietary ingredients regulated as food
1995 USDA Pathogen Reduction: Hazard Analysis and Critical Control Point (HAACP) System regulations revision	First major revision of USDA food safety regulations since 1906
1996 Food Quality Protection Act	Amendment to eliminate Delaney proviso for pesticides
1997 Food and Drug Modernization Act	Regulations for health claims established
2000 Rule on dietary supplements	Structure/function claims described
2003 Trans fat included on labels	First change in Nutrition Facts panel since 1993
2004 Food Allergy and Consumer Protection Act (FACP)	Labeling for protein (milk, eggs, fish, crustacean shellfish, tree nuts, peanuts, wheat, soybeans) allergy required
2011 Food Safety Modernization Act (FSMA)	Established food safety system and authority for enforcement

(*Continued*)

TABLE 6.9 (Continued)

Legislation	Significance
2014 Proposed updates to Nutrition Facts label	Eliminated requirement for calories from fat; changed the list of vitamins and minerals; increased the prominence of calorie information; required manufacturers to state the amount of added sugars; adjusted serving size requirements according to package size
2015 Proposed updates to Nutrition Facts label	Include %DV for added sugars

EXPANSION BOX 6.3

FOOD SAFETY MODERNIZATION ACT

FSMA was signed into law by President Obama on January 4, 2011. The primary intent of the legislation is to ensure that the US food supply is safe by shifting the focus of federal regulators from responding to cases of microbial contamination to prevention. FDA published the final rules for Preventive Controls for Human and Animal Food in 2015 with the implementation of these rules in subsequent years. Farms and businesses have between 1 and 6 years to comply, depending on the size of operation and type of food produced.

FSMA has two main components. Part 1 applies to farms that grow, harvest, pack, or hold raw produce (not including produce such as sweet corn and pumpkins, which are not eaten raw). Safety guidelines for agricultural water, biological soil amendments, health and hygiene of workers, domesticated and wild animals, and equipment, tools, and buildings are defined. The FDA will create standards for safe production and harvesting of fruits and vegetables, once they leave the farm, to further minimize the risk of food-borne illness.

Part 2 of FSMA describes preventive controls for facilities that manufacture, process, pack, or hold human food. Managers of food processing plants are required to evaluate the hazards in their operation, implement and monitor effective measures to prevent contamination, and define a plan for corrective actions. In addition to HACCP and critical control points (CCPs), food processors must have additional preventive controls especially for fruits, vegetables, and animal foods. FDA conducts inspections and food companies are accountable for prevention of contamination, and the FDA has mandatory recall authority for all food products. Food companies usually comply with FDA's requests for voluntary recalls so FDA expects it will need to invoke recalls infrequently, but this new authority is a critical improvement in the ability of FDA to protect public health. FDA has additional authority for inspection of imported foods and can block admission if a foreign country refuses FDA inspection.

FSMA directs FDA to improve training of state, local, territorial, and tribal food safety officials and authorizes grants for training of personnel and other food safety activities. Training of food industry personnel via approved courses that meet regulations under the Food Safety Preventive Control Alliance (Title 21 Code of Federal Regulations Part 117.155) are being implemented. FSMA also directed the CDC to enhance surveillance and improve identification of the causes of food-borne illness and the foods involved. Public health officials, FDA, and CDC must work together to develop new methods for detection and policies for prevention of future outbreaks.

Suggested reading: U.S. Food and Drug Administration (2016).

6.8.1 Regulations Concerning Foodborne Toxins

Many plants produce natural chemicals that protect them from insects and diseases. Some of these chemicals are also toxic to humans, such as solanine in potatoes and tomatoes. Solanine can cause gastrointestinal illness, but is rarely fatal. Mushrooms are among the most toxic foods and fatalities occur fairly regularly from accidental consumption of poisonous mushrooms. Overall, the health risk from naturally occurring toxins in foods is very small, even though some of these compounds are known carcinogens. During processing, toxins can be produced in foods. Examples that have come to the public attention in recent years include nitrosamines in bacon, heterocyclic amines and polycyclic aromatic hydrocarbons in grilled beef steaks, and acrylamide in French fries. Environmental toxins can also contaminate foods, including mercury in fish, PCBs (polychlorinated biphenyls), pesticide residues, arsenic and lead, and fungal contamination such as fumonisin in corn and patulin on apples. Growth of the mold *Aspergillus* on peanuts and grains during wet periods produces aflatoxin, a very potent carcinogen. Aflatoxin can be found in milk when cows are fed contaminated feed. Fumonisins are carcinogenic mycotoxins produced in corn in normal growing environments. Fortunately, these contaminants are routinely monitored and rarely appear at high enough levels in the US food system to affect human health.

The FDA monitors the levels of natural toxins in domestic and imported foods, assesses the potential exposure and risk, and issues recalls, market withdrawals, and safety alerts when necessary. For example, the FDA has been monitoring arsenic levels in foods for years, and noted that rice has higher levels of inorganic arsenic than other foods due to efficient absorption of arsenic from the soil during growth. Arsenic testing was expanded in 2011 when new methods of analysis became available and, in 2016, the FDA proposed a limit of 100 ppb for inorganic arsenic in infant rice cereal. FDA evaluated a large body of scientific information and determined that this level would reduce the risk of harm from arsenic consumption by infants. They also provided guidance for rice consumption by pregnant women and infant caregivers.

6.8.2 Regulations for Food Labeling

The FDA is responsible for ensuring that foods sold in the United States are properly labeled. Labels for meat and meat products are authorized by the USDA-FSIS, and the FDA and FSIS collaborate on egg product labeling. Labels for alcoholic beverages are under the authority of the US Alcohol and Tobacco Tax and Trade Bureau. The purpose of labeling on food packages is to reduce consumer confusion, help consumers make better food choices, and encourage food manufacturers to improve the nutritional profiles of foods.

A food label must contain specifically defined information. The Principal Display Panel, which is typically the front of the package, must provide the standard of identity or common name of the product and the net contents of the package. The Information Panel, which may be on the back or side, must provide the ingredient list (in decreasing order by weight), the Nutrition Facts Panel, and the name and address of the manufacturer. Including "best if used by" or "sell by" dates is voluntary. Marketing and advertising images and terms are allowed to be included on the label. However, specific terms such as *low fat, reduced, less, fewer, free, good source* (of a nutrient), and *light* or *lite* are defined by the FDA and products must meet certain standards if these terms are used.

Consumer demand for more information about how food is processed is driving manufacturers to use more terms and descriptors on their packages.

> ...Active and vocal consumers are establishing a trend that involves increasing the availability of healthy, all-natural food options with clean labels. Their advocacy and purchasing power influence how foods are labeled. As a consequence, besides ensuring that labels for packaged foods contain mandatory nutrition information, food manufacturers are adding to food labels supplementary messaging that appeals to consumers.
>
> *Tarver (2015)*

Simple, pure, safe, non-GMO, allergen-free, cage-free, cruelty-free, and *made with whole grains* are unregulated terms by the FDA but are marketing terms aimed to attract consumers who believe foods with such labels are safer, healthier, or minimally processed. Manufacturers have also responded to consumer demand for additive-free products and there are many new offerings without artificial colors, flavors, and preservatives.

Congress passed the National Organic Program in 1990, which gave the USDA authority to set national standards for organically raised crops and livestock. Farmers must apply for and meet these standards in order to be able to use the Organic label on their products. Products labeled "100% organic" must contain only organically produced ingredients, and products labeled "organic" must consist of at least 95% organically produced ingredients. Products meeting these standards may display the USDA Organic seal (Fig. 6.9) on their label.

Processed products that contain at least 70% organic ingredients may use the phrase "made with organic ingredients" and list up to three of the organic ingredients or food groups on the principal display panel. For example, soup made with at least 70% organic ingredients and only organic vegetables may be labeled either "made with organic peas, potatoes, and carrots" or "made with organic vegetables." However, for these products, the USDA Organic seal cannot be used anywhere on the package. Processed products that contain less than 70% organic ingredients cannot use the term *organic* other than to identify the specific ingredients that are organically produced in the ingredients list.

Organic animal production means that the animals received 100% organic feed, the feed contained no GMOs and was grown without synthetic fertilizers or pesticides, animals had access to the outdoors and were never confined, and received no hormones or antibiotics. The term "hormone-free" is not allowed in labeling because no animals or animal foods are without hormones. The phrase "raised without supplemental hormones" or "no hormones administered during finishing" can be used and, if these hormone claims are made with poultry and pork, the label must include the sentence "Federal regulations prohibit the use of hormones," because all poultry and pork are produced without hormone treatment.

FIGURE 6.9 The Organic Food Production Act authorized the USDA to create the Organic Standards and certification program. The Organic seal is only allowed to be used on products that contain at least 95% organically produced ingredients. *Source: USDA Agricultural Marketing Service, www.ams.usda.gov.*

An "antibiotic-free" claim is not allowed, but labels can include "no antibiotics used," "raised without antibiotics," or "livestock have never received antibiotics from birth to harvest." "Free-range" means that poultry raised for meat was allowed some access to the outdoors but the length of time and amount space are not defined. The free-range label cannot be used for laying hens. "Pasture-raised" means that animals received at least 30% dry-fed feed from pasture over the course of the grazing season. It is important to note that inspection of facilities that use these labeling terms is not mandatory, so consumers should be aware of the standards and practices of companies before assuming products meet these definitions.

Some consumers relate organic with healthier or safer foods, although there is no scientific evidence to support that organic foods are healthier or safer than conventionally grown foods. With the high consumer interest, the organic food industry has become a multibillion dollar business and some manufacturers use labeling and buzzwords to entice consumers to buy their products. Words such as "all natural," "free-range," "sustainably harvested," "sustainably raised" or "no drugs or growth hormones used" may be truthful statements, but there is no government oversight on the products labeled with terms. This produces confusion for consumers and makes comparisons with conventional foods very challenging.

The term "natural" began being used on food packages around the time the organic standard was approved. For many years the FDA refused to define the term and would allow manufacturers to use it as long as it was truthful and not misleading to consumers. Recently under increased pressure from manufacturers and consumers, the FDA has begun a process to consider setting a definition for "natural" to be used on food packages. Use of the term "natural" is not permitted in the ingredient list, with the exception of the phrase "natural flavorings." Currently, the USDA allows the use of the term "natural" to be used in meat and poultry labeling on products that contain no artificial ingredients or added color. The product also must be only minimally processed. The label must explain the use of the term "natural," for example, "no added coloring; minimally processed" and not be misleading. Natural does not refer to production methods.

The term "local" has many definitions, but no government-regulated standard. Walmart defines local as "produced within a state" and Whole Foods defines local as "products that travel less than 1 day (≤7 hours by car or truck) from farm to store." The US Congress defined local as less than 400 miles from origin or within the state in which it was produced. There is increasing interest from consumers in knowing where foods are being raised or produced and such information is being added to food packages and store shelves using smartphone technologies and other innovations.

"Sell by," "use by," and "best by" dates are often seen on food labels but, except for infant formula, are not required by federal regulation. When manufacturers started voluntarily using these dates, consumers become accustomed to them and thereby made their use expected. A fact that is often not understood by consumers is that these date labels are indicators of quality, not safety. Manufacturers decide what dates are most appropriate for their product to remain on the store shelves or suitable for home consumption. In most cases, these dates are shorter than when spoilage would occur, but consumers interpret the dates as defining when they have become unsafe to eat. Current interest in changing this system has arisen out of concern that the volume of discarded, but still safe to eat, food is becoming a major problem. Food marketers discard a large amount of food that is at or near its use by date to avoid consumer complaints, even though the food is safe to consume. Food waste contributes to greenhouse gases in landfills, is wasteful of energy used to produce and process the food, and is an overall economic loss.

6.9 CONSUMER ATTITUDES ABOUT PROCESSED FOODS

Food technology evolved along with industrialization of all other aspects of American life. New ways to process and preserve food, and conveniences in packaging and preparation, have allowed people to spend less time preparing meals. During the past 50 years, family structure, the role of women in the home and workplace, and attitudes about food have changed and will continue to change. Americans have increased the proportion of food they consume away from home from 18% in 1978 to 33% in 2010. From 2002 through 2012, the share of household food expenditures spent on "food away from home" increased from 39% to 43% (Fig. 6.10). December 2014 marked the first time that restaurant sales were higher than sales at grocery stores, which coincided with low gas prices. Food prepared away from home tends to be lower in nutrient quality, more expensive, and served in larger portions.

Every sector of the US economy has been industrialized and the food industry is no exception. Processed foods are developed by creative and innovative entrepreneurs and food scientists, and marketed because they are profitable for manufacturers and readily

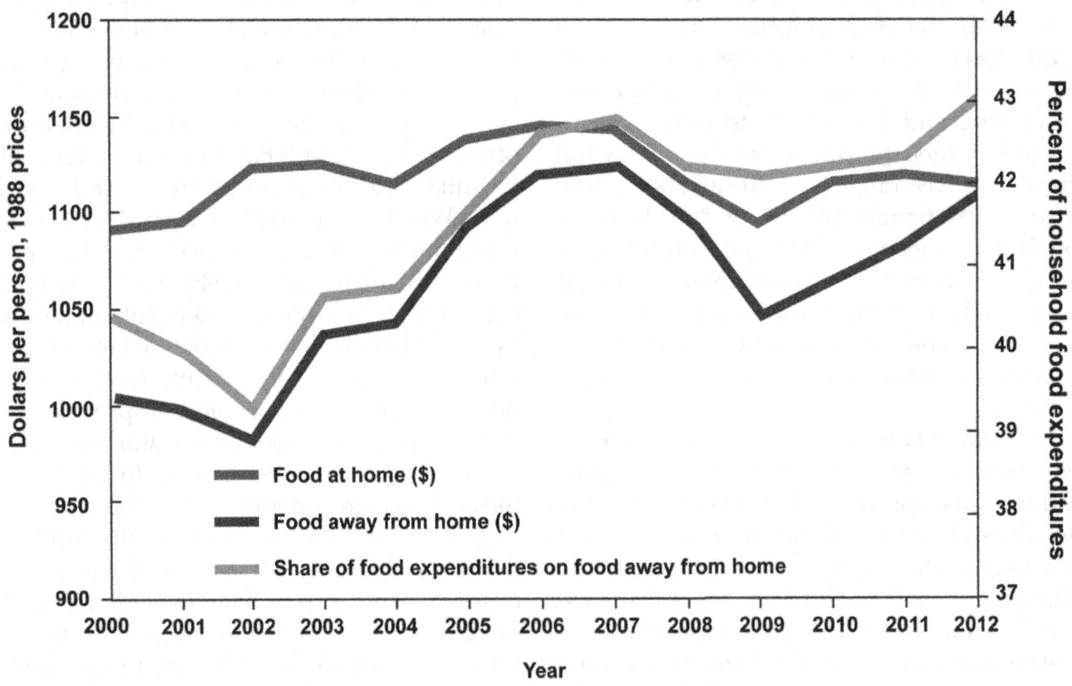

FIGURE 6.10 US consumers spent 43% of their total household expenditures on food purchased and consumed away from home in 2012. The trend between 2000 and 2007 was for consumers to spend more money on food away from home. That trend decreased during the recession years, but rebounded since 2009, indicating that people are doing less food preparation at home. *Source: USDA Economic Research Service, www.usda.ers.gov.*

accepted by consumers. Food processors follow government regulations and use approved ingredients that have been tested for safety. Yet, negative images of processed foods have arisen in the popular press and the food system is often characterized as broken in large part because of the increased reliance on processed foods. Consumers are told to only shop the perimeter of grocery stores (where the fresh food items typically are located) and to avoid the center aisles, which contain packaged foods. There is a lack of clear scientific thinking with these types of recommendations, which overlook the many positive aspects of food processing.

Some argue that safety testing of additives and food processing technologies by the government has been insufficient. In some part that is true. There has not been adequate funding for food science and nutrition research by independent scientists for several decades, which limits the knowledge base of the role of food processing in human health. Relationships between diet and health require expensive and long-term studies, and newer information can alter previously held conclusions. An example is the recent recognition that trans fats are a causative agent in cardiovascular disease and the subsequent removal of partially hydrogenated fats from the GRAS list. Because processed foods are changed from the raw form and processes may be needed to ensure safety, or additives are used to provide color, flavor, and texture attributes, many consumers are skeptical about processed foods: "...the idea that many processed foods are still inherently poor choices for nutrition is deeply ingrained in the American consumer's mind" (Fox, 2012).

According to the IFIC 2016 survey, 50% of US consumers were concerned about chemicals in food, and 57% identified food-borne illness as their most important food safety concern. These concerns are not based on scientific evidence, which supports the safety of chemical additives in foods and the very low risk of food-borne pathogens in processed foods. Many respondents were trying to replace processed foods with fresh foods because they wanted to avoid foods with sugars (52%), saturated fats (44%), artificial flavors (45%), artificial colors (43%), preservatives (50%), and HFCS (53%). A shift away from processed foods may come with higher food prices, more food waste, greater impact on the environment, and less efficient food production systems. Applying scientific thinking to define the best approach to food production that takes advantage of technology to enhance the safety, quality, and convenience of foods, while optimizing healthfulness for consumers, will be necessary.

References

Bourland, C., Kloeris, V., Rice, B. L., & Vodovotz, Y. (1999). Food systems for space and planetary flights. In H. W. Lane, & D. A. Schoeller (Eds.), *Nutrition in spaceflight and weightlessness models*. Boca Raton, FL: CRC Press, 301 p.

Floros, J. D., Newsome, R., Fisher, W., Barbosa-Canovas, G. V., Chen, H., Dunne, C. P., ... Dwyer, J. (2011). Food, fortificants, and supplements: Where do Americans get their nutrients? *Journal of Nutrition, 141* (10), 1847–1854.

Food and Drug Administration. (2016). *FDA food safety modernization act*. Available from <http://www.fda.gov/Food/GuidanceRegulation/FSMA/default.htm>.

Fox, M. (2012). Defining processed foods for the consumer. *Journal of the Academy of Nutrition and Dietetics, 112*(2), 214–221.

Sloan, E. A. (2015). The top ten food trends. *Food Technology, 69*(4), 24–43. Available from <http://www.ift.org/food-technology/past-issues/2015/april/features/the-top-ten-food-trends.aspx?page=viewall>.

Tarver, T. (2015). Food labels: Defining a new narrative. *Food Technology, 10*(1), 35–51.

Further Reading

American Chemical Society. (2015). Food safety. Available from <http://www.americanchemistry.com/Safety/ConsumerSafety/FoodSafety>.

American Crystal Sugar. (2016). Sugar processing. Available from <https://www.crystalsugar.com/sugar-processing/>.

American Meat Institute (2010). *AMI fact sheet: Carbon monoxide in meat packaging: Myths and facts*. Washington, DC: American Meat Institute. 3 p. Available from <https://www.MeatAMI.com>.

Anonymous (2000). A taste of the 20th century and a sampling of innovations, laws, and product introductions in the U.S. food industry. *Food Review, 23*(1), 30–31.

Baldwin, C. J. (Ed.), (2009). *Sustainability in the food industry*. Ames, IA: Wiley-Blackwell and IFT (Institute of Food Technologists) Press, 257 p.

Ben & Jerry's. (2015). Social and environmental assessment report. Available from <http://www.benjerry.com/about-us/sear-reports>.

Brody, A. L. (2015). Take the heat off: Minimum heat processing and packaging. *Food Technology, 68*(12), 86–87.

Brooks, C. (2007). *Beef packaging*. Centennial, CO: National Cattlemen's Beef Association. Available from <http://www.beefresearch.org/CMDocs/BeefResearch/Beef%20Packaging.pdf>.

Centers for Disease Control and Prevention. (2015). Food safety-food-borne germs and illness. Available from <www.cdc.gov>.

Clark, S., Jung, S., & Lamsal, B. (Eds.), (2014). *Food processing: Principles and applications* West Sussex, UK: John Wiley & Sons, Ltd, 578 p.

Crosby, G. (2015). The top 10 breakthroughs in food science in the past 75 years. *Food Technology, 69*(7), 120. Available from <http://www.ift.org/Food-Technology/Past-issues/2015/July/Columns/perspective.aspx?vie>.

Dunn, T. (2015). The active, smart future of packaging. *Food Technology, 69*(9), 118–121.

Economic Research Service (2014). *Processing and marketing: New products*. Washington, DC: ERS, USDA. Available from <http://www.ers.usda.gov/topics/food-markets-prices/processing-marketing/new-products.aspx>.

Eicher-Miller, H. A., Fulgoni, V. L., & Keast, D. R. (2012). Contributions of processed foods to dietary intake in the US from 2003–2008: A report of the food and nutrition science solutions joint task force of the Academy of Nutrition and Dietetics, American Society for Nutrition, Institute of Food Technologists, and International Food Information Council. *Journal of Nutrition, 142*, 2065S–2072S. Available from http://dx.doi.org/10.3945/jn.112.164442.

Environmental Protection Agency. (2016). Sustainable management of food. Available from <http://www.epa.gov/sustainable-management-food>.

Esnouf, C., Russel, M., & Bricas, N. (Eds.), (2013). *Food system sustainability* Cambridge, UK: Cambridge University Press, 303 p.

Food and Drug Administration. (2015a). *Food irradiation: What you need to know*. Available from <http://www.fda.gov/Food/ResourcesForYou/Consumers/ucm261680.htm>.

Food and Drug Administration. (2015b). *Irradiated food and packaging*. Available from <http://www.fda.gov/Food/IngredientsPackagingLabeling/IrradiatedFoodPackaging/default.htm>.

Food and Drug Administration. (2015c). *Pesticide residues in food and feed*. Available from <www.fda.gov>.

Food and Drug Administration. (2016). *Arsenic in rice and rice products*. Available from <http://www.fda.gov/Food/FoodborneIllnessContaminants/Metals/ucm319870.htm>.

FoodNet (2016). *Foodborne Diseases Active Surveillance Network (FoodNet)*. Washington, DC: U.S. Department of Health & Human Services. Available from <http://www.cdc.gov/FoodNet>.

Food Processing. (2015). *U.S. food industry*. Available from <www.FoodProcessing.com>.

Food Review. (2000). *Food processing: A history*. Available from <http://www.foodprocessing.com/articles/2010/anniversary/>.

Golan, E., & Buzby, J. C. (2015). Innovating to meet the challenge of food waste. *Food Technology, 69*(1), 21–25.

Harris, R. S., & Karmas, E. (1977). *Nutritional evaluation of food processing*. Westport, CN: The AVI Publishing Co., Inc, 670 p.

Harvard T.H. Chan School of Public Health. (2016). *The nutrition source. Salt and sodium*. Available from <www.hsph.harvard.edu/nutritionsource/salt/tasting-success-with-cutting-salt/index.html>.

Hegarty, V. (1992). *Nutrition-food and environment*. St. Paul, MN: Eagan Press, 433 p.

Hersch, D., Perdue, L., Ambroz, T., & Boucher, J. L. (2014). The impact of cooking classes on food-related preferences, attitudes, and behaviors of school-aged children: A systematic review of the evidence, 2003–2014. *Preventing Chronic Disease, 11*, 140267. Available from http://dx.doi.org/10.5888/pcd11.140267.

Institute of Food Technologists. (2015). *Food processing and packaging*. Available from <www.ift.org>.

International Food Information Council. (2008). *2008 Food and health survey: Consumer attitudes towards food, nutrition & health*. Washington, DC. Available from <http://www.foodinsight.org/Content/6/IFICFdn2008FoodandHealthSurvey.pdf>.

International Food Information Council. (2015). *2015 Food and health survey: Consumer attitudes toward food safety, nutrition, and health*. Washington, DC. Available from <http://www.foodinsight.org/2015-food-health-survey-consumer-research>.

International Food Information Council. (2016). *2016 Food and health survey, food decision 2016: The impact of a growing national food dialogue*. Washington, DC. Available from <http://www.foodinsight.org/articles/2016-food-and-health-survey-food-decision-2016-impact-growing-national-food-dialogue>.

Kuhn, M. E. (2015). Health + convenience = market opportunity. *Food Technology, 69*(11), 32–41.

Litchfield, R. (2008). *High fructose corn syrup-how sweet it is.* Ames, IA: Iowa State University.

Maga, J. A., & Tu, A. T. (Eds.), (1994). *Food additive toxicology* New York, NY: Marcel Dekker, Inc, 552 p.

Marsh, K., & Bugusu, B. (2007). Food packaging-roles, materials and environmental issues. *Journal of Food Science, 72*(3), R39–R55. Available from http://dx.doi.org/10.1111/j.1750-3841.2007.00301.x.

McHugh, T. (2015). Producing edible films. *Food Technology, 69*(4), 120–122.

McHugh, T., & Pan, Z. (2015). Innovative infrared food processing. *Food Technology, 69*(2), 79–81.

McHugh, T., & Toepfl, S. (2016). Pulsed electric field processing for fruits and vegetables. *Food Technology, 70*(1), 73–75.

Mermelstein, N. H. (2015). Pesticide analysis. *Food Technology, 69*(7), 96–100.

Mermelstein, N. H. (2016). Coloring foods and beverages. *Food Technology, 70*(1), 67–72.

Morris, J. G. (2011). How safe is our food? *Emerging Infectious Diseases, 17*(1), 126–128. Washington, DC: Centers for Disease Control and Prevention. Available from <www.cdc.gov/eid>.

NASA. (2002). *Space food.* NASA Facts. National Aeronautics and Space Administration, Lyndon B. Johnson Space Center. FS-2002-10-079-JSC. 4 p.

NASA. (2003). *Space food.* Available from <http://spaceflight.nasa.gov/living/spacefood/>.

NASA. (2005). *Cosmic cuisine.* NASA Facts. National Aeronautics and Space Administration, Lyndon B. Johnson Space Center. FS-2005-10-055-JSC. 1 p.

National Restaurant Association. (2012–2015). *Industry impact.* Available from <http://www.restaurant.org>.

NHANES (2015). *National health and nutrition examination survey.* Washington, DC: National Center for Health Statistics. Available from <http://www.cdc.gov/nchs/nhanes/index.htm>.

Paarlberg, R. (2010). *Food politics.* New York, NY: Oxford University Press, 260 p.

Painter, J. A., Hoekstra, R. M., Ayers, T., Tauxe, R. V., Braden, C. R., Angulo, F. J., & Griffin, P. M. (2013). Attribution of foodborne illnesses, hospitalizations, and deaths to food commodities by using outbreak data, United States 1998-2008. *Emerging Infectious Diseases, 19* (3). Centers for Disease Control and Prevention. Available from <http://wwwnc.cdc.gov/eid/article/19/3/11-1866_article.htm>.

Penfield, M. P., & Campbell, A. M. (1990). *Experimental food science.* San Diego, CA: Academic Press, Inc, 541 p.

Root, W., & de Rochemont, R. (1976). *Eating in America: A history.* New York, NY: The Ecco Press, 512 p.

Smith, J. S., & Pillai, S. (2004). Irradiation and food safety. *Food Technology, 58*(11), 48–55.

Tang, J., Mikhaylenko, G., & Simunovic, J. (2016). *Microwave sterilization.* Pullman, WA: Washington State University. Available from <http://www.microwaveheating.wsu.edu/factsheet/index.html>.

U.S. Bureau of Labor Statistics. (2014). *Occupational outlook handbook.* Available from <http://www.bls.gov/ooh/>.

U.S. Dept of Agriculture. (2015). *Food safety.* Available from <http://www.usda.gov/wps/portal/usda/usdahome?navid=food-safety>.

U.S. Food and Drug Administration. (2015). *2013 revision of the Grade "A" pasteurized milk ordinance.* U.S. Public Health Service and Food and Drug Administration. Available from <http://www.fda.gov/Food/GuidanceRegulation/GuidanceDocumentsRegulatoryInformation/Milk/ucm389905.htm>.

Walmart. (2016). *Sustainable food.* Available from <http://corporate.walmart.com/global-responsibility/environment-sustainability/sustainable-agriculture>.

Weaver, C. M., Dwyer, J., Fulgoni, V. L., King, J. K., Leveille, G. A., MacDonald, R. S., ... Schnakenberg, D. (2014). Processed foods: Contributions to nutrition. *American Journal of Clinical Nutrition, 99,* 1525–1542.

Williams, L. (2015). Trending down fat, sugar, sodium. *Food Technology, 69*(5), 23–31.

Winter, C. K. (2015). Chronic dietary exposure to pesticide residues in the United States. *International Journal of Food Contamination, 2*(11), SpringerOpenJournal. doi:10.1186/s40550-015-0018-y.

Ziegler, G. R. (2010). Feeding the world today and tomorrow: The importance of food science and technology. *Comprehensive Reviews in Food Science & Food Safety, 9*(5), 572–599.

CHAPTER

7

Nutrition and Food Access

7.1 DISCOVERY OF NUTRIENTS IN FOOD

At the most fundamental level, food must be consumed to provide the basic nutritional and energy needs for life. Without proper nutritional intake, growth and development are inhibited and illness occurs. The exploration of nutrient requirements established the discipline of nutritional science, and led to the elimination of some common nutrient deficiency diseases in the United States. Ensuring adequate nutritional intake for military personnel and civilians became an important use of these defined nutrient requirements. But food provides much more than nutrients to the human experience. The tastes, textures, and aromas of foods entice our senses and provide pleasurable experiences. The food industry plays a major role in the health of the population by providing foods that will meet the nutritional needs of consumers and provide the enjoyable experiences they demand. Finding ways to help consumers make wise food choices is a goal for the USDA, FDA, and CDC. These government agencies mediate between protecting the health of society, and the production, marketing, and accessibility of foods. Balancing the intake of foods that are pleasurable with those that provide the right amounts of nutrients is a daily challenge. The rise in obesity rates in the United States is indicative of the difficulties consumers have when making food choices. In contrast, food insecurity continues to be prevalent throughout the United States. These interrelationships amongst consumers, the food industry, nutrition experts, and government regulatory agencies comprise an important component of the food system.

Humans have been curious about how the body uses energy and food elements for growth and physical activity since the time of Aristotle (384–322 BC). The concept of fire, air, water, and earth as basic components of life led to the belief that four humors (substances) explained a person's health (yellow bile (dry heat), blood (wet heat), black bile (dry cold), and phlegm (wet cold)). The Greek physician Hippocrates of Cos (460–370 BC) is often credited with saying "let food be thy medicine and medicine be thy food." As the father of Western medicine, Hippocrates recognized the important connection between food consumption and the occurrence or progress of disease. An understanding of the specific components of food and their role in human metabolism evolved over many centuries and led to the establishment of the scientific discipline of nutrition. Today we know that food contains macronutrients (proteins, carbohydrates, and fats) that provide energy and functional

Understanding Food Systems.
DOI: http://dx.doi.org/10.1016/B978-0-12-804445-2.00007-7

components for hormones and enzymes, and micronutrients (vitamins and minerals) that are essential for metabolic functions.

The advent of the "chemical revolution" in France during the 1800s opened a new era in discovery of the nature of human physiology. Antoine Lavoisier, a French chemist, was among those first credited with exploration of human respiration and combustion of foods. Lavoisier found that the amount of carbon dioxide in the breath of humans correlated with physical activity and he created the first calorimeter, a device to measure the amount of heat produced by an animal. Unfortunately, Lavoisier was arrested as an enemy of the Republic and guillotined during the Reign of Terror in 1794 before he could complete his research. Jean Baptiste Boussingault continued Lavoisier's line of research and conducted the first balance (intake and output) studies. He tracked the amount of carbon (an element in carbohydrates and organic compounds) in cow feed (intake) and urine, feces, and milk (output). The results showed that carbon intake balanced carbon output in healthy animals, but input was greater than output in growing animals (because carbon was retained to build tissues). This led French chemists to determine the function of carbon and nitrogen in the human body. German scientists, including Justus von Liebig, Carl Vogt, and Max Rubner developed analytic methods for measuring organic materials and constructed a calorimeter large enough to hold a person. With these tools, they established the concepts of metabolic conversion of foods to produce body tissues and the understanding that macronutrients in food generated specific amounts of body heat. A focus of work during this period was placed on defining the amount of protein needed to maintain growth and physical strength, as protein foods were considered to be of greatest importance. There was much debate about the amount of protein needed for health and productive work, with ranges of 40–125 g/day. The amount of protein recommended for the average healthy adult today is 0.8 g/kg body weight, or about 56 g for a man and 46 for a woman, so the lower protein level was proven to be correct.

7.1.1 Nutrient Composition of Foods

The study of nutrient composition of foods in the United States began in the mid-1800s with S.W. Johnson who had studied under Liebig and became the first professor of biochemistry at Yale University. Johnson's student, Wilbur O. Atwater, received the first appropriation from Congress to conduct research in human nutrition. Atwater eventually led the Agricultural Experiment Stations within the USDA and was intrigued with the concept of the energy value of foods. One of Atwater's accomplishments was the construction of a respiration calorimeter similar to that developed in Germany. His work led to the understanding that dietary fats and carbohydrates, not only proteins, could be used for mechanical work by the body. The Atwater factors of 4 kcal/g of protein or carbohydrate, 7 kcal/g of alcohol, and 9 kcal/g of fat are still used today to estimate the amount of energy generated from foods. In 1896, Atwater created the first proximate analysis database (nitrogen, fiber, ash, ether extract, moisture, and carbohydrates) of foods to teach poor people how to obtain their protein requirements at the lowest cost. The USDA publication from his work was the first to define food within five categories of macronutrients: protein, carbohydrate, fat, energy, and water.

By 1950, the USDA had compiled a substantial database on the composition of foods and produced *Agricultural Handbook No. 8, Composition of Foods—Raw, Processed, Prepared.* This document listed the proximate analysis (protein, carbohydrate, fat, and water), five vitamins (vitamin A, thiamine, riboflavin, niacin, and vitamin C), and three minerals (calcium, phosphorus, iron) of 750 foods. In 1963, a revision of *Handbook No. 8* with added data for cholesterol, fatty acids, sodium, potassium, and magnesium, was published. Analytical methods to measure nutrients in foods continued to be developed during this time, and the demand for information about food composition was high as the role of food in health and disease was being recognized. An electronic version of *Handbook No. 8*, now called the National Nutrient Database for Standard Reference (NNDSR), was released by the USDA in 1980. The need for more data, especially for quantification of the wide range of bioactive compounds of foods, continued to grow through the 1990s. Bioactive compounds are naturally occurring chemicals in foods that may have health benefits, but are not considered essential nutrients. These bioactive compounds include antioxidants found in plants such as the flavonoid family of anthocyanidins, isoflavones, and catechins. In 1997, the USDA created the National Nutrient and Food Analysis Program to address the complexity of food composition and to coordinate and ensure the accuracy of data being collected from government labs, the food industry, and academic research. The USDA Nutrient Data Laboratory is responsible for overseeing the publicly available food composition database for the United States, which is freely accessible through their website: www.ndb.nal.usda.gov.

7.1.2 Discovery of Essential Nutrients

By the early 1900s, it was recognized that proteins were comprised of amino acids, and groundbreaking work by the team of Thomas B. Osborne and Lafayette B. Mendel led to the recognition that dietary sources of proteins differed in their ability to support growth. W.C. Rose and others established the concept of essential amino acids and defined the exact requirements for amino acids in 1937. Today, we know that humans require certain amino acids to be supplied in the diet (essential), while others can be synthesized (nonessential). Protein foods that have a balance of essential amino acids in amounts that meet human requirements are called "high-quality" proteins and come from meats, eggs, milk, and soybeans. Low-quality protein foods lack one or more essential amino acids. Most plants, including wheat, rice, nuts, and beans, are considered low-quality protein foods when consumed alone, but do meet the human amino acid requirements when consumed together in "complementary pairs." Examples of foods with complementary proteins are beans and rice or peanut butter on wheat bread. When the diet is lacking in even one essential amino acid over time, growth is impaired. Proteins from skeletal muscle and heart are broken down to provide the missing amino acid. For this reason, vegetarians should be well versed in the amino acid composition of foods to ensure their requirements are met.

To understand how the human body uses energy from food it was necessary to measure heat production. The respiration calorimeter, such as the one constructed by Atwater, was used in early nutritional research in Europe and by Atwater in the United States to understand caloric needs. From studies of the energy production during activity it was learned that humans produce energy from carbohydrates, fats, and proteins in the same amounts as when these components were combusted outside of the body. Physical activity, or work, could be quantitatively matched with energy production. Because foods contain mixtures of fats, proteins, and carbohydrates, and all of these may be used for energy, defining specific needs or requirements for these nutrients was complicated. Hamish Munro, a Scottish nutrition scientist, found in the 1940s that replacing carbohydrates with fat in the diet resulted in adequate energy, but net nitrogen loss from the body. This eventually led to the understanding that humans require a small amount of dietary carbohydrate (about 130 g/day) to maintain blood glucose levels. Glucose is essential for the functions of the brain and central nervous system and blood glucose levels must be maintained within a very narrow range to ensure these needs are met. If there is no dietary source of carbohydrate, proteins can be utilized to make glucose (the carbon structure of amino acids is used to make glucose and the nitrogen is removed and excreted). The proteins come from body tissues, mainly muscle. This explains why Munro found negative nitrogen balance in his subjects fed a diet without carbohydrates. When carbohydrates are restricted for a prolonged period of time, the body will utilize breakdown products of fatty acids (ketones) as an energy source for the central nervous system, but will continue to utilize protein to generate the minimum glucose requirement. Humans are not able to convert fatty acids to glucose. Dietary forms of carbohydrates are either simple (monosaccharides such as glucose, or disaccharides such as sucrose (glucose + fructose)) or complex (polysaccharides such as starch (chain of glucose)). Humans cannot store carbohydrates, other than for a short while in the form of liver and muscle glycogen, so a daily intake of carbohydrate is needed to prevent muscle tissue breakdown. Consumed carbohydrates that are not needed for immediate use or to replenish glycogen stores are converted to fat and stored in the body.

The most challenging macronutrient requirements to be defined were the essential fatty acids. George and Mildred Burr, working in the 1930s at the University of Minnesota, found that rats fed diets without fat failed to grow and developed scaly tails. The scientists could reverse these symptoms by adding only one polyunsaturated fatty acid, linoleic acid, to the animals' diet. But it was not clear at that time if humans also required this fatty acid. Infants fed very low-fat diets did develop skin problems, but a clear deficiency disease did not develop. George Burr and his collaborator, Arild Hansen, concluded in a paper published 1932 that "[t]his finding, however, together with various serum lipid studies, gives strong support to the theory that the human organism is unable to synthesize the unsaturated fatty acids which have been found to be essential for some animals" (Hansen & Burr, 1932). It wasn't until the 1960s when infant formulas became popular and large groups of infants were fed formulas with different amounts and sources of fat that the important role of linoleic acid became evident for humans. Additional work was required to understand that humans can synthesize most fatty acids, but must have a dietary source of either linoleic acid (C18:2 w6) or arachidonic acid (C20:4 w6) and that linoleic acid (C18:2 w3) and eicosapentaenoic acid (C20:4 w3) are also essential for humans.

The role of the different types of dietary lipids (saturated, unsaturated, and cholesterol) and classes (omega-6 and omega-3) in human health has been a major focus of nutrition research for the past 60 years. The relationship of dietary fat to disease began in earnest in the 1950s and was brought to the public's attention by the National Diet-Heart Study that was published in 1963 (Baker et al., 1963). Ancel Keys was one of the scientists that worked on the study, and he had been reporting that heart disease rates in different countries correlated with the amount and type of fat that was consumed by that population. From data collected from seven countries, Keys et al. (1986) found populations that consumed more saturated fat and cholesterol had higher rates of cardiovascular death compared to populations that consumed lower fat diets. Based on these connections between dietary fat and cholesterol, and risk of cardiovascular disease nutritionists encouraged people to limit fat and cholesterol intake. This eventually led to the fat-free craze of the 1970s, which may have initiated the obesity epidemic in the United States. Consumers avoided fat but failed to recognize that calories from carbohydrates would be converted to body fat when consumed in excess of energy needs. Debate continues today as to the ideal amount, type, and balance of lipids in the diet.

7.1.3 Nutrient Deficiency Diseases

Diseases that are now known to be due to nutrient deficiencies were common throughout history. Table 7.1 lists diseases that were widely present from the 1800s to the 1930s, usually developing as outbreaks in populations experiencing some type of dietary restriction. Early nutrition scientists used a wide range of experimental approaches to understand dietary factors, including careful documentation of food consumption and disease incidence. With limited technology and crude experimental protocols, researchers struggled to discern if the illnesses were caused by infectious agents or by toxins present in foods. As the macronutrients became better understood and purified diets could be used in animal models, it became clear that factors other than protein, carbohydrate, and fat were necessary to maintain health. The classification and characterization of these accessory factors in foods began in 1912 with Casimir Funk, who investigated the cause of beriberi and coined the term "vitamines." Funk believed beriberi, pellagra, scurvy, and rickets were caused by a deficiency of a vital amine (derived from proteins). This hypothesis turned out to be incorrect but the term vitamin was retained. In 1916, research by Elmer V. McCollum and Marguerite Davis at the University of Wisconsin led to the finding of fat-soluble (Factor A) and water-soluble (Factor B) compounds required for proper growth and development. To save research time and costs, McCollum and Davis developed protocols using rats as experimental models for dietary research, which opened a new era in nutrient discovery. Selected factors could be removed from the diet of rats until symptoms developed and then systematically returned to the diet to identify the active compound. By the end of the 1940s, the fat-soluble vitamins (A, E, D, and K) and water-soluble vitamins (niacin, riboflavin, thiamine, pantothenic acid, folic acid, vitamin B_6, vitamin B_{12}, and vitamin C) had been isolated and their functions in human biochemistry began to be understood. Simultaneously, researchers were discovering that selected minerals, such as iron, copper, zinc, iodine and calcium, sodium, phosphorus, and magnesium were also necessary for animal and human health.

TABLE 7.1 Diet-Related Diseases Common in the 1800s

Disease	Symptoms	Afflicted population	Dietary component (nutrient)
Scurvy	Bleeding from mucous membranes, weakness, death	Sailors, prisoners, potato famine	Citrus juice (vitamin C)
Beriberi	Weakness, loss of feeling in legs, heart failure	Japanese navy and soldiers, prisoners	Brown rice polishings (thiamine)
Night blindness	Loss of vision in low light, corneal ulcers, blindness	Sailors, children in orphanages	Cod-liver oil (vitamin A)
Goiter	Enlarged thyroid gland	People from the upper midwest US	Seaweed ash (iodine)
Rickets	Weak bones	Children in industrialized cities	Sunlight, cod-liver oil (vitamin D)
Pellagra	Diarrhea, dermatitis, and dementia	Poor people in southern US	Meat or dairy foods (niacin)

From Carpenter, K. J. (2003). A short history of nutritional science: Part 1 (1912–1944). Journal of Nutrition, 133, *3023–3032 (Carpenter, 2003).*

EXPANSION BOX 7.1

ANIMALS IN NUTRITION RESEARCH

A significant amount of the research that has been done to understand the role of nutrients in humans has actually been conducted using rodents. It may be curious to think that we can learn anything about human health by studying rats and mice. There are several reasons why rodents have been so widely used in nutrition research. Humans are, of course, the ideal model in which to study nutrition, but there are many reasons why human research is either impossible or impractical. In nutrition research human subjects are used when the outcome can be assessed using a biomarker (an indicator of change that can be measured such as hemoglobin to detect anemia) obtained from blood, sweat, urine, saliva, or feces. Nutrition studies can be conducted when growth, development, or bodily functions such as blood pressure, brain responses, muscle activity, digestion, or respiration are the outcome. Studies that require tissue samples or cells can be conducted in humans by obtaining biopsies, or collecting samples during surgery that is being done for a medically approved reason. Conducting nutrition research on humans has many limitations, some of which are ethical but mostly are due to the constraints of controlling and manipulating diets and lives of people over long periods of time.

Rats and mice are the most commonly used models to test and explore the ways nutrients function at the organ, cellular, and molecular level. There are many similarities between rodents and humans. Both species are mammals, which means they incubate their young *in utero* and provide milk from mammary glands. The internal organs and physiology of rats and mice are very similar to those of humans, including the circulatory, respiratory, digestive, endocrine, skeletal–muscular, and reproductive systems. These similarities allow a general correlation and suggest what occurs in the rodent is likely to also occur in humans. There are some striking differences between rodents and humans, however. Rodents have much shorter lifespans than humans, which is a positive feature for experimental research, but is a major difference in length of exposure to nutrients and or environmental factors. Rats are capable of synthesizing vitamin C, whereas humans are not. Guinea pigs are used in vitamin C research because they are unable to make the vitamin. Rats metabolize lipids somewhat differently than humans and do not develop lipid-related diseases in the same manner. By using rodent models, hypotheses can be tested and mechanisms of action identified, which then can be verified in human studies. Such exploratory work could not be done in humans. A major area of research in human nutrition currently is to understand the role of microorganisms in the intestinal tract. Rodent models are useful in these studies, for example, by treating the animals with antibiotics to kill all their gut microbes and then selectively adding them back, or by feeding very specific diets over the lifespan of the rodents to examine changes in the microorganism population. Such experiments would not be possible in humans.

The ability to manipulate the genome of rodents to develop specific genetic models has been an important component of nutrition research for the past three decades. By inserting, silencing, or overexpressing specific genes, these animal models are useful as models of human diseases including cancer, cardiovascular disease (CVD), autoimmune conditions, and genetic illnesses. By replicating the cellular events that are associated with human diseases, the role of nutrients to prevent or mitigate these

EXPANSION BOX 7.1 (*cont'd*)

diseases is possible. Conducting such studies using humans would be limited by the wide range of disease expression in people, and the inability to measure responses or collect tissues.

Since 1966 when the Animal Welfare Act was passed, all animals used in research must be handled and treated humanely. The law requires that an Institutional Animal Care and Use Committee (IACUC) be in place at any location where animals are being used. The IACUC must include researchers, veterinarians, and members of the public. IACUC members review and approval all protocols before they are implemented, inspect all research areas where animals are being used, and report any violations. IACUC standards are strictly enforced and violations are not tolerated. Treating research animals with care and respect is expected of all scientists. The use of animals in research has been essential for providing the understanding of how nutrients function in humans.

Suggested reading: USDA Animal Welfare Act: https://www.nal.usda.gov/awic/animal-welfare-act

7.1.4 Promoting Food for Health

As scientists were discovering the biochemical role of nutrients in foods in the United States, public interest in food and health was also evolving. Just as Hippocrates had connected food with health, Americans were also searching for ways to improve their health through food. In the early 1830s, Reverend Sylvester Graham, a Presbyterian minister from New Jersey, advocated for vegetarianism as a means to achieve purity of life. He recommended a diet of fruit, vegetables, and whole wheat, with no meat or spices and only sparing use of milk, egg, cheese, and butter. To encourage his followers, he created Dr. Graham's Honey Biskets, made with a special type of whole wheat flour (now called graham flour). These graham crackers were a market success and, by 1900, the several bakeries that were producing them joined together to form the National Biscuit Company, which eventually became Nabisco. "Honey Grahams" along with the popular "Teddy Grahams" continue to be sold by Nabisco (part of Mondelez International) today, although the recipe has been modified over the years.

Other popular leaders of health foods of the time were Will Keith Kellogg and his brother Dr. John Harvey Kellogg, who ran a sanitarium in Battle Creek, Michigan in the 1800s. The Kellogg brothers held the theory that the illnesses they observed among patients in the sanitarium were largely caused by bad intestinal flora. They experimented with putting patients on strict diets containing vegetables and nuts, and developed a process to make flakes from wheat and corn, resulting in a light breakfast cereal. With the concurrent introduction of pasteurized milk, the American breakfast cereal market took off. "Corn Flakes" launched the Kellogg Company in 1906, followed by "Bran Flakes" and "All Bran" cereals in 1915 and 1916. The Kellogg brothers were ahead of their time as scientists are just now starting to understand the connections between the gut microflora and health.

Candle maker William Procter and soap maker James Gamble in Cincinnati, Ohio, combined their trades to forge the Procter and Gamble Company (now P&G) in 1837. Both candles and soap were derived from the animal processing industry, so adding cooking lard (beef or pork fat) to their product line was a natural extension. In 1911, P&G introduced a new product called Crisco, which was the first all-vegetable solid fat product made using the newly developed process of hydrogenation. By adding hydrogen atoms to liquid vegetable oils, the fats became more solid at room temperature. P&G promoted Crisco as more digestible and economical than lard and touted the cleanliness and purity of the product (it was pure white) to homemakers. The company hired home economists to lead Crisco cooking schools across the country and eventually on television to demonstrate how to use the product. This new model of marketing food by demonstration ("home demonstration agents" in the Cooperative Extension Service were also effective educators for new innovations) was highly successful. The hydrogenation process used to make Crisco shortening and other solid vegetable fats, as will be discussed later, creates the unnatural *trans* form of unsaturated fatty acids, which have been linked to an increased risk of heart disease.

7.2 DEFINING NUTRIENT REQUIREMENTS FOR THE POPULATION

One of the earliest roles assumed by the USDA was to provide dietary guidance to the American people. Based on the work of Atwater, a Farmers' Bulletin was published in 1894 that suggested the amounts of protein, carbohydrate, fat, and mineral ash (vitamins and minerals had not yet been discovered) for healthy men. In 1916, the first USDA food guide, *Food for Young Children*, was developed by Caroline Hunt. By this time, foods were categorized into five groups: milk and meat, cereals, vegetables and fruits, fats and fatty foods, and sugars and sugary foods. The following year, Hunt and Atwater developed *How to Select Foods*, a guide for the general public. Slight variations of these guides, including food purchasing plans, were widely used by the public. When the United States entered the Great Depression in 1930, the USDA food guides focused on economic food plans and recommended ways to meet nutritional needs at different cost levels. The USDA has continually provided such recommendations (today called the *Thrifty Food Plan*) to provide consumers with guidance for meeting their nutritional needs within their budget allowances.

As nutrition research was uncovering that foods contained specific components necessary for growth, physical activity, and maintenance of health, quantification of the amounts of these nutrients became possible. The first attempt to define nutrient requirements occurred in 1941 following the National Nutrition Conference for Defense called by President Franklin Roosevelt. The demands of providing food for the military during WWII, combined with a loss of imports, strained the food supply. Rationing of some foods was necessary and concerns about maintaining the health of the population became an issue of national security. Three major goals were outlined by Vice President Henry Wallace during the Conference: (1) eliminate deaths caused by dietary deficiencies; (2) reduce diseases due to insufficient food; and (3) make sure everyone in the United States had an adequate diet. The first Recommended Dietary Allowances (RDA) (developed over the previous year by the Food and Nutrition Board of the National Academy of Sciences) was released at the Conference. This first publication of the RDA included recommendations for calories and nine nutrients: protein, iron, calcium, vitamins A and D, thiamine, riboflavin, niacin, and vitamin C. Recommendations were provided for different age groups, both genders, and pregnant and lactating women. The RDA was revised periodically and the number of nutrients included in the recommendations increased to 25 by 1989 when the 10th edition was completed.

TABLE 7.2 Components of the Dietary Reference Intakes

Abbreviation	Category	Definition
EAR	Estimated Average Requirement	Average daily nutrient intake level estimated to meet the requirements of 50% of healthy individuals within an age and gender group.
RDA	Recommended Dietary Allowance	Average daily dietary nutrient intake level sufficient to meet the nutrient requirements of nearly all (97.5%) healthy individuals within an age and gender group; set at 2 SD above the mean requirement (EAR).
AI	Adequate Intake	Average daily intake level by a group of apparently healthy people that is assumed to be adequate; provided when an EAR and RDA cannot be determined.
UL	Upper Limit	Highest average daily nutrient intake level likely to pose no risk of adverse health effects to almost all individuals in the general population.
EER	Estimated Energy Requirement	Average energy intake predicted to maintain energy balance in a healthy individual at a specific level of energy expenditure.
AMDR	Adequate Macronutrient Distribution Range	Range of intake of protein, fat, and carbohydrate that is associated with a reduced risk of chronic disease, yet can provide adequate amounts of essential nutrients.

From Murphy, S. P., Yates, A. A., Atkinson, S. A., Barr, S. I., & Dwyer, J. (2016). History of nutrition: The long road leading to the dietary reference intakes for the United States and Canada. Advances in Nutrition, *7, 157–168 (Murphy, Yates, Atkinson, Barr, & Dwyer, 2016).*

Defining nutrient requirements for the general population is a significant undertaking. Responsibility for the RDA was assumed by the Food and Nutrition Board within the Institute of Medicine of the National Academies of Science. The process involves convening a panel of experts, including scientists and healthcare professionals, who review the body of evidence related to each nutrient. The panel must determine the quality and accuracy of each study, and interpret the results. They take into consideration all of the variables within the studies, and draw a consensus from the evidence available at that time for the appropriate requirement for each nutrient for each age and gender category. This is a challenge for the panel members, who volunteer their time for the work. Having a clear standard of nutrient intakes is of great value for many reasons including guidance for school, military, or other congregate feeding programs; food fortification or enrichment; and clinical assessment of adequate nutrient intake.

Around the time that the 10th RDA was completed, nutrition researchers were recognizing that the concept of the RDA, which was a single value of the defined amount of each nutrient required by a specific age, gender, or pregnancy/lactation, was not sufficient for the wide range of uses for nutrient requirement information. A new concept was proposed by the Food and Nutrition Board, the Dietary Reference Intakes (DRI). The development of the DRIs began in 1994 and was a joint collaboration between the United States and Canada. It took about 10 years for all of the nutrients to be sufficiently reviewed and documented in the DRI framework. The RDA was replaced with the DRI, which is a group of values designed for specific uses, as shown in Table 7.2.

The DRIs have been effective in providing a framework for dietary guidelines and recommendations, monitoring the quality of dietary intake in the population, advising standards for school feeding programs and the military,

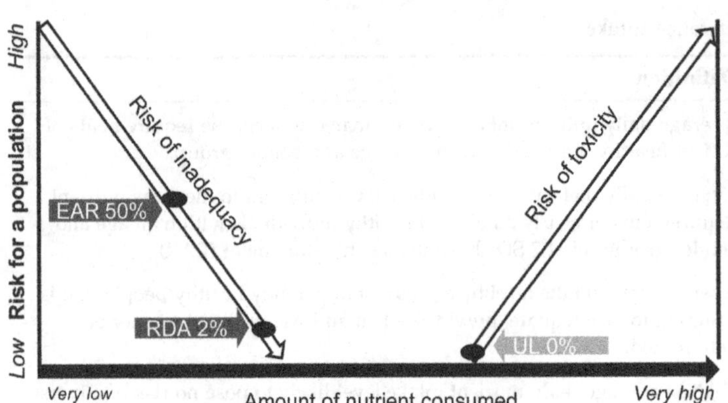

FIGURE 7.1 Nutrient requirements are defined for the population using the Estimated Adequate Requirement (EAR), Recommended Dietary Allowance (RDA), and Upper Limit (UL) values, which indicate amounts needed to prevent deficiency or toxicity. The RDA is defined to meet the nutrient needs of 98% of the population and the EAR to meet the needs for 50% of the population. The UL is set at a level that would prevent any consumer from toxicity. *Source: Illustration by Reannon Overbey.*

and for nutrition labeling of foods. The DRIs provide a means to describe the potential for inadequate or toxic levels of a nutrient for the population. A graphic illustration of three of the DRI values is shown in Fig. 7.1.

7.2.1 Nutrient Deficiency and Public Health

Hunger and malnutrition were fairly common conditions during the settling of the United States. From 1800 to 1900, the average life expectancy of white men was less than 50 years of age. The main causes of death were influenza, pneumonia, heart disease, and diarrhea. Poor sanitation and a lack of antibiotics and vaccines were the primary reasons these diseases were so common. While not recognized until much later, poor nutritional status exacerbates the risk for, and also likely contributed to, the high incidence of infectious diseases. Nutritional deficiencies including goiter, rickets, pellagra, and anemia were widespread public health concerns as well, although the link to dietary components was not yet known.

Goiter is characterized by enlargement of the thyroid gland (thyromegaly), which results in a swelling around the neck. Before 1920, regions of the United States including the Great Lakes, Appalachian Mountains, and Northwest were known as the "goiter belt" because as many as 70% of children were afflicted with the condition. During WWI, a Michigan physician noted that many local young men were being disqualified from military enlistment because of goiter. Links between iodine and goiter had been made by Jean Baptiste Boussingault in France in the 1800s. In 1917 David Marine, an Ohio physician, experimented with giving iodine to schoolgirls, which reduced the incidence of goiter in the children. Over the next few years, research confirmed that iodine supplementation prevented goiter. This led the Michigan State Medical Society to launch the first food fortification campaign to address a public health problem. Salt was identified as the best vehicle to widely administer iodine to the public because everyone used salt. Although the salt industry balked at first, stating that they were not in the business of pharmaceuticals, the public health campaign soon overcame their resistance. The first iodized salt was available in grocery stores in 1924. No federal legislation was ever passed requiring salt to be iodized, and the only current ruling is that any salt sold in the United States must be labeled as "supplying" or "not supplying" "iodide, a necessary nutrient."

The bone deformation condition known as rickets was likely rare in early civilizations that arose in regions with abundant sunlight. When industrialization began, people moved from farms to crowded cities with high levels of air pollution and children went to work in factories, and the incidence of rickets increased. Children with rickets are characterized by small size, head and chest deformations, and bowed legs associated with weakened bone structure. Cod liver oil was effective in preventing rickets, as was exposure to sunlight. It was also found that feeding children irradiated milk (ultraviolet light converts sterols to vitamin D) stimulated bone health and scientists at the University of Wisconsin developed an industrial-size milk irradiator to provide this product (although it was never widely used). Eventually, the fat-soluble vitamin D was identified as a key component of the condition, and since 1932, milk has been fortified with vitamin D to prevent rickets in children. Vitamin D facilitates the absorption of calcium from foods in the intestine, so an adequate source of both calcium and vitamin D are needed to prevent rickets, thereby making milk the ideal source for fortification. As with iodized salt, the movement to add vitamin D to milk came from the medical community, not the government. The dairy industry embraced the process as a means of connecting with mothers and physicians to promote the health benefits of their product.

Pellagra is characterized by diarrhea, dermatitis, dementia, and death (known as the 4 Ds). The incidence of pellagra increased rapidly across the United States between 1907 and 1940, and because it was considered an infectious disease, was raising public concern. To understand and curtail the outbreak, the US Public Health Service sent Dr. Joseph Goldberger to Mississippi to investigate the problem. From his observations, Dr. Goldberger proposed that pellagra was not caused by an infectious agent, but rather was related to poverty and a poor diet. He conducted experiments, including exposing himself and his assistants to affected patients, and showed that the illness was not transmitted through personal contact or body fluids. He observed that families with diets lacking protein foods, such as milk and meat, were more likely to develop the illness and it could be prevented by providing these foods. Pellagra was eventually found to be caused by inadequate intake of niacin (a B vitamin). From the 1930s, bakers had begun voluntarily adding yeast mixtures (rich in B vitamins) to bread in response to appeals from nutrition professionals to improve vitamin intake of the public. The American Medical Association (AMA) had also raised the idea of enriching wheat flour with B vitamins and iron because Americans were increasingly consuming refined flour. Eliminating the bran and germ fractions from wheat during the refining process removes the majority of the vitamins and minerals. Standardized flour enrichment did not occur until 1943 when the War Foods Administration issued War Food Order No. 1 making enriched bread required by law, and the Institute of Medicine (IOM) adopted a resolution encouraging the enrichment of flour and bread with B vitamins. Today the FDA sets the standards for flour enrichment to contain defined amounts of niacin, thiamine, riboflavin, folic acid, and iron.

These examples of enrichment and fortification of foods to prevent public health problems established the association between food and disease mitigation in the minds of Americans. Many more foods, in addition to salt, milk, and wheat flour, are fortified with nutrients. These include breakfast cereals (vitamins and minerals), pastas (vitamins and minerals), rice (vitamins and minerals), breads (vitamins and minerals), and orange juice (vitamin D and calcium). Food manufacturers see market benefit of making their products "healthier" by adding nutrients, and consumers logically assume more nutrients make the product better. The FDA provides guidance to the food industry as to which nutrients may be added to foods under the FDA Fortification Policy. The policy specifies that essential nutrients, associated with a public health concern, may be used for enrichment or fortification. The FDA sets guidelines for the concentration of added nutrients per caloric intake in attempt to ensure overall intake of each nutrient will be within the safe and effective range for the majority of consumers.

In recent years, nutrients and other compounds, including caffeine and botanical extracts, have been added to energy drinks and shots (such as 5-hour ENERGY), which are especially popular among youth. The FDA does not define or regulate the levels of nonessential compounds added to supplemented foods and beverages. Some nutritionists and health professionals have raised concern that the widespread addition of nutrients, even those that are FDA regulated, to so many foods may potentially create intakes that exceed the DRI upper limits for safety. The risk of excessive nutrient intakes would be exacerbated by the use of vitamin and other dietary supplements in combination with consumption of enriched, fortified, or supplemented foods. As shown in Fig. 7.1, toxicity can occur at doses of nutrients above a defined threshold. There is increasing awareness that high intakes of both essential nutrients and supplemented compounds (e.g., caffeine) may become a problem for some segments of the population, particularly children.

7.2.2 Public Policy and Nutrition Programs

A policy is "a course of action or principle adopted or proposed by a government, party, business, or individual." There are two components of policies: (1) statement or goal of what is to be accomplished and; (2) set of practical rules, guidelines, or regulations to accomplish the goal or program. The tenets of US nutrition policy are to provide adequate food at affordable cost; ensure quality, safety, and wholesomeness of the food supply; ensure food access to people in need; provide information and education to support informed food choices; support science/research base in food and nutrition; and integrate nutrition into preventative and medical care.

The primary federal agencies involved in administering nutrition policy include the USDA Food and Nutrition Service, Centers for Disease Control and Prevention (CDC), and the Food and Drug Administration (FDA) (for nutrition labeling).

Policy makers involved in nutrition programs include those in county government, city government, school districts, school boards, and corporate managers. The legislative process follows discussions and revisions of the policy agenda, as well as policy formulation, adoption, implementation, evaluation, and termination. The process involved in federal legislative actions starts when a bill is developed and introduced, then referred to a committee (Senate or House) for hearings and modifications. The committee will take action (vote in favor or against), refer the bill to a conference committee, then for a floor vote. When the bill has been approved by both chambers (House and Senate) it is forwarded to the president for approval or rejection. A bill that makes it through to presidential action becomes a law. The law is assigned to a federal program for administration, where policy is defined and published in the Federal Register (www.federalregister.gov).

Within the USDA, there are seven undersecretaries. Three secretaries oversee nutrition programs: Food Nutrition and Consumer Services (FNCS); Research, Education, and Economics (REE); and Food Safety. Under the Food and Nutrition Service are 15 federal nutrition assistance programs:

- National School Lunch Program (NSLP)
- Supplemental Nutrition Assistance Program (SNAP)
- Special Supplemental Nutrition Program for Women, Infants & Children (WIC)
- School Breakfast Program (SBP)
- Child and Adult Care Food Program (CACFP)
- Commodity Supplemental Food Program (CSFP)
- Food Assistance for Disaster Relief (FADR)
- Food Distribution Program on Indian Reservations (FDPIR)
- Farmers' Market Nutrition Program (FMNP)
- Nutrition Assistance Block Grants (NABG)

- Senior Farmers' Market Nutrition Program (SFMNP)
- Summer Food Service Program (SFSP)
- Special Milk Program (SMP)
- The Emergency Food Assistance Program (TEFAP)

Funding for three of these programs constitute the majority of money allocated for the Farm Bill: SNAP (Supplemental Nutrition Assistance Program) to improve diets of low income people, WIC (Supplemental Nutrition Program for Women, Infants, and Children), and the National School Lunch (NSLP) and Breakfast (SBP) programs. In 2014 the USDA provided almost $74 billion for SNAP and much less for WIC, NSLP, and SBP (Fig. 7.2).

7.2.3 Federal Food Assistance Programs

In the United States, public food assistance programs began during the Great Depression. The Agricultural Adjustment Act (AAA) of 1933 introduced programs that gave the USDA the authority to purchase excess commodities from farmers and distribute the food to people in need (Chapter 2: History of US Agriculture and Food Production). The legacy of this effort continues in the federal food assistance programs in effect today. A timeline of food assistance programs that have been implemented are listed in Table 7.3.

7.2.3.1 The Food Stamp Program

The Food Stamp Program, begun by Secretary of Agriculture Henry Wallace, was designed such that people would buy orange stamps with money they would ordinarily use for food and, as a bonus, would receive 50¢ worth of blue stamps for every dollar spent. They could buy any food with the orange stamps but the blue stamps could only be used to buy surplus commodity

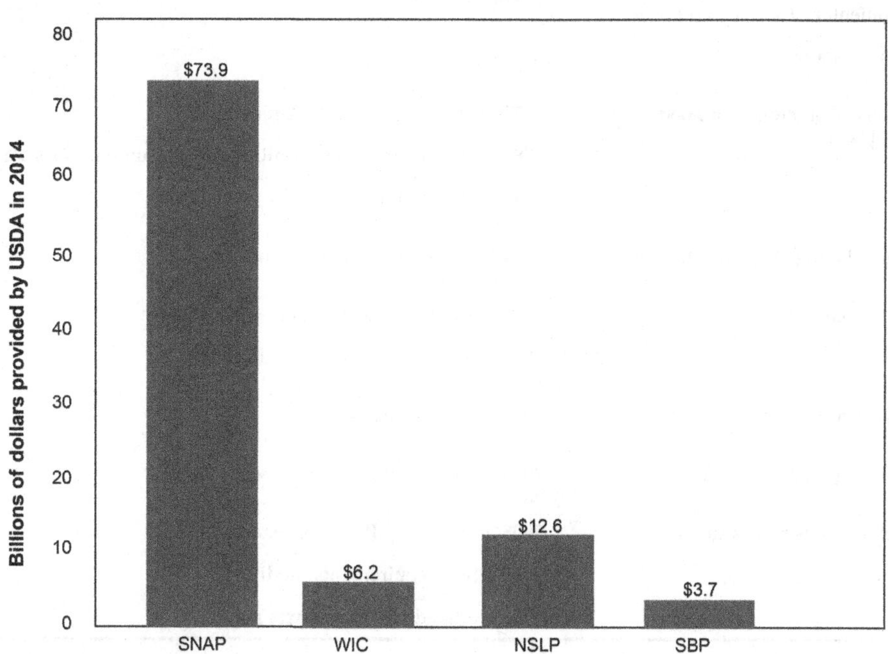

FIGURE 7.2 The majority of funding provided in the 2014 Farm Bill for nutrition assistance is committed to the Supplemental Nutrition Assistance Program (SNAP). The National School Lunch Program (NSLP); Special Supplemental Nutrition Program for Women, Infants, and Children (WIC); and School Breakfast Program (SBP) receive smaller allotments of funding. *Source: USDA Economic Research Service, www.usda.ers.gov.*

TABLE 7.3 Timeline of Major Events and Federal Food Assistance Programs

Major events	Food assistance programs
1930–39 Great Depression and Dust Bowl	1933 Agricultural Adjustment Act
	1935 Commodity Food Distribution
	1939–43 Food Stamp Program
	1938 Food Drug and Cosmetic Act
1941–45 World War II	1941 Recommended Dietary Allowances
	1942 Emergency Price Control Act (food rationing)
	1943 Wheat enrichment with B vitamins
	1946 National School Lunch Act
1962 Michael Harrington *The Other America* published	1961 Food Stamp pilot program
	1964 Food Stamp Act
1960 Presidential campaign Kennedy vs Nixon	1972 Special Supplemental Food Program for Women, Infants, and Children
1964 Lyndon Johnson War on Poverty	1972 Congregate Meals Program and Home Delivered Meals Program (meals on wheels)
1968 CBS documentary *Hunger in America*	
1969 March on Washington	
1969 White House Conference on Food, Nutrition, and Health	1977 Dietary Goals for the United States
	1980 Nutrition and Your Health: Dietary Guidelines for Americans
	1989 DRI introduced
1990 Nutrition Labeling and Education Act	1994 Nutrition Facts panel required
2000 Healthy People 2000	1994 FDA allows health claims on packages
	2000 Rates of obesity become alarming
2010 Affordable Care Act	2007 CFBAI formed
2010 Healthy Hunger Free Kids Act	2008 USDA identifies food deserts
2013 AMA defines obesity as a disease	2008 Food Stamp Program renamed SNAP
	2015 NSLP requirements modified
	2018 Updated Nutrition Facts panel

From USDA, www.usda.gov; FDA, www.fda.gov; CDC, www.cdc.gov.

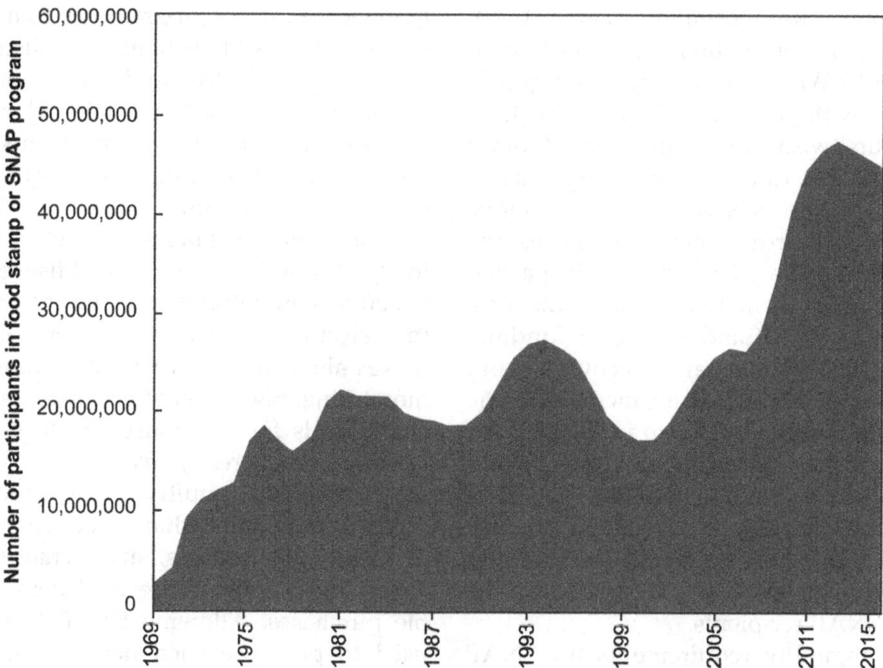

FIGURE 7.3 The number of people that utilized food stamps or Supplemental Nutrition Assistance Program (SNAP) benefits increased quickly through the 1980s. Changes in government-regulated eligibility created periods of decreased or increased access through the 1990s. The economic downturn during the mid-2000s created high need for food support among Americans. *Source: USDA Economic Research Service, www.usda.ers.gov.*

foods (that had been purchased from farmers by the USDA). This first food stamp program lasted 4 years, but served 20 million people at a cost of $262 million. The program was considered to be successful because it kept farmers from bankruptcy and saved people from starvation. By 1943, the economy had recovered and the Food Stamp Program was discontinued.

President John Kennedy restarted the Food Stamp Program in 1961 in response to the hunger and poverty he observed during his presidential campaign. This pilot program did not require participants to purchase surplus commodities, but did require them to purchase some stamps to obtain free ones. The program reached over 380,000 people within 3 years. Congress passed the Food Stamp Act of 1964, which made the program permanent, in support of President Lyndon Johnson's War on Poverty. Growth in the use of the program was steady; by 1974 over 15 million people were participating. During the 1970s, program reforms included establishing national eligibility standards to replace the state-by-state regulations. The Food Stamp Act of 1977 increased access and established income eligibility based on the poverty threshold and eliminated the requirement that participants had to purchase stamps to obtain free ones. These changes increased participation significantly and over 20 million people were accessing food stamps by 1980 (Fig. 7.3). Fiscal cutbacks in the 1980s, in response to the national recession, caused reduction in funding and participation remained fairly flat over the decade. There was a small surge in participation through the 1990s because Congress restored funding to address hunger issues in the United States. Then in 1996, President Bill Clinton promoted the Personal Responsibility and Work Opportunities Reconciliation Act, also known as "Welfare to Work." This program required participants without dependents to work at least 20 hours/week and implemented other requirements that caused food stamp participation to decline steadily. The economic downturn of 2000 created new demand for the program and generated resurgence in participation. The 2008 Farm Bill changed the program name to SNAP and increased funding, and the 2009 American Recovery and Reinvestment Act provided an increase in the allotment per family. As shown in Fig. 7.3, the number of participants climbed from 20 million in 2000 to 45 million in 2016. The 2014 Farm Bill approved by President Barack Obama cut SNAP funding by $800 million per year for 10 years, which will likely reduce the number of SNAP recipients.

Current eligibility requirements for SNAP benefits are less than $2250 in resources and an income of less than $26,124 a year for a family of three. Able-bodied adults between 16 and 60 years of age, without dependent children, can access benefits for only 3 months if they are not working. SNAP benefits would be about $511 per month for a family of 3 or about $1.90 per person per meal. It is very difficult to meet nutrient requirements on this amount of money and SNAP participants frequently access community food pantries and other services to secure sufficient food.

The Food Stamp/SNAP program has been called "the cornerstone of the nation's nutrition safety net." Criticism of the SNAP program includes concerns over the government subsidizing food assistance and potential misuse of the funds by participants. Some argue that participating in the program discourages individual work ethic and motivation. As with any large government program, abuses occur, but the great majority of people receiving SNAP benefits use them appropriately. Evidence from the program has found that the majority of SNAP benefits are distributed to families with children and about 4 in 10 SNAP recipients live in a household where at least one person works. Improvement in school performance and health of children are documented benefits of the program. The SNAP program has an economic impact by increasing food sales and creating jobs. Misuse of SNAP benefits is monitored and the introduction of the electronic benefit transfer (EBT) card makes abuse more difficult and allows careful monitoring. SNAP benefits can be used to purchase foods for the household to eat, such as breads and cereals, fruits and vegetables, meats, fish and poultry, dairy products, and seeds and plants that produce food. Soft drinks, candy, cookies, snack crackers, bakery cakes, and ice cream are food items and eligible purchases. Allowing SNAP benefits to be used to purchase such items is controversial considering the need to prevent obesity and focus on healthier food choices. In some areas, restaurants can be authorized to accept SNAP benefits from qualified homeless, elderly, or disabled people in exchange for low-cost meals. SNAP benefits cannot be used to buy beer, wine, liquor, cigarettes or tobacco, or any nonfood items, such as pet foods, soaps, paper products, household supplies, vitamins and medicines, food that will be eaten in the store, or hot foods.

7.2.4 National School Lunch Program

Massachusetts was the first state to enact a compulsory education law in 1852 requiring all children to attend school. Other states soon followed, with the intent of ensuring all citizens had access to a basic education and to discourage children from being exploited to work in factories. Providing meals to school children was the responsibility of parents, and a disparity between rich and poor was apparent. Recognition that hungry children are not able to achieve their learning potential was brought to the public's attention in the book *Poverty* (1904) by Robert Hunter, based on his observation of hungry school children in New York City. In response, many schools attempted to meet the needs of children by partnering with private groups and organizations that provided meals to students. It wasn't until the Great Depression when child nutrition became a national crisis that the federal government began providing assistance for school meals. Surplus commodities purchased by the USDA from farmers were given to schools starting around 1935. In addition, the Works Progress Administration provided funding for school lunch workers, giving jobs to needy women to help support their families. These programs were of great benefit and likely prevented malnutrition and starvation of millions of children.

The advent of the United States' entry into WWII changed the economic balance of the country and surplus food and workers were no longer readily available. As a result, schools were not able to provide low-cost meals to students. To address this problem, Congress in 1943 extended funding to continue school lunch and milk programs and provided cash subsidies to schools, which became a year-by-year appropriation. By 1946, it was apparent that a more consistent funding model was needed and the National School Lunch Act was passed (Table 7.3). This legislation defined how federal funds would be distributed to states, based on the number of school-aged children and the per capita income of the state. States with larger populations of low-income families received more funding. Schools received a federal allotment based on the number of meals provided to children who were not able to pay for their own meals. Specific criteria for the quality of the meals were defined within the act, as were reporting and monitoring expectations.

The school lunch program expanded over the next 20 years as did the federal investment. The growth of the program created challenges for school and brought public scrutiny. Schools objected to having to cover the costs and manage the overhead for the program, and some people disagreed with this type of federal welfare program. In 1968, a national debate about hunger in America was reopened when the Citizen's Board of Inquiry into Hunger and Malnutrition in the United States released its report *Hunger USA* (1968), and CBS aired a documentary, *Hunger in America* (1968), that highlighted the extent of poverty among families and children. It became apparent that many children who were eligible for free or reduced-cost school lunches were not receiving them and schools were not adequately funded to provide meals. Economic and racial discrimination was a major problem during this period in American history and public schools were at the forefront of the controversy. These issues were debated and discussed during the White House Conference on Food, Nutrition, and Health, and led President Richard Nixon to call for a new agency to be established within the USDA, the Food and Nutrition Service, which would be responsible for overseeing all federal food programs, including school lunch. Additionally, Congress enacted amendments to the National School Lunch and Child Nutrition Acts to ensure access to breakfast and lunch for needy children and to maintain confidentiality for recipients of the program.

The NSLP has been the target of criticism and controversy throughout its history. Over time, efforts to make meals more acceptable to students and to reduce costs allowed items such as French fries, hot dogs, pizza, and chicken nuggets to be served. In the 1980s, schools were allowed to sell foods of minimal nutritional value (including from vending machines that dispersed candy, cookies, and soda) that competed with school lunch menus.

Funding for nutrition education and training was cut during difficult financial times. In some communities, fast food restaurants opened near schools and students opted to buy lunch off-campus rather than eat the school lunch offerings. At the same time, public health experts were raising the alarm about increasing rates of childhood obesity. The release of *Healthy People 2000* (1990) by the Department of Health and Human Services (DHHS) called on school meal programs to meet the Dietary Guidelines for Americans (DGA) by 2000. The USDA responded by implementing new regulations for school meals, including reducing the amount of fat and adding more fruits, vegetables, and whole grains. Further changes came with the Healthy Hunger-Free Kids Act of 2010, which was approved by President Barack Obama. This sweeping legislation required schools to develop wellness policies, specifically defined nutrition standards for foods sold in schools (including vending machines), reinstated training and education for school food service directors, increased breakfast and after-school food service, and created the Farm-to-School program. New standards for meals ensured students were offered both fruits and vegetables every day of the week, substantially increased offerings of whole grain–rich foods (Fig. 7.4), allowed only fat-free or low-fat milk varieties, limited calories based on the age of children being served to ensure proper portion size, and increased the focus on reducing the amounts of saturated fat, *trans* fats, and sodium.

Today, the NSLP is operating in over 100,000 US schools and 31 million children receive low-cost or free meals daily. Children are eligible for free meals if their family has an income at or below 130% of the poverty line and children are eligible for reduced-price meals (students charged no more than 40¢) if their family has an income between 130% and 185% of the poverty level. A family of 4 with an income of $44,863 or less is eligible for reduced-price meals and families with incomes less than $31,525 are eligible for free meals. Children whose families have an income above 185% of the poverty level pay the full price, which is set by the local school food authorities, but the meal service operation must run as a nonprofit program. The USDA provides schools with cash reimbursement for the free and reduced meals they serve. Free lunch rebates to schools equal $2.86 with a bonus of

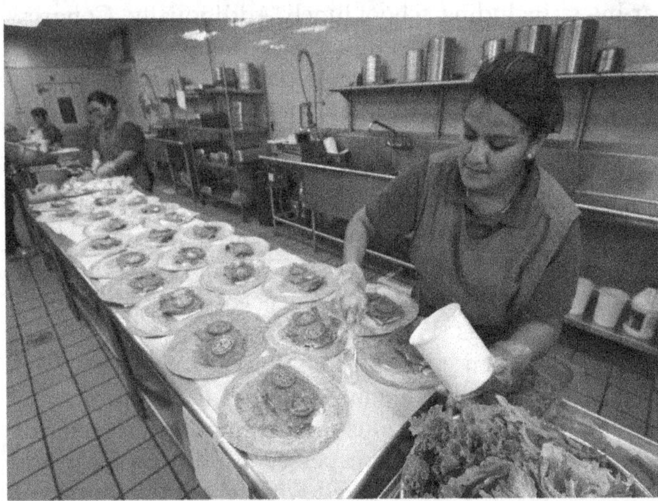

FIGURE 7.4 The current standards for the National School Lunch Program (NSLP) require meals to provide whole grains, vegetables, and lean meats. Almost 73% of school children in the United States qualify for free or reduced price meals. Schools are challenged to provide healthy meals that meet these standards within the cost structure of the NSLP. *Source: Photo from the USDA image gallery.*

6¢ per meal if the meals are certified to be in compliance with nutritional requirements. This level of reimbursement provides a challenge for schools to provide high-quality, nutritious meals.

7.2.5 Women, Infants, and Children

Widespread poverty in the United States during the 1960s came to the attention of pediatricians working with inner city populations, notably Dr. David Paige at Johns Hopkins School of Public Health. Nutrition research at the time was demonstrating that diets lacking micronutrients while *in utero* and during early life resulted in poor growth and development, increased illness, and reduced the ability to learn. Public health workers were aware that infant malnutrition among the poor was becoming a serious issue. To address this problem, Paige and the Maryland Board of Health developed a program to provide vouchers for iron-fortified infant formulas to low-income women. The pilot program was such a success in improving the health of the children that other programs including one at St. Jude's Children's Hospital in Memphis, Tennessee undertook similar programs and documented their impact. Based on the scientific evidence that providing adequate foods to mothers and their children resulted in better pregnancy outcomes and healthier babies, leading to lower healthcare costs overall, Senator Hubert Humphrey became an advocate for a legislative program to expand food supplementation for women and young children. The program was signed into law in 1972, to be implemented by the USDA (Table 7.3). It took several years for the USDA to develop a mechanism to deliver the program but the first Women, Infants, and Children (WIC) clinic opened in Pineville, Kentucky in 1974.

WIC programs are administered locally within each state using federal grants to provide supplemental foods, healthcare referrals, and nutrition education for low-income pregnant women, mothers, infants, and children who are at nutritional risk. Unlike the SNAP program, only selected foods are included in the WIC food packages. Children from 1 to 4 years of age, and pregnant, postpartum, and breastfeeding women may receive milk, yogurt, cheese, juice, cereal, eggs, peanut butter and/or dried beans, fish, whole wheat bread, and fruits and vegetables. Infants receive iron-fortified formula, baby foods, and plain cereal. WIC participants receive education and support that encourages breastfeeding and balanced nutrition for the first years of life. The impact of WIC has been significant in reducing malnutrition of mothers and children and increasing access to healthcare. In 2012, 51% of infants and 30% of pregnant women in the United States participated in the WIC program. California and Texas account for over 25% of all WIC participants nationwide. Southern states, including Mississippi, Arkansas, Kentucky, Louisiana, South Carolina, and Alabama have high enrollment with over 60% of all infants in the state receiving WIC benefits. On average, women receive about $35–$50 and infants $37–$50 per month in benefits from WIC, which can only be used to purchase the approved food items.

7.2.6 Nutrition Education for the Public

Conveying nutrition information to the public in a simple, clear manner is challenging. During the 1940s, the USDA created a visual tool, *A Guide to Good Eating*, which classified foods into the Basic Seven food groups as a means of describing the type of diet that would ensure nutrient adequacy (Fig. 7.5). The guide was presented as a circle with pictures and simple messages with the goal of helping make nutrition information accessible. Over the years, the guides were modified as

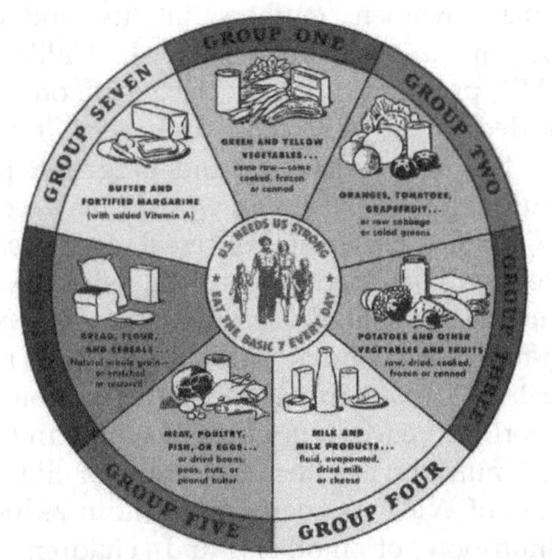

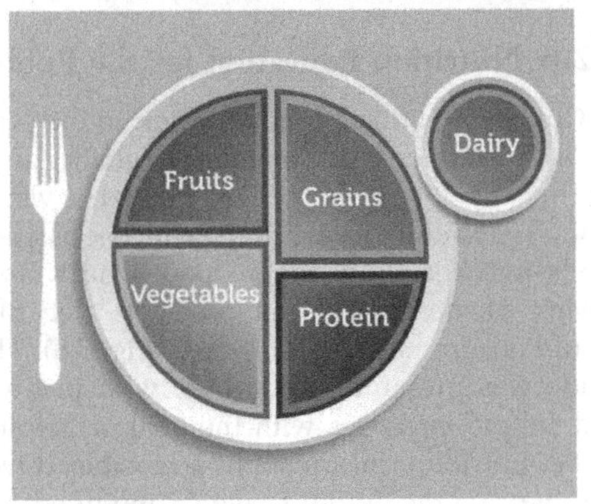

nutrition information changed. The number of food groups was reduced to four in the 1950s to include milk, meat, bread and cereal, and vegetable and fruit. A dramatic shift occurred with the 1992 Food Guide Pyramid, which transitioned from the circular image to a pyramid. The pyramid shape was intended to visually illustrate the amounts of each food group that should be consumed from the wide base (cereals and grains) to the narrow top (fats, oils, and sweets). In 2005, the MyPyramid Food Guidance System was introduced and, for the first time, included the message that exercise was a core component of a healthy lifestyle by adding steps with the image of a person (Fig. 7.5). The MyPyramid system included a website that provided more detailed and individualized information. In 2011, after several years of consumer research, the USDA introduced MyPlate to replace MyPyramid (Fig. 7.5). The MyPlate image was thought to be closer to how consumers view food, i.e., sections of a plate divided into the four food groups with dairy on the side. MyPlate has fostered the concept of "make half your plate fruits and vegetables" among consumers. But the release of MyPlate was not without controversy. The food groups included are fruits, vegetables, grains, dairy, and protein. Labeling a group by a nutrient, protein, rather than by foods, had not been done before and raised concerns. The meat and poultry industries feared that people might not understand the options for that category and would assume meat and poultry

FIGURE 7.5 The USDA has provided images to convey ways to meet nutritional needs through food choices since the 1940s. The Good Eating Guide from the 1940s showed seven food groups and encouraged consuming foods from all groups. The Food Pyramid from the 1990s showed five food groups with relative intake defined by the width of the triangle, and also encouraged physical activity. MyPlate from the 2000s includes five food groups using a plate to show recommended relative proportions. *Source: USDA Economic Research Service, www.usda.ers.gov.*

should not be consumed. The protein category of MyPlate includes meat, poultry, seafood, beans and peas, eggs, processed soy products, and nuts and seeds. The Choose MyPlate website has many additional tools for consumers to personalize their nutrition and fitness routines. For example, the SuperTracker system can be used by individuals or groups to monitor diet and physical activity, set goals, and create challenges for diet and exercise improvements.

7.2.7 Dietary Guidelines

In 1968, the US Senate Select Committee on Nutrition and Human Needs, led by Senator George McGovern, was created to study the problems of hunger and malnutrition in the United States. Over the next several years, the scope of the committee expanded as hearings were held with health and nutrition experts, schools, and medical professionals. It was becoming clear during this time that chronic diseases such as heart disease, diabetes, and obesity were closely linked to dietary habits. In 1977, the committee released a report, *Dietary Goals for the United States*, which summarized this information and promoted reduced intakes of fat, cholesterol, and sugar and increased consumption of fruit, vegetables, and whole grains. This report, although highly controversial at the time, began a public dialogue about nutrition, food, and health interrelationships. The USDA and the DHHS were called upon to verify and document the scientific evidence in the report and, in 1980, released *Nutrition and Your Health: Dietary Guidelines for Americans*. This publication had seven recommendations:

1. Eat a variety of foods
2. Maintain ideal weight
3. Avoid too much fat, saturated fat, and cholesterol
4. Eat foods with adequate starch and fiber
5. Avoid too much sugar
6. Avoid too much sodium
7. If you drink alcohol, do so in moderation

The Secretaries of the USDA and DHHS have jointly issued revisions of the *Dietary Guidelines for Americans* every 5 years since 1980 (Table 7.4). The process for development of the DGA is for the USDA and DHHS to convene an advisory committee comprised of leading nutrition and health professionals and researchers. The advisory committee reviews the scientific literature, hears testimony from other experts and convenes public meetings, then compiles a report. The report is opened to the public and other federal agencies for comment. The Secretaries of the DHHS and USDA review the report and the comments and issue the new version of the DGA. Because each advisory committee interprets information differently, the organization, structure, and details of the Guidelines have changed over the years. The most recent DGA advisory committee (2015) took a very different approach than any of the previous committees. They focused on eating patterns as a means of helping consumers choose foods, rather than addressing nutrients as the main target of the guidelines. The five main themes of the 2015 DGA are:

1. Follow a healthy eating pattern across the lifespan
2. Focus on variety, nutrient density, and amount
3. Limit calories from added sugars and saturated fats and reduce sodium intake
4. Shift to healthier food and beverage choices
5. Support healthy eating patterns for all

The core messages of the DGA have remained fairly consistent since the 1980 version. A healthy eating pattern includes a variety of vegetables, fruits, whole grains, dairy, and a variety of protein foods. Limiting consumption

TABLE 7.4 Historical Comparison of the United States Dietary Guidelines

1980 and 1985	1990	1995	2000
Eat a variety of foods	Same	Same	
Maintain ideal weight	Same	Balance the food you eat with physical activity—maintain or improve your weight	Aim for fitness—aim for a healthy weight—be physically active each day
Avoid too much fat, saturated fat, and cholesterol	Same	Choose a diet low in fat, saturated fat, and cholesterol	Choose a diet that is low in saturated fat and cholesterol and moderate in total fat
Eat foods with adequate starch and fiber	Choose a diet with plenty of vegetables, fruits, and grain products	Choose a diet with plenty of grain products, vegetables, and fruit	Choose a variety of grains daily, especially whole grains
			Choose a variety of fruits and vegetables daily
Avoid too much sugar	Use sugars only in moderation	Choose a diet moderate in sugars	Choose beverages that limit your intake of sugars
Avoid too much sodium	Use salt and sodium only in moderation	Choose a diet moderate in salt and sodium	Choose and prepare foods with less salt
If you drink alcohol, do so in moderation	Same	Same	Same
			Keep food safe to eat

The Format of the Dietary Guidelines Changed in 2005 to Focus on Eating Patterns

2005	2010	2015
	Building a healthy eating pattern	
Meet recommended intakes within energy needs by adopting a balanced eating pattern, such as the USDA Food Guide or the DASH Eating Plan	Balancing calories to manage weight	
Consume <10% of calories from saturated fats, <300 mg cholesterol, and keep *trans* fats as low as possible	Foods and food components to reduce:	A healthy eating pattern limits:
	Sodium <2300 mg	Consume <10% of calories from added sugars
	Less than 10% of calories from saturated fats	Consume <10% of calories from saturated fats
Total fat to 20%–35% of calories	Less than 300 mg cholesterol	Consume <2300 mg of sodium per day
	Trans fats as low as possible	

(Continued)

TABLE 7.4 (Continued)

The Format of the Dietary Guidelines Changed in 2005 to Focus on Eating Patterns

2005	2010	2015
Reduce sugar and starch-containing foods and beverages	Reduce calories from solid fats and added sugars	
	Limit refined grains	
2 cups of fruit and 2.5 cups of vegetables per day	Foods and nutrients to increase:	Consume a healthy eating pattern:
	Vegetables and fruits	A variety of vegetables
Variety of fruits and vegetables	Half of all grains whole grains	Fruits, especially whole fruits
	Increase fat-free or low fat milk and dairy	Grains, at least half of which are whole grains
Consume 3 or more ounces of whole grains per day	Variety of protein foods	Fat-free or low-fat dairy
	Increase seafood	A variety of protein foods
Consume three cups per day of fat-free or low-fat milk	Oils to replace solid fats	Oils
	Foods with potassium, fiber, calcium, and vitamin D	
Alcohol in moderation	Alcohol in moderation	Alcohol in moderation
Avoid microbial foodborne illness	Follow food safety recommendations to reduce risk of foodborne illness	

From USDA, www.usda.gov.

of saturated fats, *trans* fats, added sugars and sodium and moderate alcohol intake continue to be recommended. An outcome of the 2015 DGA has been the recommendation to list "added sugars" on the Nutrition Facts panel of food labels. The FDA announced this change in 2016; it will become required in 2018.

7.2.8 Nutrition Labeling

Food marketers have always been interested in using nutrition and health claims to sell their products. The FDA is responsible for assessing the efficacy and safety of drugs to treat and manage disease on one hand and overseeing food labeling and safety on the other. The distinction between drugs and foods was clear in the legislation but less clear in practice. During the 1930s as vitamins and minerals were being discovered and linked to disease, foods containing these nutrients were advertised for their preventative and curative effects. And when scientists began linking diets high in fat and cholesterol with heart disease in the 1950s, new challenges to connect disease with foods arose. Throughout its history, the FDA strongly discouraged advertising claims about the health effects of foods and threatened to reclassify foods as drugs if such claims were made.

By the mid-1970s, there was convincing evidence from nutrition research that diets rich in fiber were associated with lower risks of colon cancer. The National Cancer Institute (NCI) began reviewing diet and cancer relationships during that time and released recommendations in 1979 that included, among other things, a generous intake of dietary fiber. In this scenario, consumers were receiving information about foods and health from these reliable sources, but were unable to connect that information directly with food products. The Kellogg Corporation saw the potential benefit of linking fiber to cancer prevention for their "All-Bran" cereal and launched a marketing campaign in 1984 promoting the cereal as a way to reduce the risk of cancer. The NCI actually cooperated with Kellogg's on the campaign by reviewing the ads for accuracy and providing a toll-free telephone number for consumers to contact the NCI for information about diet and cancer. The FDA had not approved the campaign, however, and voiced its objection. With encouragement from the NCI, the FDA eventually allowed the campaign to continue and announced a review of its policy about health claims on foods. This opened the door for additional foods and claims and much public and legal debate about the process. Finding the proper balance between consumer protection and access to information was a challenge and the FDA was rapidly overwhelmed with trying to regulate and manage this new environment.

In 1990, in response to the challenges faced by the FDA and increasing pressure to provide consumers with information about food composition and its role in health, Congress passed an amendment to the Federal Food, Drug, and Cosmetic Act. The Nutrition Labeling and Education Act (NLEA) required that nutrition information be included on the food package and defined the regulatory process for health claims. By 1994, the Nutrition Facts panel was developed and standardized by the FDA and became required for all foods sold in the United States. The purpose of the Nutrition Facts panel was to provide consumers with the serving size; the amount of calories, fat, and fiber in the serving; and a way to determine how well the product met nutritional needs (Fig. 7.6). The latter was not a simple task and the FDA struggled with finding a clear and understandable way to post information on a package about nutrient content. The decision was made to use the new concept of % Daily Value (DV). As described previously, the nutrient requirement (RDA) for each of the vitamins and minerals is defined by age and gender. It would not be possible to list all of these possible recommended levels on a food label. As a compromise, a DV was selected for each essential nutrient to be used on the Nutrition Facts panel. A DV was defined for components that did not have an RDA value but were required to be included on the Nutrition Facts panel including total fat, saturated fat, cholesterol, sodium, total carbohydrate, and dietary fiber. The DV is designed to ensure adequate nutrient intake when consuming a diet of 2000 kcal/day. A food item with a %DV of less than 5 is defined as a low source and a %DV of 20 as a high source for that particular nutrient.

Finding a more effective way to convey nutrition information on the food package continues to be a challenge. For the past few years, consumers, scientists, and government experts have been evaluating and debating the efficacy of the Nutrition Facts panel. An outcome of these discussions was that in 2016 the FDA announced changes to the Nutrition Facts panel for implementation by 2018 (Fig. 7.6). Eight changes have been implemented for the new Nutrition Facts panel including:

1. Larger type for serving size
2. Larger type for calories
3. Updated serving sizes to match consumer expectations for the packaged food
4. Updated Daily Values
5. Change in the nutrients required to be on the panel (vitamin D, calcium, iron, and potassium)

Original Label

Nutrition Facts
Serving Size 2/3 cup (55g)
Servings Per Container About 8

Amount Per Serving	
Calories 230	Calories from Fat 72
	% Daily Value*
Total Fat 8g	**12%**
Saturated Fat 1g	**5%**
Trans Fat 0g	
Cholesterol 0mg	**0%**
Sodium 160mg	**7%**
Total Carbohydrate 37g	**12%**
Dietary Fiber 4g	**16%**
Sugars 1g	
Protein 3g	
Vitamin A	10%
Vitamin C	8%
Calcium	20%
Iron	45%

* Percent Daily Values are based on a 2,000 calorie diet. Your daily value may be higher or lower depending on your calorie needs.

	Calories:	2,000	2,500
Total Fat	Less than	65g	80g
Sat Fat	Less than	20g	25g
Cholesterol	Less than	300mg	300mg
Sodium	Less than	2,400mg	2,400mg
Total Carbohydrate		300g	375g
Dietary Fiber		25g	30g

New Label

Nutrition Facts
8 servings per container
Serving size 2/3 cup (55g)

Amount per serving	
Calories	**230**
	% Daily Value*
Total Fat 8g	**10%**
Saturated Fat 1g	**5%**
Trans Fat 0g	
Cholesterol 0mg	**0%**
Sodium 160mg	**7%**
Total Carbohydrate 37g	**13%**
Dietary Fiber 4g	**14%**
Total Sugars 12g	
Includes 10g Added Sugars	**20%**
Protein 3g	
Vitamin D 2mcg	10%
Calcium 260mg	20%
Iron 8mg	45%
Potassium 235mg	6%

* The % Daily Value (DV) tells you how much a nutrient in a serving of food contributes to a daily diet. 2,000 calories a day is used for general nutrition advice.

FIGURE 7.6 The Nutrition Facts panel has been required on food packages since 1994. The composition and required information on the panel has remained the same since then and is shown on the left panel. Changes to the panel, as shown on the right, have been approved to be implemented by 2018. The main changes are to increase the type of serving size and calories, include added sugars, and to change the required nutrients from vitamins A and C, calcium, and iron to vitamin D, calcium, iron, and potassium to better reflect the nutrients that consumers are not consuming adequately. *Source: USDA Economic Research Service, www.usda.ers.gov.*

6. Amounts of nutrients listed with the %DV
7. Added sugars listed
8. Footnote to explain %DV

The Nutrition Facts panel provides information to consumers about the nutritional value of the food in a variety of ways. The caloric value of the food for a single serving, the serving size, and servings per container are listed. The serving size was defined by the manufacturer but will be standardized for food categories by the FDA when the new panel goes into effect. This change was made because manufacturers were tempted to reduce the serving size of their product to keep the calorie value low. Consumers' perception of a serving size may not be in agreement with the label. A prime example is breakfast cereals, which typically list a serving size as ³/₄ cup. When consumers pour a bowl of cereal the amount is closer to 1.5–2 cups of cereal. Making the consumer expectation for a serving more in agreement with the Nutrition Facts panel is a goal of the new regulations. The middle section of the Nutrition Facts panel includes information about nutrients of concern, fats, cholesterol, sodium, sugars, and proteins listed in gram quantity and %DV. Listing added sugars will be required in the new regulation. The bottom panel lists the amount and %DV of selected micronutrients. The micronutrients required by the FDA to be listed will change with the 2018 regulations to include vitamin D, calcium, iron, and potassium. These nutrients are considered to be the ones most commonly consumed in inadequate amounts by the population. Manufacturers may list other micronutrients on the panel if they choose. The panel includes the %DV measure to allow consumers a way to quickly estimate the nutritive value of the food, but it is important to remember that the value is based on a 2000 calorie diet.

NLEA gave FDA clear authority to define health claims on food labels. A health claim is defined by the FDA as a statement that "describe(s) a relationship between a food substance (a food, food component, or dietary ingredient), and reduced risk of a disease or health-related condition." NLEA required the food industry to demonstrate scientific evidence for health claims and for the FDA to approve them before they could be used on a label or advertising campaign. The basic premises of the law are that claim statements must follow FDA guidelines and may not claim to directly cure or treat disease. The FDA has approved a wide range of health claims and numerous "qualified" health claims that have emerging but not definitive scientific evidence. An example of a health claim that is approved by the FDA would be "a diet low in total fat may reduce the risk of some cancers." The FDA also approves structure/function claims that describe the role of a nutrient or dietary ingredient that affects the normal structure or function of the body. An example of an approved structure/function claim would be "calcium builds strong bones." The FDA accepts petitions for health claims and posts the petition on its website to invite public comment. The scientific evidence for the claim is then assessed by the FDA and they determine if it will be approved or rejected.

Four years after NLEA was signed into law, the Dietary Supplement, Health, and Education Act (DSHEA) was approved. The dietary supplement industry was growing rapidly at this time and had substantial lobbying influence. The 1994 DSHEA amendment to the Food, Drug, and Cosmetic Act created a separate system for the management of dietary supplements from the established processes for either food or drugs. A *dietary supplement* is defined as a vitamin, mineral, herb or botanical, amino acid, or any supplement to the diet that is not a conventional food or sole item of a meal. The law allows manufacturers to market any products that were already on the market before 1994 without FDA approval. New products entering the market after 1994 must receive FDA approval. The law was quite liberal in allowing products to be marketing without the rigorous testing that would be required for drugs. Under the law, the FDA may restrict the sale of products only if they can show they are unsafe or ineffective. DSHEA requires the packages of all dietary supplements to include a Supplement Facts panel similar to the Nutrition Facts panel, a list of ingredients, and information about the source and manufacturer. Unlike drugs, dietary supplements are not subject to purity or quality standards. If the label makes any claim about the health benefits of the product, it must include a disclaimer that "this statement has not been evaluated by the Food and Drug Administration" but the FDA does not have to approve the claim.

7.3 ASSESSING AMERICANS' NUTRITIONAL STATUS

In response *Hunger USA* (1968), the US Congress commissioned the Ten-State Nutrition Survey to measure the incidence and locations of hunger and malnutrition. A total of 86,352 people living in Texas, Louisiana, New York, Kentucky, West Virginia, Michigan, California, Massachusetts, Washington State, and South Carolina were given dental and nutritional evaluations. A smaller group, 11,337 people, were asked to provide a 24-hour recall of their food intake as well. The amount of data collected was significant and so overwhelmed the capacity and available funding that it took years for reports to be compiled. One report that was issued in 1975 found that while malnutrition and poverty were of concern, the rate was not as high as had been depicted in *Hunger, USA*. The survey found that socioeconomic status impacted overall growth and development, noting that higher body fatness correlated with greater height and bone structure of children. The report documented, perhaps for the first time, that obesity and overweight were present among children, occurred more frequently in low income populations, and tended to occur in families. The nutritional assessment of the population did not find any specific nutrient deficiencies in the diet. The cause of malnutrition was determined to be insufficient access to enough food to maintain health, primarily due to poverty.

7.3.1 National Health and Nutrition Examination Survey

In 1956, the National Health Survey Act provided for the collection of clinical data from the US population. Between 1960 and 1970, the National Center for Health Statistics and the US Public Health Service carried out three national health examination surveys (NHES). NHES I measured chronic diseases of adults, NHES II assessed growth and development of children aged 6–11, and NHES III focused on children aged 12–17. In 1970, the important role of diet and nutrition to disease risk was becoming evident, so the National Nutrition Surveillance System was proposed and added to the NHES. The new program was named the National Health and Nutrition Examination Surveys (NHANES) and three main surveys were conducted as shown in Table 7.5.

Since 1999, NHANES has been conducted on a continual basis using mobile medical trailers that are located, for a period of time, in selected communities. People come to the trailers where the assessments are performed. The process has been compared to space missions because of the complexity and logistical challenges of bringing the self-contained assessment trailers to sometimes remote locations. Approximately 7000 randomly selected people in the United States are surveyed annually.

The NHANES program has contributed significantly to monitoring the health of the nation and understanding the role of food, nutrition, and lifestyle on the incidence of health and disease. NHANES data have been used to develop the DRIs and growth charts of children. A major impact of NHANES was the

TABLE 7.5 National Health and Nutrition Examination Surveys

Survey	Years	Population
NHANES I	1971–75	28,000 people; ages 1–74
NHANES II	1976–80	28,000 people; ages 1–74
NHANES III	1988–94	40,0000 people; ages 2 months and above
HHANES	1982–84	16,000 people; Mexican, Cuban, and Puerto Rican descent

From USDA, www.usda.gov; FDA, www.fda.gov.

FIGURE 7.7 An assessment of the nutritional status of the population conducted by the USDA in 1941 found that people living on farms had better diets than those living in villages or cities. This survey was one of the first to assess the diets of populations as a means of determining the health of Americans. *Source: Illustration created by Reannon Overbey, from Are we well fed? USDA Bureau of Home Economics Publication Number 430, 1941.*

observation of high levels of lead in children, which provided evidence to support the ban on lead-based gasoline and paints. In 1998 NHANES data showed a high incidence of low folate levels in women of child-bearing age. Previous research had correlated low folate levels in pregnant women with increased risk of neural tube defects in their babies. Based on the NHANES findings, the FDA decided to add folate to the nutrients added to enriched bread, cornmeal, rice and other grains to improve the folate intake of the population. This has led to reduced incidence of neural tube defects in the US.

7.3.2 Food Consumption Surveys

As the overseer of agriculture and farm production, the USDA assesses food production and markets. The Economic Research Service (ERS) provides free access to this wide range of data on its website (www.ers.usda.gov). The USDA has conducted surveys and collected data on food consumption since the 1930s. One of these early surveys reported that farm families in Pennsylvania and Ohio in 1940, comprised of a husband, wife, and two children, had on average an annual income of $1000–$1249. On average, families had $435 worth of food with $298 of that food generated on their farm and $155 worth of food purchased. A comparison of farm to city families in 1941 found that more people living in cities had fair or poor diets compared to those living on farms (Fig. 7.7).

The USDA data collected between 1942 and 1948 found that the consumption of iron, thiamine, riboflavin, and niacin had improved due to the enrichment of flour with these nutrients, but also because of increased consumption of milk and dairy products. Income was then and continues to be a significant determinant of nutritional adequacy. In 1955, 62% of families with an annual income of $6000 had adequate dietary intake compared to only 34% of families with incomes less than $2000, with families in the southern United States at higher risk than those in northern states.

The USDA began conducted the Continuing Survey of Food Intakes of Individuals (CSFII) in 1985 to determine the food and nutrient content of diets consumed by Americans. The first surveys consisted of personal interviews of women to gather 1-day food and nutrient intakes for themselves and their children. In 2001, the CSFII was combined with dietary surveys being collected in NHANES. The new platform was renamed the What We Eat in America (WWEA) survey. WWEA is conducted annually, collecting dietary intake information from almost 10,000 people. The value of these data are significant for understanding changes in food intake patterns, identifying nutrient deficiencies and excesses, and finding correlations between dietary habits and health outcomes.

7.4 FOOD INSECURITY

Widespread hunger and limited access to food first occurred in US history during the Great Depression. Prior to that time, people certainly experienced hunger linked to poverty or social status and during crop failures. The culminating events of the 1930s, including economic collapse and environmental disaster, were extreme and put millions of lives at risk. As described in Chapter 2, History of US Agriculture and Food Production, the New Deal policies implemented by President Franklin Roosevelt were effective in getting food from farmers to those in need. Hunger issues were not because of food shortages during that time, but rather distribution. In the spring of 1942, soon after the United States entered WWII, that situation changed and rationing of food, as well as many other materials, became necessary due to increased demand for the war efforts and reduced imports. The Food Rationing Program issued ration books to families based on the number of family members, which specified the amount of food that could be purchased. Red Stamps covered meats, butter, fat, oils, and some cheeses. Blue Stamps rationed canned, bottled, and frozen fruits and vegetables; juices; dry beans; canned soups; and baby food. Sugar Buying Cards were also issued. Public service messages were used to encourage people to conserve and limit their consumption of certain products to support the troops and war effort (Fig. 7.8). People were encouraged to plant "victory gardens" wherever land was available to supplement their food needs. The Agricultural Extension Service

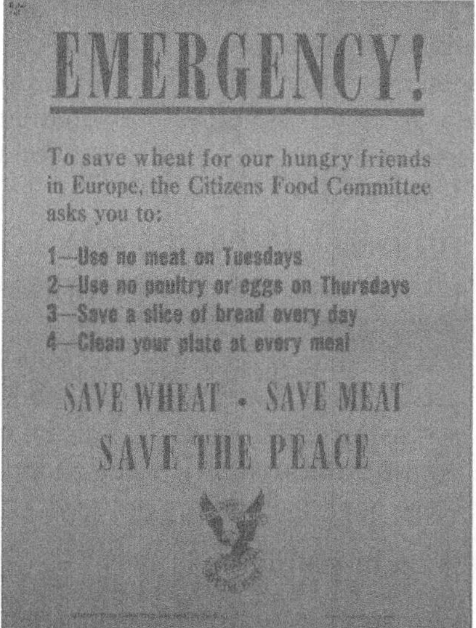

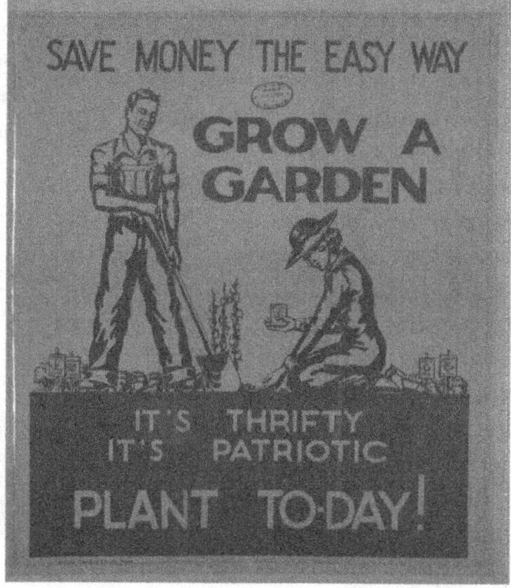

FIGURE 7.8 During WWII, rationing of certain foods was needed because of shortages in production and disruption of imports. Posters were used to encourage the public to plant gardens and conserve food so that adequate food could be provided to the military and general population. *Source: Images from the National Agricultural Library.*

was activated to teach home canning courses and to provide nutritional guidance to consumers. These efforts were successful and malnutrition was not a major problem during the war years in the United States.

When the war ended and through the 1950s, there was general economic growth and prosperity, the Baby Boom generation began, and funding for food assistance programs was largely discontinued. But hunger and poverty had not been eliminated. The tumultuous years of the 1960s, with racial unrest and increased inner city populations, created a new environment for poverty and food insecurity. The poverty rate in 1959 was 22% and was three times higher among blacks compared to Caucasians. There was also a significant increase in the number of poor people living in families headed by women, which was as high as 50% in poor minority populations. The Civil Rights Movement was active in seeking political action to improve economic and social equality for minorities and, in 1963, organized the March on Washington. Over 200,000 people participated in the rally that included the "I Have a Dream" speech given by Dr. Martin Luther King. As discussed in Chapter 2, History of US Agriculture and Food Production, President John Kennedy restarted the food stamp program and his successor Lyndon Johnson launched his War on Poverty. The demands of the Vietnam War left little time or money for domestic food assistance programs but by the end of his term, Johnson had made important progress by passing the Food Stamp Act of 1964 and expanding the National School Lunch Program (NSLP). President Richard Nixon continued efforts to address hunger and poverty during his term. He called for the White House Conference on Food, Nutrition, and Health, which brought together leaders in nutrition, public health, food, and agriculture; federal, state, and local governments; and consumer groups for the first time to discuss food and nutrition issues. The outcomes of the conference were enhancement of the food stamp, school lunch, and nutrition education programs, and creation of the Supplemental Food Program for Women, Infants, and Children, and Congregate Meals and Home Delivered Meals programs to address hunger among the elderly. These programs, along with a strong economy during the 1970s, resulted in reduction of poverty rates to 11% by 1973.

7.4.1 Food Insecurity in the US

The concept of "hunger" proved hard to quantify from an economic or access perspective. A working group from the American Institute of Nutrition (now the American Society for Nutrition) introduced the terms "food security" and "food insecurity" in 1990. As a means of tracking information about food security, the USDA and the National Center for Health Statistics developed the Food Security Scale (FSS) consisting of a set of 18 questions that were included in the annual Census Bureau Current Population Survey beginning in 1995. These questions provided quantitative measures, collected on an annual basis, of the amount of food insecurity experienced by Americans. The USDA uses the FSS data as well as other information on food access to define food security as follows:

Food Security:

- High food security—Households had no problems, or anxiety about, consistently accessing adequate food.
- Marginal food security—Households had problems at times, or anxiety about, accessing adequate food, but the quality, variety, and quantity of their food intake were not substantially reduced.

Food Insecurity:

- Low food security—Households reduced the quality, variety, and desirability of their diets, but the quantity of food intake and normal eating patterns were not substantially disrupted.

- Very low food security—At times during the year, eating patterns of one or more household members were disrupted and food intake reduced because the household lacked money and other resources for food.

Food insecurity rates in 1995 were about 12% of the total US population and remained fairly steady until 2007 when the Great Recession began. Rates of food insecurity jumped between 2007 and 2009 to 15% and have only declined slightly (Fig. 7.9). The rates of food insecurity in the United States have been proportionally higher among black and Hispanic households compared to other race/ethnicity or white households. A significant increase in rates among all households occurred during the 2007–09 recession. Among Hispanic households, food insecurity reached 27% in 2009 and was still at 22% by 2014. Among black households, the rate of food insecurity rose to 26% in 2009 and stayed above 25% through 2014. The rates for white households rose from about 8% to 11% during that time period.

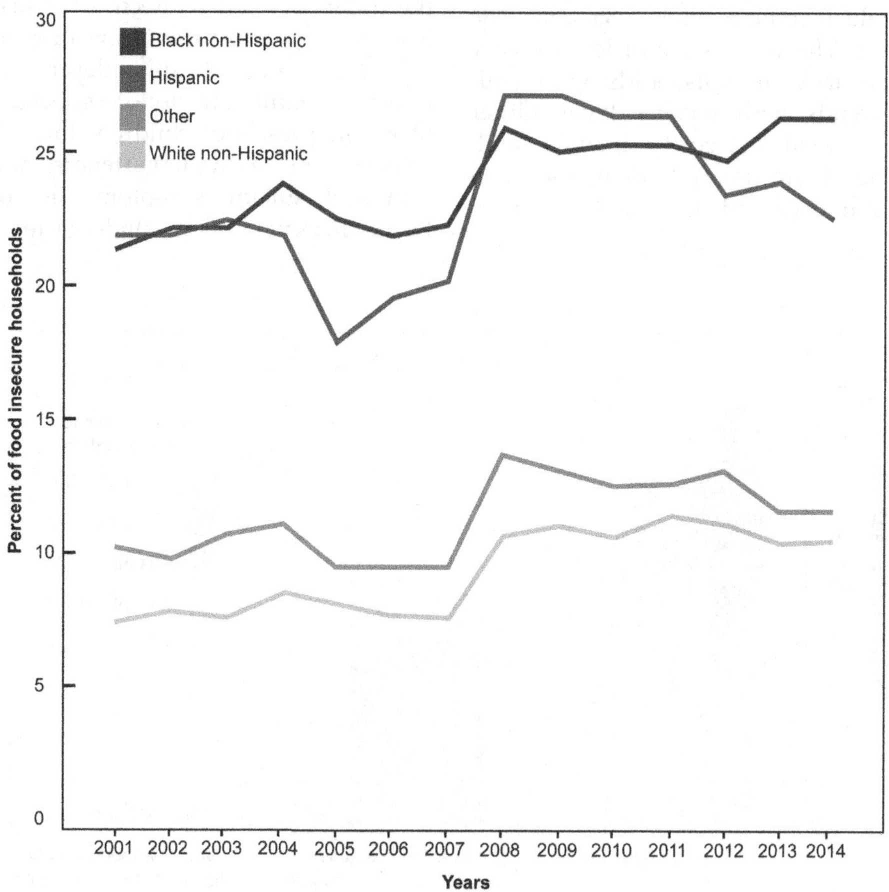

FIGURE 7.9 Food insecurity in the United States population increased during the 2007–09 recession especially among black and Hispanic households. The rate of food insecurity among all populations has not returned to the levels before the recession. *Source: USDA Economic Research Service, www.usda.ers.gov.*

Food insecurity is strongly linked to economic and social conditions. Of all food-insecure households in 2014, 35% were headed by a single woman, 26% were black, 22% were Hispanic, 22% were headed by a single man, and 20% had children under 6 years of age. Very low food insecurity is greatest among households with children headed by a single woman, and black households. Income is a primary driver of food insecurity. Households with incomes at or below the poverty threshold (<1.0) have high rates of food insecurity (Fig. 7.10). Food insecurity continues to be high until the household income reaches 1.85 times the poverty level. The prevalence of low or very low food security in households with children also trends with income level. Other predictors of food insecurity include high housing rents, high unemployment rate, residential instability, and low use of SNAP benefits.

7.4.2 Impact of Food Insecurity on Children

The 2015 USDA report on food insecurity reported that 17% of households with children were food insecure at some time during the year. Families experiencing food insecurity reported that children were hungry, skipped a meal, or did not eat for a whole day. Inadequate nutrient intake during childhood has severe consequences. Diets lacking adequate calories, proteins, essential fatty acids, and micronutrients affect brain development, bone structure and dental health and reduce the ability to resist infectious diseases. Food insecurity has been found to cause poor physical and mental health, depression, health issues in infants and toddlers, behavioral problems in preschool children, low educational achievement in kindergarteners, and depression and suicide symptoms in adolescents. Food insecurity in kindergarten children

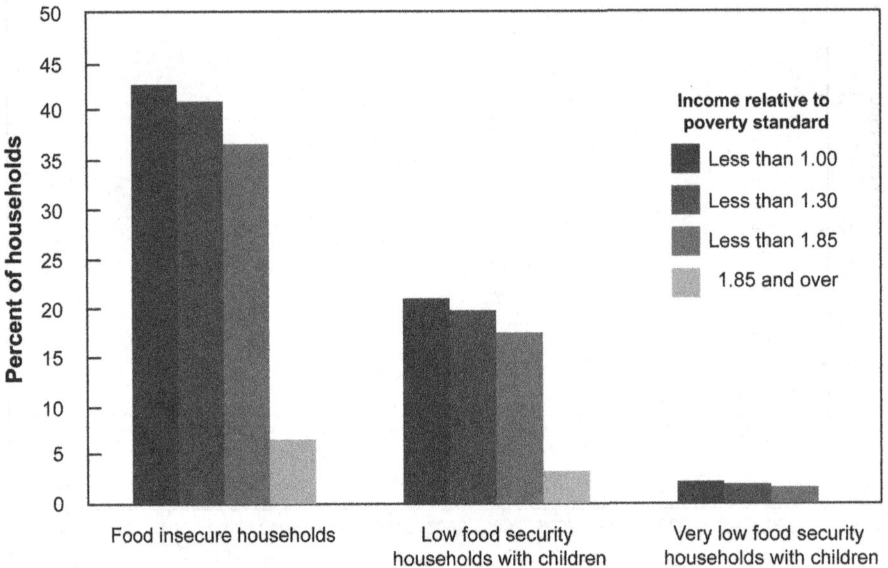

FIGURE 7.10 Food insecurity is strongly associated with poverty. Households with incomes that are at the poverty level (1.00) or at 85% of the poverty level (1.85) have higher prevalence of food insecurity compared to those with incomes above that level. *Source: USDA Economic Research Service, www.usda.ers.gov.*

predicted impaired academic performance in reading and math for boys and girls, greater decline in social skills for boys, and greater weight and body mass index (BMI) gains for girls. Participation in SNAP, which potentially reduced food insecurity, was found to result in higher math and reading scores in children in kindergarten to third grade compared to those that were not accessing SNAP benefits. An association has also been made between body weight and marginal and low food security households due to the reduced quality of food consumed to avoid hunger. Households may rely on high energy, low nutrient-dense foods when access to food and/or money is low. The result may be increased rates of obesity, which would seem to be paradoxical in a food insecure environment.

Adequate nutrition during pregnancy is also important for the health of the mother and the *in-utero* development of the infant. Low birth-weight and preterm deliveries are more common in women with poor diets immediately prior to and during pregnancy. Low iron and folate intakes are of particular concern as deficiency of these nutrients can cause preterm births, fetal growth retardation, or neural tube defects. The ability to adequately breastfeed an infant due to poor maternal nutrition is also a major concern. Infants gain significant benefits from being breastfed for the first several months of life, including better growth rates, ability to resist infection, and fewer chronic health problems.

7.4.3 Food Deserts

Having adequate money and the ability to access wholesome food are both necessary to prevent food insecurity. American development since the 1800s has been characterized by a shift of population away from rural agricultural locations to urban and suburban communities. The introduction of cars and highways allowed people to move miles away from towns where stores were located and the development of supermarkets reduced the need for frequent food shopping. As a consequence, small local grocery stores closed and were replaced with regional shopping areas. The concept of a food desert was coined in 2008 when it became apparent that the built environment in some communities had created areas where people had limited access to a variety of healthy and affordable food. *Built environment* refers to the infrastructure of a community, such as the location and types of roads, sidewalks, types of buildings and stores, and access to green areas or playgrounds. The USDA began quantifying food deserts as an additional measure of food security. A "low access" area is defined as one where 500 people or 33% of the population resides more than 1 mile from a supermarket or large grocery store in urban areas, or greater than 10 miles in rural areas. There are over 6500 food desert tracts as defined by the USDA in the United States. Food deserts are likely to exist in locations with a population of minority ethnic groups; high poverty rates; the rural areas of the West, Midwest, and South; dense urban areas of the West, Midwest, and South; and areas with a high amount of vacant property. The consequences of living in a food desert are a reliance on convenience stores for food purchases, which typically supply packaged foods and beverages and do not include fruits, vegetables, dairy, meats, or eggs. Such foods have been associated with high calorie intakes and low nutritive value, thereby contributing to obesity risks.

While serving as First Lady, Michelle Obama took on the issue of food deserts as part of her Partnership for a Healthier America (PHA). Convenience store companies, after partnering with PHA, began to offer healthier items. For example, KwikTrip convenience stores have added fruit and vegetables options, whole grains, and low-fat dairy products in their stores, and the vending machine supplier Sodexo has added healthier options. Walmart, which is the largest food marketer in the United States (Walmart had 68 supercenters with grocery stores in 1994 and 3288 in 2015), began opening smaller stores in urban and rural food desert areas to provide more options for local residents. Walgreens, which is primarily a pharmacy, expanded its food offerings in urban stores to make more foods available in these areas. The USDA has funded grants to encourage increased farm-to-market programs that would expand food options in food desert areas. Other initiatives have been introduced by states and cities to address the food desert issue including the Pennsylvania Fresh Food Initiative and the Stable Food Ordinance in Minneapolis, which required retailers selling small amounts of food to offer more fruits, vegetables, milk, and whole-grain products. There are also programs to support existing and new grocery stores in rural areas.

When the concept of a food desert was introduced, the types of foods available locally were thought to be a significant factor in food purchasing decisions. This aspect of purchasing may not be as important as was initially thought. In 2016, the USDA completed a study in which food purchases of consumers living in food desert areas were tracked. When consumers living in low food access areas shopped locally, they purchased slightly less fruit, vegetables, and milk products but more red meats and diet drinks than people not living in food deserts. When the people from food desert areas traveled farther to a store, presumably to have better food choices, their food purchased changed only slightly compared to what they purchased closer to home. Similarly, when a new grocery store was opened in a community, only about 25% of residents changed their shopping habits to use the new store, and the impact on consumption of fruits and vegetables was very small. These findings illustrate that there may be multiple factors that influence food purchasing behavior beyond simply having access to food markets. Food prices, knowledge about food preparation, eating habits, and cultural preferences all play important roles in food buying decisions.

7.4.4 Nongovernmental Food Assistance Programs

There is a great need for food assistance in the United States and this need is not fully met by federal programs. Consequently, a wide range of nonprofit volunteer organizations have stepped up to fill the gap. These organizations secure funding from a network of federal, state, and local grants, seek donations from individuals and corporations, and rely on volunteers to carry out most of the activities. An important legislative act that assisted the growth in nongovernmental food assistance programs was the Bill Emerson Good Samaritan Food Donation Act. This act was signed into law by President Bill Clinton in 1996, and protects donors from criminal liability when donating food to a nonprofit organization if the food should cause illness or harm. It includes a provision that says providing food close to the recommended date of sale is not grounds for gross negligence.

John Arnold van Hengel started the first food bank in Phoenix, Arizona in 1967. Working as a volunteer in the Immaculate Heart Church, he saw the great need of people coming to the soup kitchen for food. He also noticed the large amount of food being discarded by stores and restaurants and he began to collect and distribute these foods to homeless people. Van Hengel opened St. Mary's Food Bank with the support of St. Mary's Catholic Church, which provided funding and a warehouse where he could store and distribute food. Van Hengel continued to expand his reach in 1976 by launching America's Second Harvest (now Feeding America www.feedingamerica.

org) and garnering support and donations from large corporations and foundations. In 1983, van Hengel established Food Banking Inc. (now The Global Food Banking Network www.foodbanking.org), which is an international nonprofit organization to create, support, and strengthen food banks. Today, every state has numerous food banks and pantries run by faith-based organizations, local volunteer groups, and communities that provide food and necessities to those in need. Food drives to collect food are organized by local groups, schools, athletic teams, and even the US Postal Service. A great deal of food is also donated by the food industry. Products that are safe to eat but may have a crooked label or that have been overproduced for the market may be donated to a food pantry. There has been some concern that the types of foods available through food banks are not always the most healthful options; specifically fresh fruits and vegetables, dairy, and meats tend to be difficult to provide to food pantries. Some recent efforts to improve this include creating "plant an acre" programs where farmers plant extra crops to be donated to food pantries, creating community gardens specifically for food pantry donations, and encouraging hunters and wild game processing centers to donate their products.

Meals on Wheels America is a national, public–private organization that oversees a network of community programs that deliver meals to senior citizens who are unable to leave their homes. Federal funding provides some support for Meals on Wheels through the Older Americans Act (OAA), which is administered by the USDA, but most programs rely on donations and other support to meet the local demand. Over 9% of elderly people living alone in 2015 were food insecure. States with the highest food insecurity among senior citizens include Mississippi, South Carolina, Arkansas, Texas, and New Mexico. Funding from the OAA is also used to provide congregate meals to elderly persons who are able to go to a local community center for the meals. These environments provide daily social interactions and opportunities for educational programs for seniors in addition to the meals.

Many faith-based organizations provide meals and food assistance to low-income families. These organizations may partner with USDA nutrition assistance programs or other federal, state, and local organizations for part or all of the funding. They may provide food to day care facilities, food banks, pantries and soup kitchens, schools, after-school programs, homeless shelters, health clinics, and summer activity programs.

EXPANSION BOX 7.2

FOOD ASSISTANCE PROGRAMS ON COLLEGE CAMPUSES

College campuses feed thousands of students every day. These large food service operations are run efficiently, but often overproduce food that is not consumed. Food recovery programs unite students at colleges and universities to fight food waste by recovering surplus perishable food that would otherwise be discarded from their campuses and donating it to people in need. The *Food Recovery Network* was first implemented by students at the University of Maryland–College Park. The Sodexo Foundation provided funding to grow the program and today there are 191 chapters on college campuses around the United States. A similar program, Campus Kitchens Project, began in 2001 to connect schools with community groups to provide food assistance. Students work with the dining staff of their

EXPANSION BOX 7.2 *(cont'd)*

institutions to collect and distribute the edible, unneeded food to food pantries and congregate meal sites in their local communities. On some campuses, the students use the campus kitchens in the summer, when the facilities would otherwise not be operating, to prepare foods for the community. Part of the mission of these student groups is to raise awareness among their fellow students about hunger and food insecurity and to encourage them to not waste food.

Gleaning is another method of obtaining food for distribution to food-insecure people. Gleaning programs link farmers who have crops that are edible, but not marketable, with those who distribute food to the needy. These programs rely on volunteers, often college or high school students, to pick and deliver the produce. The Stanford Gleaning Project is an example of a student-run program to harvest food that is given away to those in need.

Food insecurity is an increasing problem for college students. A study of students attending colleges in Illinois found that 35% of students were food insecure. Higher costs of education, and the challenges of working while going to school, have put students in a situation where they may not be able to afford food. To meet these needs, many schools are providing student-run food pantries on their campuses. Students majoring in dietetics at Iowa State University were researching food insecurity for a class project and learned about the challenges some college students face with regard to food. Their research found that students who are single parents, or paying for their college tuition on their own, were having difficulty meeting their nutritional needs. To address the problem, the dietetic students created a food pantry, called the SHOP (Students Helping Our Peers). Members of the SHOP come from many different majors and contribute their skills to promote the program. Students hold food donation drives on campus and collect food donations from local farmers, merchants, and restaurants; advertise the SHOP on campus; and organize and distribute the food. Other college campuses, and K–12 schools, have similar programs to provide food to students in need.

Suggested websites: www.foodrecoverynetwork.org, www.campuskitchens.org, www.bewell.stanford.edu, www.theshop.stuorg.iastate.edu

7.5 DIET AND CHRONIC DISEASE

The role of nutrients in disease began with the identification of essential nutrients that when deficient in the diet caused illness. By the 1950s, vitamins and minerals, essential amino acids, and essential fatty acids were isolated, chemically characterized, and purified, and standards of intake required to prevent disease were defined. After these scientific accomplishments, nutrition researchers began to determine the role of diet in chronic disease. Chronic diseases are those that are not cured in a short time but require life-long treatment and management and have an impact on quality of life. Many chronic diseases are strongly linked to dietary habits and nutritional status. Analysis of two types of chronic conditions, colon cancer and heart disease, began to direct nutrition science and chronic disease research.

7.5.1 Colon Cancer

An observation made by Denis Burkitt, a British physician working in Africa, led to the connection between colon cancer and diet.

Burkitt observed that colon cancer was very rare among African populations and developed the hypothesis that diets high in complex carbohydrates and fiber were a primary reason for the lower rates of colon cancer in Africans compared to rates in the United Kingdom and United States. He also postulated that high rates of colon cancer in developed countries could be explained by the consumption of refined carbohydrates that lacked fiber. In 1971 he presented his findings at a national conference in the United States. This led to an intense research agenda by many scientists to explain the connection between diet and colon cancer risk.

Cancer is a complex disease with many possible causative factors, including genetic familial risks, environmental components, viral and bacterial agents, and dietary compounds, most of which have not been fully defined. The nature of the colonic environment includes a cell population with rapid turnover and a rich bacterial concentration that generates acids, bases, and other signaling compounds. Diet composition directly and indirectly affects the colonic cells and interacts with the microbial environment as well. These factors play a role in regulation of cell growth and possibly the development of cancer. The complexities make understanding the causes of and influences on colon cancer risk challenging.

The number of research papers focused on dietary fiber and colon cancer exploded after Burkitt's hypothesis was presented. Medical groups and healthcare providers encouraged people to consume more fiber and the food industry responded with hundreds of fiber-enriched products on store shelves. Colon cancer incidence rates and deaths have decreased steadily in the United States since 1985. Dietary fiber intake may have played a role in the decrease, but increased emphasis on screening colonoscopies for earlier detection, better treatment protocols, and reduction in smoking also contributed to the decline. Current research suggests that a diet high in dietary fiber is likely protective of colon cancer risk, but family history, the colonic microbial population, dietary components, and lifestyle (obesity and a sedentary lifestyle are risk factors) play important roles.

7.5.2 Cardiovascular Disease

As with the connection between dietary fiber and colon cancer, a relationship between dietary fat and CVD arose from comparisons of populations. CVD includes conditions related to the heart and blood vessels, such as hypertension, atherosclerosis, heart attack, and stroke. From the 1950s Ancel Keys from the University of Minnesota, along with research colleagues from around the world, collected data on the incidence of heart disease and dietary intakes of populations in different countries. This came to be known as the Seven Countries Study and was published a report entitled *The diet and 15-year death rate in the seven countries study* (Keys and others, 1986). The report correlated causes of death with dietary intake within populations and included a graph that showed higher deaths from CVD in populations that had higher intakes of saturated fats compared to the lower death rate in populations with lower saturated fat intake. These correlations, along with research findings in animal studies launched a decades-long investigation of the role of dietary fat in CVD.

It had been known since the 1950s that high-fat and high-cholesterol diets could induce heart disease in animal models. Plaque buildup (comprised mainly of cholesterol) in arteries had been found during autopsies of young soldiers from the Korean War, which suggested CVD started early in life. These and other lines of evidence created a credible connection between dietary saturated fat and cholesterol with higher risk of CVD, leading nutrition researchers to focus on defining this relationship. The first DGA (1980) included recommendations to reduce fat intake, especially saturated fats, and to limit cholesterol intake. Consequently, many consumers switched from eating red meat to chicken and consumed fewer eggs and whole milk. Vegetable oils, hydrogenated margarine and shortening were substituted for butter and animal fats. The food industry again responded by developing "fat-free" and "low-fat" foods. Consumers embraced these fat-free products but fat-free foods were certainly not calorie-free. To replace the mouthfeel and creaminess that fats provide, gums, carbohydrates, and sugar were often added to the products. There has been some recent discussion among nutrition scientists that removal of fat from foods may not have been the best public health approach to reducing CVD, and the consumption of high-carbohydrate foods may have been one of the factors that led to the obesity epidemic.

The two primary dietary targets for CVD have been saturated fats and cholesterol. Structurally, fatty acids contain carbon, hydrogen, and oxygen molecules. The carbons are linked by bonds that are either saturated (a single carbon–carbon bond and two hydrogens per carbon) or unsaturated (a double bond is formed between the two carbons, with one hydrogen per carbon; Fig. 7.11). Saturated fats are characterized by being solid, whereas unsaturated fats are liquid, at room temperature. Saturated fats are found mainly in animal products, such as meat and butter, and unsaturated fats are plant-based, such as vegetables oils. One other group of fatty acids found in foods is monounsaturated fatty acids. These are characterized by having only one double bond in the hydrocarbon chain and are typically present in plant oils. Within each class of fats, there are specific types of fatty acids, defined by the number of carbons for saturated fats and the location and number of double bonds for unsaturated fatty acids. There are many ways to classify and categorize fatty acids, which can be somewhat confusing. Short-chain fatty acids are those with fewer

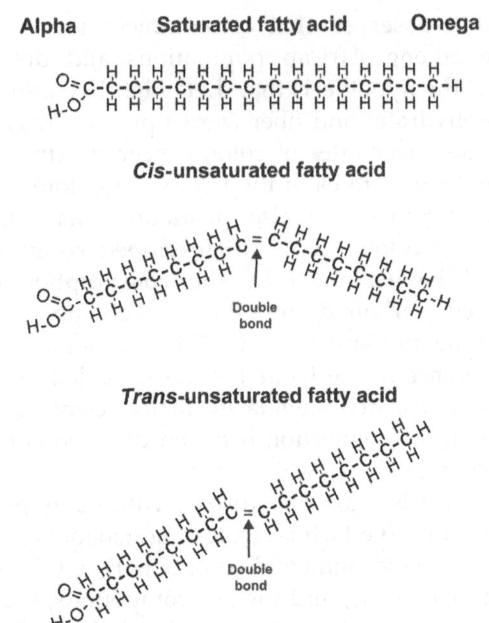

FIGURE 7.11 Fatty acids are classified as saturated (without double bonds) or unsaturated (with double bonds). *Cis*-fatty acids are the natural form for unsaturated fats, whereas *trans*-fatty acids are formed during the hydrogenation process. The location of the first double bond from the omega end of a fatty acid is used as a classification system for fatty acids. *Source: Illustration by Reannon Overbey.*

than 10 carbons, and long-chain fatty acids have more than 20 carbons.

A clever way to categorize the unsaturated fatty acids is the use of families based on the location of the first double bond from the omega end. Fatty acids have an alpha (α symbol meaning the beginning) and omega (ω symbol meaning the end) side. The alpha side contains oxygen molecules bound to the carbon while the omega end has only hydrogens (Fig. 7.11). Adding carbons to a fatty acid can occur only from the alpha side, so the family remains constant. There are three families of unsaturated fatty acids of importance to human biology; the omega 3, 6, and 9 families (designated as *n*-3, *n*-6, or *n*-9; or ω3, ω6, or ω9). Humans are not able to synthesize the ω3 or ω6 families of fatty acids, so these must be consumed from food (and are called "essential" fatty acids) while the ω9 family can be synthesized (nonessential fatty acids). The ω6 fats are present in vegetable oils and plants and the ω3 fats are in high concentration in fatty fish such as salmon. Both these fatty acid families provide precursors for important regulatory compounds called prostaglandins (PGs), but the structures of the PG made from the two sources have different biological effects. PGs from the ω3 fatty acids tend to be more effective in reducing inflammatory processes while PGs from the ω6 promote inflammation. The ω3 fatty acids are also very important for development of tissues in the brain and nervous system and therefore have been added to infant formulas in recent years. There is increasing recognition that Americans have been consuming too much ω6 fatty acids and not enough ω3 fatty acids, especially in populations that do not regularly consume fish. With this knowledge, fish oil supplements became popular and dietary recommendations to increase fish consumption have been made to reduce CVD risk.

An additional feature to the structure of unsaturated fatty acids is the orientation of the carbon chains around the double bond. In the human body, all of the double bonds within fatty acids are in the *cis* orientation, or in the same plane (Fig. 7.11). The process of hydrogenation, in which plant oils are exposed to hydrogen gas to add hydrogen molecules to some of the double bonds, can create *trans* fatty acids, which have the carbon chains oriented in opposite planes. *Trans* fatty acids are taken up by the body normally, but once inside the cell, they cannot be broken down because the enzyme systems in humans cannot access the bonds in this orientation. *Trans* fats have been the target of recent research that has discovered a strong correlation between consumption of these types of fats and CVD risk. The FDA has taken the decisive move to remove partially hydrogenated oils from the GRAS list because of the concerns about the role of *trans* fats in human health.

The process for determining which types of fatty acids are most involved in CVD risk has not been straightforward. The Keys study (1986) targeted saturated fats as being strongly linked to deaths from CVD. Research over the past two decades has found that not all saturated fatty acids may be bad for health. Initially, recommendations were made to avoid all sources of saturated fats, including those from dairy foods and meats. More recent evidence suggests that some saturated fats, especially those from dairy foods, may be beneficial to health. There is also evidence that recommending a low-fat diet may not have been the best for overall health. As noted, dietary fats are important for providing essential fatty acids, but fats provide satiety because they are digested more slowly than proteins and carbohydrates. Fats carry flavors and improve the absorption of fat-soluble nutrients. The 2015 DGA makes the recommendation to consume diets with a balance of healthy fats.

The role of cholesterol in CVD is complicated because humans require cholesterol as a component of cellular membranes and as a precursor of many important regulatory compounds in the body. Humans synthesize cholesterol from acetyl-CoA, which is a molecule normally generated during fatty acid metabolism. In healthy people, cholesterol synthesis is tightly regulated and the absorption of dietary cholesterol from food sources is balanced with production. Dietary cholesterol circulates in the blood as part of lipid droplets, called chylomicrons, which deliver fats and cholesterol from the intestine to the liver. In the liver, cholesterol and fats from the diet and those synthesized endogenously are repackaged into lipoproteins that leave the liver and circulate in the blood. As the lipoproteins pass through the body, fats and cholesterol from the lipoproteins are released and taken up by cells within organs and tissues. The fats and cholesterol are used by the cells to make membranes and to synthesize many important compounds. Some of the cholesterol and fats may also be deposited in arteries, forming plaque which can eventually cause CVD. Eventually, most of the fat and cholesterol is released from the lipoprotein leaving a smaller particle called a low-density lipoprotein (LDL). These LDL particles are cleared from the blood by being taken up the liver. Another type of particle, a high-density lipoprotein or HDL, is made by and released from the liver. HDLs begin as empty vesicles and collect and transport excess cholesterol and fats from tissues back to the liver. In this system, LDLs deliver cholesterol and fats to tissues and HDLs take them away, so having high LDL in the blood increases risk of CVD while having high HDL is protective. In some people, cholesterol synthesis, production, and/or clearance of LDL or production of HDL are not well regulated causing what is known as dyslipidemia and increased risk of CVD. In these individuals, reducing dietary cholesterol intake has no effect and medications are needed to correct blood lipid levels. More recent research has found that dietary cholesterol intake in most people does not cause high cholesterol levels in the blood, which is in stark contrast to the belief that dietary cholesterol was a primary cause of CVD risk. With this newer understanding of cholesterol regulation, the 2015 DGA removed the recommendation to limit the consumption of eggs, which had been a recommendation in previous DGAs to reduce cholesterol intake. Some people will need to reduce their dietary cholesterol intake, especially those taking medications to control their blood lipids or if they have CVD.

A major function of cholesterol in the body is as a precursor of bile acids. Cholesterol is converted to bile acids in the liver and stored in the gallbladder. When dietary fats are consumed, the gallbladder is signaled to release the bile acids into the small intestine. The bile acids serve to *emulsify* dietary fat, allowing digestive enzymes to convert the fats into their basic components (glycerol and fatty acids) that are absorbed from the intestine. Emulsification is the process of forming a coat of nonpolar molecules around polar molecules. In the intestine, bile acids are the nonpolar compounds that form a coat around the polar lipids. This is the same principle used to make homogenized milk (proteins coat the lipids) or salad dressings (proteins or carbohydrates coat the vegetable oils). After the fat is digested, most of the bile acids are reabsorbed from the intestine and returned to the gallbladder. But a small amount of them are lost into the feces. This is the only route through which cholesterol can be excreted from the body. Diets containing more fiber are thought to increase the amount of bile acids that are not reabsorbed and therefore reduce cholesterol levels in the body. Adding certain sterols to the diet and some medications also target this pathway to reduce serum cholesterol levels.

7.5.3 Obesity

Colon cancer and CVD are illustrative of the complicated role that dietary factors play in chronic disease and the challenges of recommending dietary guidelines for the general population. In recent years, the attention of the nutrition research community has turned to the issue of obesity. Obesity is a chronic condition, although it has not typically been classified as a disease.

Early attempts to define healthy body composition used height and weight charts as a standard comparison. The life insurance industry developed charts of ideal body weights based on death rates of their clients. These charts, for males and females, were used for years as guides for recommending a healthy body weight. They were limited in applicability because they were based a small number of people, mainly Caucasians between 25 and 59 years of age (typical policy holders).

The concept of using the ratio of body weight to height squared was introduced in the 1800s by a mathematician. This ratio was first applied to human nutrition in 1972 by Ancel Keys, who called the ratio the BMI. BMI is calculated as (weight (kg)/height (m)^2).

BMI replaced the height and weight tables as a standard reference tool and is used to define stages of obesity. The CDC is careful to note that BMI is useful as a screening tool but should not be diagnostic of body fatness or health. BMI does not take into consideration muscle mass or bone density. For example, BMI measurements on athletes can be misleading because they tend to have high muscle mass and bone density. For the general population of adults, BMI provides a useful indicator of obesity. BMI defines stages of obesity, as shown in Table 7.6.

Other tools used to define nutritional status and body composition include anthropometric measurements. Weight circumference is a useful tool to define amount of fat within the abdomen. A waist circumference of greater than 40 in. in men and 35 in. in women would be considered in the obese category. Skinfold measurements taken at three or four sites are also useful to define overall body composition. Typically, these are taken at the triceps (upper back of arm), biceps (above elbow), subscapular (upper back), and suprailiac (hip) using a caliper.

TABLE 7.6 Classification of Obesity Based on Body Mass Index (BMI)

BMI	Weight status
Below 18.5	Underweight
18.5–24.9	Normal or healthy weight
25.0–29.9	Overweight
30.0–34.9	Class I obesity
35.0–39.9	Class II obesity
40.0–49.9	Class III obesity

From CDC, www.cdc.gov.

More direct measures of body fatness include underwater weighing (water displacement), air displacement (Bod Pod), bioelectrical impedance, or dual-energy X-ray absorptiometry (DEXA). Each of these measurements provides a reliable value for total body fat. DEXA has the advantage of also showing where body fat is located on the body.

EXPANSION BOX 7.3

OBESITY BECOMING AN EPIDEMIC

Life expectancy has improved greatly in the United States. In 1900, the average life expectancy for men was 46 and for women 48 years of age (Fig. 7.12). Today, a man is expected to live to 76 and a woman to 81 years of age. Improved nutrition and reduced incidence of childhood diseases have been credited with improving growth rates of children over the last century. Recent trends in body composition have changed the outlook for the health of the nation. The average body weight of a 20- to 29-year-old American male increased 25 pounds between 1900 and 2000 (Fig. 7.13). Women were 10 pounds heavier in 1900 compared to 1955, but were 30 pounds heavier in 2010 compared to 1955.

Between 1960 and 2010 body weights of 20- to 29-year-old men increased 20 pounds; women increased 34 pounds. Average heights for this age group increased only a half of an inch during that time (Fig. 7.14). The gain in body weight has largely been from a gain in body fat.

EXPANSION BOX 7.3 (cont'd)

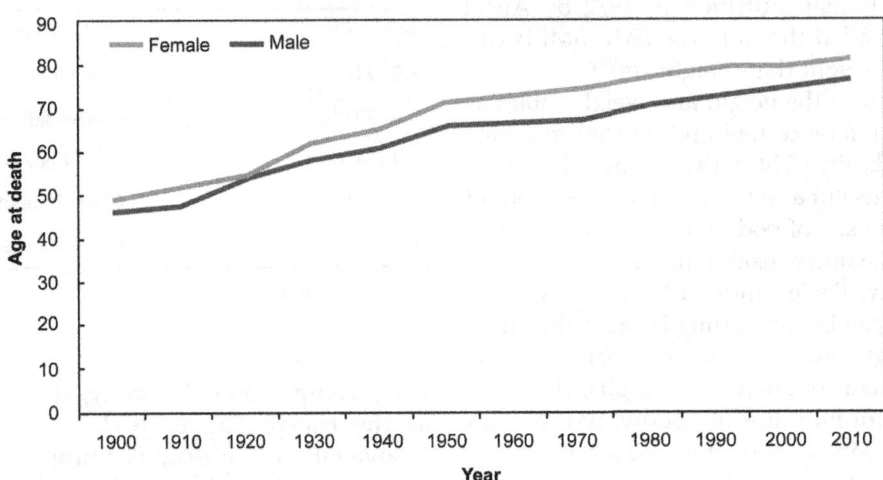

FIGURE 7.12 Life expectancy for men and women in the 1900s was less than 50 years. Due to improved nutrition and healthcare, life expectancy has increased to around 76 years for men and 81 for women. *Source: CDC FastStats, www.cdc.gov.*

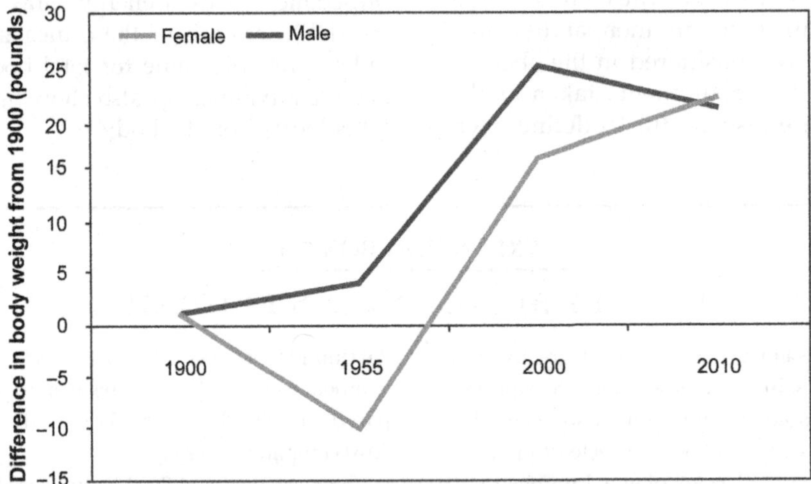

FIGURE 7.13 The average body weight of men and women in 2010 was 24 pounds greater than in 1900. Women weighed less between 1900 and 1950 possibly because they were more likely to be living in cities and not on farms doing manual labor. *Source: CDC, www.cdc.gov.*

The incidence of obesity within the US population increased from 10%–15% in 1960 to 35%–40% in 2013 (Fig. 7.15). The fastest rate of increase occurred between 1976 and 1990. An increase in the percent of people with extreme obesity, a BMI greater than 40, has increased steadily during this time as well.

EXPANSION BOX 7.3 (*cont'd*)

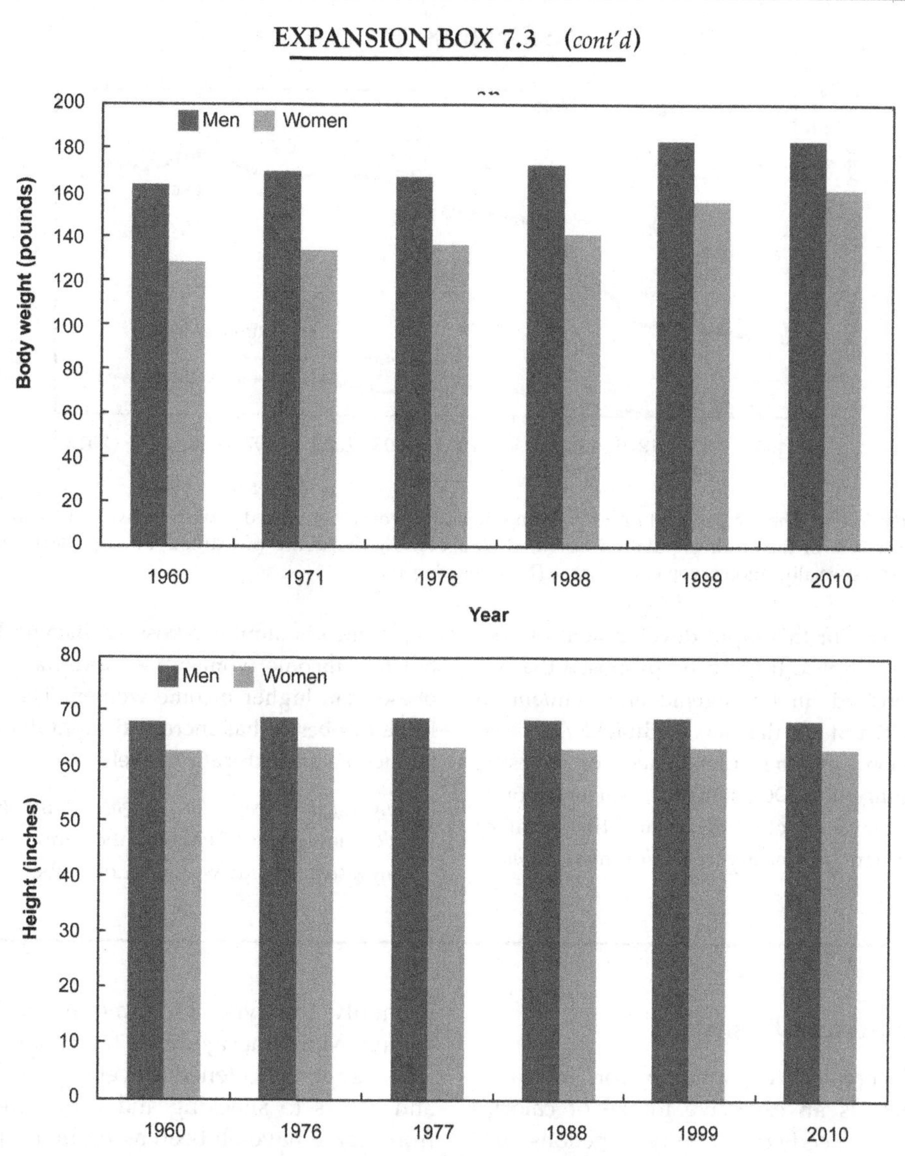

FIGURE 7.14 The average body weight of men and women (top panel) has increased since 1960, whereas the average height (bottom panel) has remained constant. The greater body weight is suggestive of a gain in body fatness that is not beneficial to overall health. *Source: CDC, www.cdc.gov.*

The prevalence of obesity within states has been tracked by the CDC since 1985 via NHANES data. In 1990, only 10 states had obesity rates as high as 10%. By 2000, no state had obesity rate less than 10% and 23 states had rates of 20%–24%. By 2010 there were no states reporting obesity prevalence less than 10% and 12 states with prevalence above 30%.

EXPANSION BOX 7.3 (*cont'd*)

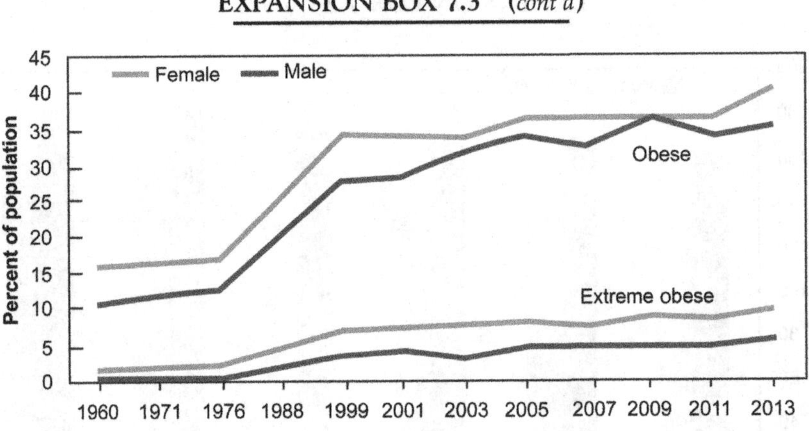

FIGURE 7.15 The incidence of obesity among men and women increased rapidly between 1976 and 2000 with over 35% of the adult population classified as obese. The incidence of extreme obesity (BMI > 40) has increased, especially among women. *Source: CDC, www.cdc.gov.*

The causes for this rapid development of obesity are complex. It has been proposed that we have created an obesogenic environment in the United States that has facilitated the rapid expansion in the prevalence of obesity. According to CDC statistics, socioeconomic status plays a role but is not the primary determinant of obesity risk. For men, obesity prevalence is similar across all income levels, but low income women are more likely to be obese than higher income women. The prevalence of obesity has increased in adults across all income and education levels.

Suggested video: The Weight of the Nation, http://theweightofthenation.hbo.com/

Suggested website: www.cdc.org/obesity

7.5.4 Causes of Obesity

The biochemical explanation for increased body fatness is an excessive intake of calories beyond those needed for body functions and physical activity. The human body is highly efficient at storing excess energy from dietary fat, carbohydrate, and protein by converting it to fat. Humans have an essentially unlimited capacity to build body fat stores. Factors that increase energy intake and/or reduce energy expenditure will contribute to obesity risk. The types of factors associated with increased energy intake are primarily the type and amount of food consumed. More energy-dense foods, larger portion sizes, sugar-sweetened beverages, convenience and access to snacking and eating meals away from home have all been associated with obesity risk. Portion sizes on all types of foods from restaurants to packaged foods to sodas have increased leading to higher calorie intakes (Fig. 7.16). A trend for larger portions began in the 1970s. Consequently, the average calorie intake has correspondingly also increased. From CSFII data, the average calorie intake of children increased between 1977 and 2003 by about

FIGURE 7.16 Portion sizes of common foods have increased over the past 20 years leading to higher caloric intake. These larger portion sizes may be contributing to the higher rates of obesity. *Source: Illustration by Reannon Overbey.*

180 cal/day, primarily from larger portion sizes and higher intakes of snack foods. Assuming that the amount of energy expenditure of children did not increase over this time period, this higher caloric intake would be expected to cause increased body fat.

On the energy expenditure side, less physical activity in daily life, built environments that inhibit walking and playing outside, more sedentary work and increased screen time (computer, television, smart phones), less leisure time, air-conditioned homes and workplaces, and more time spent in cars have all been linked to increased obesity risk. The move to suburbs during the 1950s meant that children rode buses to school and adults drove to work and shopping. The time students were allowed for recess in school decreased and children increasingly participated in scheduled sports rather than free-play. These factors have been proposed as having an impact on the rise in obesity.

Over the past decades, Americans changed their views of, and participation in, exercise. It was fairly uncommon for adults to participate in daily exercise in the 1950s and going to a gym to exercise was rare. An early leader in popularizing fitness was Jack LaLanne, who hosted a popular television show from 1953 to 1985. LaLanne promoted exercising in the home and demonstrated mostly calisthenic types of exercises. Jane Fonda used the newer medium of video in the 1980s to provide workout routines that women could do in their homes. Richard Simmons, having undergone a significant personal weight loss, became a popular fitness and motivational celebrity during the 1990s. Fitness became electronic and personal with the introduction of smart phones. Geospatial sensors that detect movement and global positioning systems (GPS) that track location created the opportunity to monitor individual physical activity. The FitBit, a wearable fitness tracker, launched in 2009 and was instantly popular. The Android Wear and Apple Watch are similar systems that record heart rate, steps taken, distance traveled, sleep patterns, and calories burned. These devices can even send reminders to help achieve goals.

The concept of a fitness center for the average person may have evolved from the activities of The Young Men's Christian Association (YMCA). The first YMCA building with a gymnasium opened in 1869 and began offering group fitness classes, swimming lessons, and team sports. The game of basketball was invented by James Naismith at a YMCA in Massachusetts. During the 1970s, people became more interested in fitness and leisure sports and venues were developed to offer these opportunities. In 1977, there were 2700 fitness clubs in the United States and by 2015, there were over 34,000. The fitness industry has grown to be a major part of the US economy, generating over $24 billion in 2014. High-tech equipment in sleek fitness centers are commonly found in hotels, on college campuses, in nursing homes and retirement centers, and in the workplace. The types of clubs and workout programs available are almost unlimited including yoga, cycling, aerobics, kickboxing, swimming, running, and weight training. Group events such as marathons and triathlons, bike treks, kayaking, and rock climbing are widely available. This fitness craze has occurred simultaneously with the rise in obesity rates, which suggests different segments of the population are accessing these programs.

7.5.5 Childhood Obesity

The large number of obese adults in the United States is a serious concern, but the high prevalence of obesity in children is a greater risk to the nation's health. Obese children are more likely to continue to be obese into adulthood and obesity during development affects physical, mental, and social well-being. Obese children have higher rates of school absences, which affects their learning, and have a more difficult time fitting in socially leading to more depression and social isolation than normal weight children. Defining obesity in children is more complicated than in adults because children are growing. In addition, there may be risk of stigmatization by labeling children as obese or overweight at a certain period of growth, as they may transition out of that phase into a normal range. There is the need for a clinical assessment tool to determine when children may be at an unhealthy weight. For children, a "BMI-for-age" measurement, which takes into account age- and gender-specific percentile, is used as the guide. The BMI-for-age growth charts allow comparison to define children who might be at risk of overweight or obesity (Table 7.7).

TABLE 7.7 Definition of BMI-for-Age for Children

Weight status category	Percentile of the CDC growth charts
Underweight	Less than the 5th percentile
Normal or healthy weight	5th percentile to less than the 85th percentile
Overweight	85th to less than the 95th percentile
Obese	95th percentile or greater

From CDC, www.cdc.gov.

Rates of obesity have increased steadily in children over 6 years of age since 1963, but the rate slowed for children aged 2–5 since 2003 (Fig. 7.17). This may be reflective of a greater awareness about childhood obesity among mothers and pediatricians. Despite this positive trend, the overall prevalence of childhood obesity is nearly 20%.

7.5.6 Consequences and Costs of Obesity

There has been significant debate as to whether obesity should be classified as a disease, a lifestyle condition, or a risk factor for other diseases. This distinction is of importance because medical care, treatments, and cost reimbursements are defined by the specific terminology of diagnosis. When obesity is not classified as a disease, treatment is focused on the other diseases that occur in obese patients rather than the root cause of those diseases. A significant change occurred in 2013 when the AMA adopted the policy that defined obesity as a disease. By defining obesity as a disease and not a condition, medical treatments will be developed and covered by insurance. The AMA policy uses BMI to

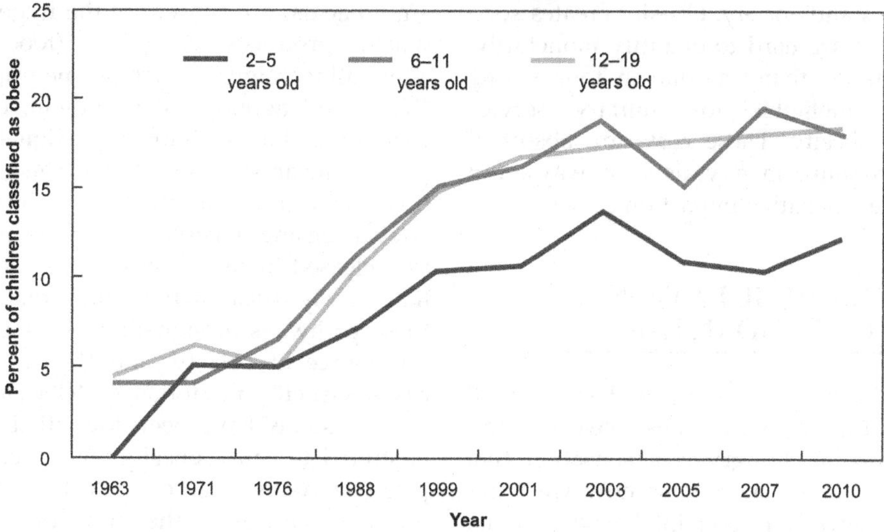

FIGURE 7.17 The prevalence of obesity among children of all ages has increased since 1960. Increased awareness of the problem and implementation of programs to address childhood obesity may be helping to keep the incidence from continuing to increase in recent years. *Source: CDC, www.cdc.gov.*

diagnose obesity and provides guidelines for physicians and healthcare providers to treat and manage the disease. The guidelines include three main concepts:

1. Healthcare providers should screen patients annually and determine their BMI to identify those who would benefit from weight loss.
2. Patients should be enrolled in medically supervised weight loss programs for at least 6 months in order to learn how to manage their weight.
3. Weight loss surgery should be considered for extremely obese patients who have one or more obesity-related health conditions.

It has long been recognized that obesity increases the risk for many severe conditions, called comorbidities, including hypertension, dyslipidemia, type 2 diabetes, coronary heart disease, stroke, gallbladder disease, osteoarthritis, sleep apnea, respiratory problems, reproductive problems, and some types of cancer. Obese people have a higher rate of disability and premature death than nonobese people.

The costs of obesity are both direct and indirect. Direct costs include those associated with medical care such as office visits, laboratory tests, physical therapy, drug therapy, and hospitalizations. By some estimations, obesity-related medical care may represent as much as 20% of total medical spending in the United States with increases every year. Hidden costs such as having to replace hospital beds, wheelchairs, and toilets with larger versions and more sturdy construction, or reconfiguring ambulances to hold larger people must also be considered. More gasoline and airplane fuel is used to transport heavier people than for transportation of lighter people.

The indirect costs of obesity include the impact on the productivity of the individual. Obese workers are more likely to miss work, experience periods of short-term or long-term disability, and may not work at full capacity. Lower worker productivity is an indirect cost to employers and society. Obesity creates societal costs that are hard to quantify monetarily. For example, as many as one in four young adults are ineligible for military service because of obesity. These costs are absorbed into the economy in a variety of ways, but overall have a negative impact on society.

7.6 REGULATION OF FOOD INTAKE

First and foremost, eating food is essential for life. Eating also provides pleasure and enjoyment. Being hungry is a sensation that everyone has experienced, but how we manage and balance our food intake with actual physiological needs is complicated. Definitions of hunger include a feeling of discomfort with a desire to eat. The English language includes a wide range of words to describe a desire to eat including *craving, famished, starving, ravenous, yen, munchies,* and *appetite.* What and when people eat depends on many factors. We eat when we are happy and sad, bored and excited, and celebrating and mourning. The tastes and texture of foods influence the amount we consume, and these preferences vary widely between people and among cultures. Defining how the body regulates food intake and energy stores (body fat) has been somewhat explained through research, but not to the level where it can be manipulated on a long-term basis.

It has been known from studies of patients with head injuries that areas of the brain participate in food intake regulation. Damage to some areas of the hypothalamus cause uncontrolled eating (*hyperphagia*) whereas damage to other areas causes extreme *anorexia* (refusal to eat). The hypothalamus serves as a central receiving center for signals in the blood that monitor food intake and energy stores. Circulating hormones (insulin, leptin, and ghrelin) and nutrients (glucose, fatty acids, amino acids) are sensed by the hypothalamus, which produces *orexigenic* (food intake–stimulating) signals such as neuropeptide Y (NPY) and agouti-related protein, or *anorexigenic* (food intake–inhibiting) signals such as proopiomelanocortin (POMC). The gastrointestinal tract also monitors food intake using stretch-sensing receptors and hormones that are released in response to meal size and contents. Early researchers thought that defects in these pathways or signaling systems might be the source of obesity, and they are in some cases. Genetic mutations in NPY, leptin, and other systems have been identified and demonstrated to cause obesity, and specific therapies targeting these defects have been effective. However, the hope of finding the "obesity gene" has not materialized. The causes of dysfunctional eating and understanding the reasons some people are able to balance their food intake to maintain normal body weight over the lifespan while others cannot, has not yet been achieved.

The decision about when and what to eat involves the cognitive, emotional, attention, and reward systems within the brain as well as sensory signals, such as the sight, smell, and taste of food. Current research using brain imaging technologies is just beginning to uncover how the brain reacts to these multiple stimuli. Some early experiments using these tools generated public attention when it was reported that eating or even just looking at pictures of sugary foods created a similar brain response as occurs when using highly addictive drugs. This led to the idea that some types of foods are addictive and accusations that food companies made foods addictive in order to increase sales. There is little evidence that people become addicted to foods, although cravings and overconsumption of highly palatable foods can be a problem for some people. Preliminary evidence suggests that there may be complex changes that develop in the brain of obese individuals that make them less responsive to regulatory signals that inhibit eating.

A primary factor in energy management of the body is glucose. Glucose levels in the blood must be constrained within a fairly narrow range in order for the brain and central nervous system to function properly. *Hypoglycemia* (low blood sugar) or *hyperglycemia* (high blood sugar) are both dangerous and potentially life-threatening. Two hormones work together to ensure blood glucose levels are maintained within this narrow range. When blood glucose levels drop, the hormone glucagon is released from the pancreas and acts to stimulate glucose production by the liver. When glucose levels increase, the pancreas releases insulin, which acts to move glucose from the blood into tissues where it is used for energy, stored as glycogen, or converted to fat. At the cellular level, there is a complex system stimulated by insulin for glucose uptake involving hormone receptors and carrier proteins. For many reasons, there may be a breakdown in the communication between insulin and these regulatory systems that leads to insulin resistance. Type 1 diabetes occurs when the body cannot make sufficient insulin. The only treatment for type 1 diabetes is to provide insulin via regular injection or implanted pump. Type 2 diabetes is characterized by insulin resistance, that is, insulin is produced but not able to properly maintain blood glucose levels.

Obesity and sedentary lifestyles have been found to increase the risk of type 2 diabetes.

When insulin resistance occurs, blood glucose levels may not be maintained in a normal range and a state of hyperglycemia occurs, causing glucose to be excreted into the urine (not a normal event). Hyperglycemia can cause symptoms of excessive thirst, headache, blurred vision, and fatigue. Prolonged hyperglycemia leads to a process called glycation in which cellular proteins and lipids become bound to glucose, leading to tissue damage. One test that is used to define the presence of hyperglycemia is the amount of hemoglobin A1C (or glycosylated hemoglobin) present in the blood. If left untreated, hyperglycemia increases the risk of atherosclerosis (precursor of heart disease, kidney disease, stroke, and blindness) and may lead to ketoacidosis (high level of blood ketones and lowered blood pH), which may be a life-threatening condition. Type 2 diabetics may experience infections that don't heal, especially on the feet, which may progress to require amputations. Treatment of type 2 diabetes includes lifestyle changes such as management of diet and increased exercise, weight loss, and medications. There has been a steady increase in the number of people with type 2 diabetes in the United States, from about 5.5 million people in 1980 to over 22 million in 2014. A person with type 2 diabetes accrues 2.3 times more medical care than someone without diabetes, which puts additional burden on the healthcare costs of the nation.

Type 2 diabetes was considered a condition that affected adults, in fact, it was first termed adult-onset diabetes (type 1 diabetes was termed juvenile diabetes because it occurred most frequently before the age of 20). Type 2 diabetes in children is still uncommon, but the incidence increased over 30.5% between 2001 and 2009, and it is projected that as much as a fourfold increase in incidence is possible into the future. Because of the severe long-term effects of type 2 diabetes, developing the condition during childhood is of great concern.

The role of carbohydrates in obesity and type 2 diabetes risk has received increased attention. Refined carbohydrates are sugars (from sugar cane or sugar beets), caloric sweeteners (honey, molasses, agave nectar, high fructose corn syrup (HFCS)) and starches, which are rapidly digested and enter the blood as glucose. Many foods contain natural sugars, including fruits and vegetables and whole grains. Added sugars are commonly found in processed foods. The 2015 DGA has recommended limiting added sugars to no more than 10% of energy intake. So that consumers can monitor their intake, the FDA will require added sugars to be listed on the Nutrition Facts panel in 2018. One type of added sugar that has received a great deal of attention is HFCS. As described in Chapter 6, Food Processing, HFCS is made from corn and is structurally similar to table sugar. The scientific consensus is that HFCS is digested and metabolized similarly to sugar and does not pose a unique risk to health. Currently, there is substantial debate as to the direct effect of added sugar on risk of chronic disease. This is complicated by the fact that foods containing added sugars tend to be high in calories and low in nutrients, so they create a calorie-to-nutrient imbalance. There is no evidence that added sugars directly affect chronic disease risk, but a diet pattern with high intakes of added sugars would likely provide excessive calories and insufficient nutrients.

7.6.1 Fad Diets and Weight Loss

Obesity does not develop overnight and cannot be corrected overnight, but Americans tend to want quick solutions to losing weight. We have a long history of fad diets that have been interesting, but ineffective, and probably have added to health risks. Dieting is a national pastime and, on any given day, more than half of Americans are trying to lose weight. Bookstore shelves are full of programs and weight loss gimmicks that promise rapid success, and consumers spend billions of dollars on them. The most effective short-term diet plans are those that take away all decision-making by providing fixed meals or strict schedules. Some versions of these diets make whole categories of foods off limits, such as carbohydrates or fats. The Atkins Diet and South Beach Diets promote strict limitations on carbohydrates for example. Other programs provide meal substitutes or prescribe detailed menus for periods of time. Newer approaches include fasting for 2 or 3 days/week. Such programs that take away all choice from the dieter make them easier to follow. Unfortunately, once the dieter returns to their normal eating habits, the weight usually returns because they have not learned how to manage their intake over the long term. More effective programs engage in teaching the dieter how to assess their calorie intake (keep a food diary), read and understand food labels, change the way they prepare foods and develop strategies for eating out. Weight Watchers (WW) started in 1975 and has been effective in using these types of approaches. In addition, WW provides personal support and encouragement through group meetings or online advice, which are also key components of an effective weight loss program. Mindful eating and training people to understand their motivations with food are important strategies. Evidence suggests that changing eating habits and regular exercise are the key elements required for effective long-term weight loss.

The search for medications that will induce weight loss has been ongoing; however, no drug has yet been developed that effectively treats obesity. In the 1940s, amphetamines were prescribed for weight loss, but the side effects included sleeplessness, anxiety, and depression, and these drugs generated only modest weight loss. During the 1990s, drugs that targeted the appetite-regulating hormones serotonin, norepinephrine, and dopamine were approved for weight loss by the FDA. In 1996, Redux, a combination of the drugs fenfluramine and phentermine (fen-phen), was approved and given to millions of people. Unfortunately, pulmonary vascular damage developed in many of the patients, causing the FDA to withdraw fen-phen from the market the next year. A similar type of drug, sibutramine (Meridia), was approved soon after

Redux, but following reports of high blood pressure and increased heart rate, it was withdrawn by the FDA in 2010. Currently, there are five appetite-suppressing drugs approved by the FDA, Belviq (lorcaserin), Contrave (combination of bupropion and naltrexone), phentermine, and Qsymia (phentermine and topiramate), and one drug that affects blood glucose regulation, Saxenda (liraglutide). The effectiveness of these drugs is generally modest and management of diet and exercise are required in order for weight loss to occur.

A different drug approach to weight loss was to reduce the body's ability to absorb dietary fat. Orlistat (also called Xenical and Alli) acts by inhibiting the enzymes called lipases that cleave fatty acids from dietary triglycerides in the small intestine. This step is necessary in order for the fats to be absorbed by the body. The undigested fats stay in the intestine and are excreted in the feces, which reduces calorie intake. Orlistat continues to be available as both a prescription and over-the-counter medication, but it has some negative side effects that limit its acceptability. If a very low-fat diet is not adhered to, the unabsorbed fat can cause diarrhea, oily stools, and urgent bowel movements, which are unpleasant.

A large number of dietary supplements have been promoted to facilitate weight loss. Among these are green tea extract, guar gum, bitter orange, and green coffee extract. Dietary supplements are not regulated or studied for efficacy, unlike prescription drugs, which have to be shown to be effective. There is limited scientific evidence that any dietary supplement can affect weight loss. Some products may contain dangerous compounds. Ma huang is a naturally occurring plant that has been marketed for its ability to help lose weight. Ma huang contains ephedra, which is similar to the hormone epinephrine (adrenaline). Several cases of death from heart attack and stroke were reported in people taking ma huang, leading the FDA to ban supplements that contain ephedra because of these serious risks.

Surgical treatments to reverse obesity have been developed and can be effective in some cases. The earliest surgical approach was the Roux-en-Y gastric bypass procedure in which a small pouch of the stomach is created and connected to the lower part of the small intestine. This reduces the amount of food that can be consumed and, by bypassing the majority of the small intestine, reduces the amount of nutrients that are absorbed from the food. This procedure has been associated with nutrient deficiencies and other complications including vomiting and diarrhea. Two other approaches use restriction of the stomach to reduce the amount of food that can be consumed. Gastric sleeve surgery involves removing about 75% of the stomach and stapling the remaining section to form a tube. Patients are limited in the size of meals they can consume, which facilitates weight loss. A less invasive procedure, gastric banding, has been used in which a silicone band is placed around the upper part of the stomach and inflated to create a smaller pouch to reduce capacity. The size of the pouch can be adjusted as needed and there are no complications with nutrient absorption. Candidates for weight loss surgery typically must have a BMI greater than 40 or have a serious health problem related to obesity. Surgery has been found to be effective in mitigating some of the comorbidities associated with severe obesity, but management of weight loss remains a life-long struggle.

Prevention of obesity is a far better strategy than trying to lose weight. Obesity-related treatments are not covered by Medicare and policies vary greatly among private insurance companies for nutrition counseling or other obesity-related therapies. In addition to limitations on financial access, there is a significant social barrier to seeking obesity treatment. The belief that obesity is caused by a character flaw or personal weakness may cause people to not seek medical help or for medical professionals to offer words of advice rather than effective treatment. Given the enormity of this problem in the United States, more focus on medical and societal support is needed.

7.6.2 Obesity Prevention Programs

The USDA, CDC, and most major medical organizations have developed programs and messages to promote healthy dietary habits and encourage more physical activity. One attempt to use a mass-marketing approach to change eating behavior was the Five a Day Campaign for Better Health promoted by the NCI and the Produce for Better Health Foundation, and then later, the CDC. The program, which ran from 1991 to 2006, was created by researchers based on behavior theory. The simple message was to eat five fruits and vegetables a day. Signage in stores and schools, printed materials, and public advertisements to deliver the message were used. In 2007, the message was changed to "Fruits & Veggies—More Matters." Promotion of fruits and vegetables is also the goal of the FNV (an acronym for fruits 'n vegetables) campaign, which is part of the Partnership for a Healthier America. FNV has recruited top celebrities and sports figures to donate their names and image to promote the healthfulness of fruits and vegetables. These programs logically contend that healthful foods, such as fruits and vegetables, do not benefit from the major advertising campaigns that other foods such as breakfast cereals, potato chips, and cookies receive from the food industry. Marketing of foods is big business and it is nearly impossible for government agencies to compete with food industry advertisements.

Many other programs aimed at prevention of obesity have been implemented through the public schools, especially through the NSLP. Team Nutrition is an integrated, behavior-based, comprehensive plan for promoting the nutritional health of children. It involves schools, parents, and the community in efforts to continuously improve school meals and to promote the health and education of 50 million school children in more than 96,000 schools nationwide.

A major effort launched in 2010 by First Lady Michelle Obama was the Partnership for a Healthier America (PHA). PHA brings together the private and public sectors to develop strategies to solve the childhood obesity problem. PHA corporate partners pledge to make a commitment to change some aspect of their business to reduce obesity risk. Walmart was one of the first corporations to join PHA and, in 2011, made commitments to modify product formulations, open stores in food deserts, and market healthier food options. With the wide reach that Walmart has in food sales, these changes should make an impact.

According to requirements of the 2010 Affordable Care Act, large restaurant chains must provide nutrition information on their menus and vending machine suppliers must make calorie content visible before purchase. -Ready-to-eat foods in grocery and convenience stores are required to have nutrition information posted on the package or in the display case. The intent is to make it easy for consumers to make healthy choices. It is not fully clear yet if having more nutrition information will have an impact on consumer choices. A positive impact may be that restaurants and chefs will reduce portion sizes or provide healthier options in response to the required labeling.

7.7 FOOD MARKETING

Marketing, which is a targeted, research-based strategy to promote a product, is effective in creating demand for products.

Advertising forms the visible component of marketing that develops the brand image and perceived value to the consumer. Consumers are influenced by marketing programs that use celebrities and sports icons, but also by the behaviors that friends and neighbors are adopting. During childhood, skills in consumerism develop from interactions within the family and social groups, but the broad access to media has a significant impact. From the earliest days of food marketing, the methods used to market products to children have been controversial. In the 1970s, the Federal Trade Commission attempted to restrict advertisement to children, based on evidence that children are unable to distinguish between persuasion and information, but Congress refused to pass any regulations. Children are an important target market for companies because they represent lifelong customers. Children also directly and indirectly influence purchases of their parents. Some have argued that bans on magazine, radio, and television advertising, similar to those implemented for the tobacco industry, should be imposed for unhealthy foods targeted to children. Such a regulatory system would be complicated and hard to regulate and some believe it impinges on the First Amendment right to free speech. Efforts have been focused on a voluntary approach to encourage companies to refocus their marketing strategies for children. These strategies have had some success. Since 2007, Mars Inc., the makers of M&Ms and other candies, has curtailed its advertising to children younger than 12 years of age. Other confectionery companies have joined the Children's Confection Advertising Initiative to be more responsible in their promotion practices; these include The Hershey Company, Mondelez International, and Nestle.

The amount of money spent on advertising by the food industry is substantial. According to the Yale Rudd Center for Food Policy and Obesity, in 2012, McDonald's spent $972 million in advertisements. That amount was 2.7 times more than all the fruit, vegetable, bottled water, and milk advertisers combined ($367 million). Over $1 billion was spent in 2014 to advertise products through online video games alone. The great majority of marketing to children occurs through television, but online marketing is rapidly outpacing television. Television and online commercials attract children by using characters, bright colors, play, fun, and happy scenes. Children lack the critical facility to distinguish commercials from program content. By the age of 8, this skill is usually developed but can be suppressed when messages are cloaked as entertainment, information, or public service announcements. The use of cartoon characters and celebrities to target both children and parents has been shown to increase preference for advertised foods, consumption of advertised foods, overall calorie consumption, and requests for parents to buy advertised foods. The types of foods most commonly advertised to children are high in calories, fat, sugar, and sodium(breakfast cereals, cookies and snacks, soda, and fast-food restaurant meals). In addition to television advertising, food companies include brand logos on toys, games, collectibles, company-sponsored magazines and clubs, celebrity endorsements, product placement in movies, counting books using foods (Cheerios, Froot Loops, M&Ms), and clothing with logos found in specialty stores.

The IOM convened two special reviews of marketing to children in 2006 and 2013. Following the release of the 2006 report, *Food Marketing to Children and Youth: Threat or Opportunity*, in which they recommended that the food and beverage, and media and entertainment industries apply their marketing skills to promote more healthful foods to children, the Children's Food and Beverage Advertising Initiative (CFBAI) was established. CFBAI includes the Council of Better Business Bureaus and leading food companies that represent 80% of child-directed TV food advertising and food products that are consumed by children. CFBAI members developed uniform nutrition standards for child-targeted foods and pledged to modify their advertisements toward children in alignment with a set of core principles.

General Mills made the following pledge to join CFBAI in 2007:

> General Mills is pleased to submit its Pledge to The Children's Food and Beverage Advertising Initiative (the CFBAI). The CFBAI is a voluntary self-regulation program currently comprised of eleven of the largest food and beverage companies in the United States. By advancing and hastening a shift in the mix of messaging to encourage healthier dietary choices and healthy lifestyles in advertising to children under 12, the CFBAI represents a significant step forward.
>
> General Mills is proud to be a charter member of the CFBAI and is fully supportive of its goals. As an industry leader, General Mills is committed to maintaining the highest standards for responsible advertising to children. We have a long history of advocating increased levels of physical activity and support of fitness programs, particularly for children, in public policy arenas and through various private sector initiatives. Company initiatives, such as the General Mills Foundation's Champions for Healthy Kids program, reflect our decades of continuing support for youth nutrition and fitness initiatives of many kinds, further underscoring that commitment.
>
> The scope and breadth of our Pledge to the CFBAI demonstrates General Mills' continuing commitment to high standards and to providing clear leadership on this important issue.

Additional goals of the CFBAI are to modify the formulations of foods within set calorie limits, reduce targeted nutrients (saturated fat, *trans* fat, sodium, sugars), and enhance healthier food options. Some of these changes are shown in Table 7.8.

According to the 2013 IOM report summary of the impact of CFBAI, over 100 food product formulations to meet nutrition standards had been made by the participating companies, advertising of healthier options had occurred and some products were no longer being advertised to children. But marketing of snacks, cookies, candies, and sodas continues to reach children and youth. New venues in social media such as Facebook, YouTube, and Snapchat offer ways to market products that have far-reaching impacts. For example, during the 100th anniversary of Oreo cookies, the Kraft Company launched a social media campaign to encourage people to share their experiences with Oreos and provided daily ads on their Facebook page. The number of likes increased over 110% in two months and likely increased sales. Most food companies

TABLE 7.8 Changes to Food Products as Part of the Children's Food and Beverage Advertising Initiative

Company	Product	Change
Dannon	Danimals Smoothie	25% reduction in sugar
ConAgra Foods	Chef Boyardee canned pastas	Reductions of 8% in sodium
ConAgra Foods	Kid Cuisine frozen meals	Reductions of 28% and 52% in the sugar content in two and the addition of a new meal with 40% less sugar than a prior comparable meal
Campbell Soup Company	Pepperidge Farm Goldfish Grahams	Increases in the whole grain content
Mondelēz	Graham crackers	8 g of whole grains per serving

From Children's Food and Beverage Advertising Initiative, www.bbb.org/council/the-national-partner-program/national-advertising-review-services/childrens-food-and-beverage-advertising-initiative.

have websites and use social media that include games and offers associated with their products. Some have referred to this as stealth marketing that induces brand loyalty and stimulates purchases. Increasingly, advertisements are individually tailored and reach youth through social media. These specific and personal messages create new challenges for parents, policy makers, and others concerned about controlling messages about food and health. In contrast, for teens and young adults, social media has been used as a means to promote health and wellness. There are a plethora of smart phone apps that track activities and food intake, provide reminders and encouragement to eat right or exercise, and offer recipes and food preparation ideas.

7.7.1 Marketing in Schools

It is unclear exactly when vending machines first appeared in public schools, but likely they were present in teachers' lunchrooms soon after they were developed in 1888. During the 1960s, vending machines dispensing lunch items became popular on college campuses and companies made the push to enter public schools as well. Because of the legal aspects of the NSLP, schools were not allowed to offer meals via vending machines, only snacks and drinks. In 1970, the USDA made modifications to the NSLP that allowed alternative and competitive foods to be sold in schools, which opened the door for vending machines to be accessible to students. Over the next four decades, vending machine contracts with schools provided extra funds for school activities. Beverage companies contracted exclusive pouring rights for specific soft drinks with public schools and snack offerings increased. The substantial financial return to schools became integrated into the school budget to fund programs and, in turn, overt marketing pressure for increased sales occurred. In 2001, the size of sodas sold in vending machines was increased from 12-ounce cans to 20-ounce bottles, resulting in an increase of 250 calories per serving. During this time period, childhood obesity rates were increasing and the food choices available in schools became a target of concern. Banning vending machines was resisted by school administrators who had become dependent on the funding stream and by companies that argued schools and students should have the freedom to choose what they consumed. Eventually, under pressure from government and health advocacy groups, beverage manufacturers agreed to only offer low-sugar or sugar-free drinks and juices, and water in school vending machines. Stricter regulations (the "Smart Snacks in School" rule) were implemented on all foods and beverages sold in vending machines in schools as part of the Healthy Hunger Free Kids Act in 2010 (Table 7.9). All food items sold in schools, including vending machines, must meet these standards.

It is yet to be determined if these programs and regulations will have an impact on the rates of obesity in children, but there is little doubt that many approaches are needed. More focus on physical activity during and after the school day has also been implemented. Several organizations have launched public campaigns

TABLE 7.9 Smart Snacks in School Criteria

Be a whole-grain product OR
Have as the first ingredient a fruit, vegetable, dairy, or protein food OR
Contain 10% of the Daily Value of calcium, potassium, vitamin D, or dietary fiber
Must meet nutrient requirements:
Calories: <200 for a snack or <350 for an entrée
Fat: <230 mg for a snack or <480 for an entrée and no *trans* fat
Sugar: <35% sugar by weight

From Healthy Hunger Free Kids Act, www.fns.usda.gov/school-meals/healthy-hunger-free-kids-act.

to encourage children to be more active. These include the Let's Move campaign promoted by First Lady Michelle Obama, the Fuel up to Play 60 program sponsored by the National Dairy Council, and the NFL Play 60 program offered by the National Football League. Reducing obesity rates in children should be a national priority because the consequences of not reversing the rates of childhood obesity are dire for the health, economy, and prosperity of the United States.

References

Carpenter, K. J. (2003). A short history of nutritional science: Part 1 (1912-1944). *Journal of Nutrition, 133*, 3023–3032.

Baker, B. M., Frantz, I. D., Keys, A., Kinsell, L. W., Page, I. H., ... Stare, F. J. (1963). The national diet-heart study. *Journal of the American Medical Association, 185*(2), 105–106.

Department of Health and Human Services (1990). *Healthy people 2000: National health promotion and disease prevention objectives for the year 2000*. Hyattsville, MD: Department of Health and Human Services, Center for Disease Control.

Hansen, A. E., & Burr, G. O. (1932). Essential fatty acids and human nutrition. *Journal of the American Medical Association, 132*(14), 855–859.

Hunger in America. (1968). *Documentary*. CBS Reports [YouTube]. New York, NY: Carousel Films.

Institute of Medicine (2006). *Food marketing to children and youth: Threat or opportunity?* Washington, DC: The National Academies Press.

Keys, A., Menotti, A., Karvonen, J. M., Aravanis, C., Blackburn, H., Buzina, R., ... Toshima, H. (1986). The diet and 15-year death rate in the seven countries study. *American Journal of Epidemiology, 124*(6), 903–915.

Murphy, S. P., Yates, A. A., Atkinson, S. A., Barr, S. I., & Dwyer, J. (2016). History of nutrition: The long road leading to the dietary reference intakes for the United States and Canada. *Advances in Nutrition, 7*, 157–168.

United States Senate Select Committee on Nutrition and Human Needs (1977). *Dietary goals for the United States* (2nd ed.). Washington, DC: U.S. Government Printing Office.

Further Reading

Alaimo, K., Olson, C. M., & Frongillo, E. (2001). Food insufficiency and American school-aged children's cognitive, academic, and psychosocial development. *Pediatrics, 108*(1), 44–53.

Bishai, D., & Nalubola, R. (2002). The history of food fortification in the United States: Its relevance for current fortification efforts in developing countries. *Economic Development and Cultural Change, 51*(1), 37–53.

Branum, A. M., Rossen, L. M., & Schoendort, K. C. (2014). Trends in caffeine intake among U.S. children and adolescents. *Pediatrics, 133*, 386–393.

Bray, G. A. (2004). Medical consequences of obesity. *Journal of Clinical Endocrinology & Metabolism, 89*(6), 2583–2589.

Burkhalter, T. M., & Hillman, C. H. (2011). A narrative review of physical activity, nutrition, and obesity to cognition and scholastic performance across the human lifespan. *Advances in Nutrition, 2*, 201S–206S.

Burkitt, D. P. (January 7–9, 1971). Epidemiology of cancer of the colon and rectum. Cancer. Presented at the *National Conference on Cancer of the Colon and Rectum*. San Diego, CA.

Carpenter, K. J. (2003a). A short history of nutritional science: Part 1 (1785-1885). *Journal of Nutrition, 133*, 638–645.

Carpenter, K. J. (2003b). A short history of nutritional science: Part 2 (1885-1912). *Journal of Nutrition, 133*, 975–984.

Carpenter, K. J. (2003c). A short history of nutritional science: Part 1 (1945-1985). *Journal of Nutrition, 133*, 3331–3342.

Cawley, J. (2015). An economy of scales: A selective review of obesity's economic causes, consequences, and solutions. *Journal of Health Economics, 43*, 244–268.

Chick, H. (1975). The discovery of vitamins. *Progress in Food and Nutrition Science, 1*(1), 1–20.

Chow, P. K. H., Ng, R. T. H., & Ogden, B. E. (2007). *Using animal models in biomedical research: A primer for the investigator*. Hackensack, NJ: World Scientific Publishing Co. Pte. Ltd, 308 p.

Citizen's Board of Inquiry Into Hunger and Malnutrition in the United States (1968). *Hunger USA*. Boston, MA: Beacon Press, 96 p.

Coleman-Jensen, A., Rabbitt, M. P., Gregory, C. A., & Singh, A. (2016). Household food security in the United States in 2015. *Economic Research Report Number 215*. Washington, DC: Economic Research Service, U.S. Department of Agriculture.

Cook, J. T., Black, M., Chilton, M., Cutts, D., Ettinger de Cuba, S., Heeren, T. C., ... Frank, D. A. (2013). Are food insecurity's health impacts underestimated in the United States population? Marginal food security also predicts adverse health outcomes in young U.S. children and mothers. *Advances in Nutrition, 4*, 41–51.

Council for a Strong America (2010). *Too fat to fight*. Washington, DC: Mission: Readiness. Available from <www.missionreadiness.org>.

Crider, K. S., Bailey, L. B., & Berry, R. J. (2011). Folic acid food fortification: Its history, effect, concerns and future directions. *Nutrients, 3*, 370–384.

Dabelea, D., Mayer-Davis, E. J., Saydah, S., Imperatore, G., Linder, B., Divers, J., ... Hamman, R. F. (2014). Prevalence of type 1 and type 2 diabetes among children and adolescents from 2001 to 2009. *Journal of the American Medical Association, 311*(17), 1778–1786.

Department of Health and Human Services (2015). *Healthy people 2000*. Washington, DC: National Center for Health Statistics, Centers for Disease Control and Prevention. Available from <http://www.cdc.gov/nchs/healthy_people/hp2000.htm>.

Farr, O. M., Li, C.-S., & Mantzoros, C. S. (2016). Central nervous system regulation of eating: Insights from human brain imaging. *Metabolism, 65*, 699–713.

Fishman, A. P. (1999). Aminorex to fen/phen: An epidemic foretold. *Circulation, 99*, 156–161.

Flegal, K. M., Kruszon-Moran, D., Carroll, M. D., Fryar, C. D., & Ogden, C. L. (2016). Trends in obesity among adults in the United States, 2005 to 2014. *Journal of the American Medical Association, 315*(2), 2284–2291.

Fryar, C. D., Gu, Q., & Ogden, C. L. (2012). Anthropometric reference data for children and adults: United States, 2007–2010. National Center for Health Statistics. *Vital Health Statistics, 11*(252), 1–40.

Fryar, C. D., Gu, Q., Ogden, C. L., & Flegal, K. M. (2016). *Anthropometric reference data for children and adults: United States, 2011–2014*. Vital and Health Statistics, Series 3, Number 39. U.S. Department of Health and Human Services, Centers for Disease Control and Prevention, and National Center for Health Statistics.

Gabe, T. (2015). *Poverty in the United States: 2013*. Washington, DC: Congressional Research Service, 76 p.

Garn, S. M., & Clark, D. C. (1976). Trends in fatness and the origins of obesity. *Pediatrics, 57*(4), 443–456.

Geiger, C. J. (1998). Health claims: History, current regulatory status, and consumer research. *Journal of the American Dietetic Association, 98*, 1312–1322.

Greenwald, I. (1960). The significance of the history of goiter for the etiology of the disease. *American Journal of Clinical Nutrition, 8*, 801–807.

Hacker, J. D. (2010). Decennial life tables for the white population of the United States, 1790–1900. *History Methods, 43*(2), 45–79.

Harper, A. E. (1985). Origins of recommended dietary allowances: An historic overview. *American Journal of Clinical Nutrition, 41*, 141–148.

Harris, J. L., LoDolce, M., Dembek, C., & Schwartz, M. B. (2015). Sweet promises: Candy advertising to children and implications for industry self-regulation. *Appetite, 95*, 585–592.

Hoisington, A., Butkus, S. N., Garrett, S., & Beerman, K. (2001). Field gleaning as a tool for addressing food security at the local level: Case study. *Journal of Nutrition Education, 33*, 43–48.

Hopkins, L. C., & Gunther, C. (2015). A historical review of changes in nutrition standards of USDA child meal programs relative to research findings on the nutritional adequacy of program meals and the diet and nutritional health of participants: Implications for future research and the summer food service program. *Nutrients, 7*, 10145–10167.

Hunter, R. (1905). *Poverty* (p. 405). London, UK: MacMillan Company.

Hurwitz, L. B., Montague, H., & Wartella, E. (2016). Food marketing to children online: A content analysis of food company websites. *Health Communication*, 1–6. Available from http://dx.doi.org/10.1080/10410236.2016.1138386.

Institute of Medicine (2013). *Challenges and opportunities for change in food marketing to children and youth: Workshop summary*. Washington, DC: The National Academies Press.

International Food Information Council Foundation. (2016). *Food and health survey*. Available from <http://www.foodinsight.org/articles/2016-food-and-health-survey-food-decision-2016-impact-growing-national-food-dialogue>. Accessed 13.08.16.

Kahan, S., & Zvenyach, T. (2016). Obesity as a disease: Current policies and implications for the future. *Current Obesity Report, 5*, 291–297.

Kraak, V. I., & Story, M. (2015). An accountability evaluation for the industry's responsible use of brand mascots and licensed media characters to market a healthy diet to American children. *Obesity Reviews, 16*, 433–453.

Landers, P. S. (2007). The food stamp program: History, nutrition education, and impact. *Journal of the American Dietetic Association, 107*, 1945–1951.

Lear, S. A., Gasevic, D., & Schuurman, N. (2013). Association of supermarket characteristics with the body mass index of their shoppers. *Nutrition Journal, 12* (117), 3–8.

Lehnert, T., Sonntag, D., Konnopka, A., Riedel-Heller, S., & König, H.-H. (2013). Economic costs of obesity and overweight. *Best Practice and Research Endocrinology and Metabolism, 27*, 105–115.

Leitch, I., Baines, A. H. J., & Hollingsworth, D. F. (1963). Diets of working class families with children before and after the second world war. *Nutrition Abstracts and Reviews, 33*(3), 653–668.

Leung, A. M., Braverman, L. E., & Pearce, E. N. (2012). History of U.S. iodine fortification and supplementation. *Nutrients, 4*, 1740–1746.

McDowell, M. A., Fryar, C. D., Ogden, C. L., & Flegal, K. M. (2008). *Anthropometric reference data for children and adults: United States, 2003–2006*. U.S. Department of Health and Human Services, Centers for Disease Control and Prevention, and National Center for Health Statistics, Number 10.

Messinger, W. J., Porosowska, Y., & Steele, J. M. (1950). Effect of feeding egg yolk and cholesterol on cholesterol levels. *Archives of Internal Medicine, 86*(2), 189–195.

Morris, L. M., Smith, S., Davis, J., & Null, D. B. (2016). The prevalence of food security and insecurity among Illinois university students. *Journal of Nutrition Education and Behavior, 48*(6), 376–382.

Mozaffarian, D. (2016). Dietary and policy priorities for cardiovascular disease, diabetes, and obesity: A comprehensive review. *Circulation, 133*, 187–225.

National Health Survey, Vital and Health Statistics (1966). *Weight by height and age of adults United States, 1960-1962*. Washington, DC: Center for Disease Control.

Ng, S. W., Slining, M. M., & Popkin, B. M. (2014). The Healthy Weight Commitment Foundation pledge calories sold from U.S. consumer packaged goods, 2007–2012. *American Journal of Preventative Medicine, 47*(4), 508–519.

Nixon, L., Mejia, P., Cheyne, A., Wilking, C., Dorfman, L., & Daynard, R. (2015). "We're part of the solution": Evolution of the food and beverage industry's framing of obesity concerns between 2000 and 2012. *American Journal of Public Health, 105*, 2228–2236.

Ogden, C. L., Carroll, M. D., Lawman, H. G., Fryar, C. D., Kruszon-Moran, D., Kit, B. K., & Flegal, K. M. (2016). Trends in obesity prevalence among children and adolescents in the United States, 1988-1994 through 2013-2014. *Journal of the American Medical Association, 315*(2), 2292–2299.

Ogden, C. L., Fryar, C. D., Carroll, M. D., & Flegal, K. M. (2004). *Mean body weight, height, and body mass index, United States 1960–2002*. U.S. Department of Health and Human Services, Centers for Disease Control and Prevention, and National Center for Health Statistics. Advanced Data Number 347.

Oliveira, V. (2015). The food assistance landscape. *FY 2014 Annual Report. ERS, Economic Information Bulletin Number 137*. Washington, DC: Economic Research Service, U.S. Department of Agriculture.

Osborne, T. B., & Mendel, L. B. (1911). The role of different proteins in nutrition and growth. *American Association for the Advancement Science, 34*(882), 722–732.

Piernas, C., & Popkin, B. M. (2011). Food portion patterns and trends among U.S. children and the relationship to total eating occasion size, 1977–2006. *Journal of Nutrition, 141*, 1159–1164.

Pimpin, L., Wu, J. H. Y., Haskelberg, H., Del Gobbo, L., & Mozaffarian, D. (2016). Is butter back? A systematic review and meta-analysis of butter consumption and risk of cardiovascular disease, diabetes, and total mortality. *PLoS One, 11*(6), e0158118.

Piro, A., Tagarelli, G., Lagonia, P., Tagarelli, A., & Quattrone, A. (2010). Casimir Funk: His discovery of the vitamins and their deficiency disorders. *Annals of Nutrition and Metabolism, 57*, 85–88.

Poole-De Salvo, E., Silver, E. J., & Stein, R. E. K. (2016). Household food insecurity and mental health problems among adolescents: What do parents report? *Academic Pediatrics, 16*, 90–96.

Popkin, B. M. (2011). Agricultural policies, food and public health. *EMBO Reports, 12*(1), 11–18.

Rippel, J. M., & Angelopoulos, T. J. (2016). Added sugars and risk factors for obesity, diabetes and heart disease. *International Journal of Obesity, 40*, S22–S27.

Rodriguez, J. E. (2016). Past, present and future of pharmacologic therapy in obesity. *Primary Care, 43*(1), 61–67.

Roh, E., Song, D. K., & Kim, M.-S. (2016). Emerging role of the brain in the homeostatic regulation of energy and glucose metabolism. *Experimental Molecular Medicine, 48*, 1–12.

Sacco, J. E., Dodd, K. W., & Tarasuk, V. (2013). Voluntary food fortification in the United States: potential for excessive intakes. *European Journal of Clinical Nutrition, 67*, 592–597.

Schwerin, H. S., Stanton, J. L., Riley, A. M., Schaefer, A. E., Levelle, G. A., Elliott, J. G., ... Brett, B. E. (1981). Food eating patterns and health: A reexamination of the Ten-State and HANES I surveys. *American Journal of Clinical Nutrition, 34*, 568–580.

Semba, R. D. (2012). The discovery of the vitamins. *International Journal of Nutrition Research, 82*(5), 310–315.

Spahlholz, J., Baer, N., Konig, H.-H., Riedel-Heller, S. G., & Luck-Sikorski, C. (2016). Obesity and discrimination: A systematic review and meta-analysis of observational studies. *Obesity Reviews, 17*, 43–55.

Spector, A. A., & Kim, H.-Y. (2015). Discovery of the essential fatty acids. *Journal of Lipid Research, 56*, 11–21.

Stare, F. J. (1947). Ideal intake of calories and specific nutrients. *American Journal of Public Health, 37*, 515–520.

Stern, M. (2008). The fitness movement and the fitness center industry, 1960-2000. *Business and Economic History, 6*, 1–26. Available from <http://www.thebhc.org/publications/BEHonline/2008/stern.pdf>.

Stiebeling, H. K., Monroe, D., Coons, C. M., Phipard, E. F., & Clark, F. (1941). Family food consumption and dietary levels, five regions. *USDA Miscellaneous Publication Number 405*. Washington, DC: U.S. Department of Agriculture.

Strong, J. P. (1986). Coronary atherosclerosis in soldiers: A clue to natural history of atherosclerosis in the young. *Journal of the American Medical Association, 256*(20), 2863–2866.

Tallie, L. S., Ng, S. W., & Popkin, B. M. (2016). Walmart and other food retail chains: Trends and disparities in the nutritional profile of packaged food purchases. *American Journal of Preventative Medicine, 50*(2), 171–179.

U.S. Department of Health and Human Services and U.S. Department of Agriculture. (December 2015). *2015–2020 Dietary guidelines for Americans* (8th ed.). Washington, DC: Office of Disease Prevention and Health Promotion. Available from <http://health.gov/dietaryguidelines/2015/guidelines/>.

Van Sycykle, C. (1945). Some pictures of food consumption in the United States Part II: 1860-1940. *Journal of the American Dietetic Association, 21*, 690–695.

Vikraman, S., Fryar, C. D., & Ogden, C. L. (2015). Caloric intake from fast food among children and adolescents in the United States, 2011–2012. *National Center for Health Statistics, NCHS Data Brief Number 213*. Washington, DC: U.S. Department of Health and Human Services, Centers for Disease Control and Prevention.

Vitiello, D., Grisso, J. A., Whiteside, K. L., & Fischman, R. (2005). From commodity surplus to food justice: Food banks and local agriculture in the United States. *Agriculture and Human Values, 32*, 419–430.

Wiese, H. F., Hansen, A. E., & Adam, D. J. D. (1958). Essential fatty acids in infant nutrition: I. Linoleic acid requirement in terms of serum di-, tri- and tetraenoic acid levels. *Journal of Nutrition, 66*, 345–360.

Woods, R. (1948). The essential fatty acids. *Borden's Review of Nutrition Research, 9*(4), 1–12.

Young, L. R., & Nestle, M. (2002). The contribution of expanding portion sizes to the United States obesity epidemic. *Public Health, 92*, 246–249.

CHAPTER

8

Sustainability of the Food System

8.1 DEFINING SUSTAINABLE FOOD SYSTEMS

Sustainable agriculture, sustainable farming, sustainable harvesting, and *sustainable food systems* are terms that have become part of everyday conversations. When asked to define sustainability within food and agriculture, people will respond with many different interpretations, depending on their frame of reference. For farmers, sustainable may mean they can make enough money each year to protect their investment over the long term. For environmentalists, sustainable may mean farming should be done using methods that protect natural resources. For consumers, sustainable may mean less packaging or fewer chemicals are used in foods. For rural communities, sustainable may mean returning to a pre-WWI version of agriculture with many farmers working and raising their families on small farms. For humanitarians, sustainable may mean the right balance of nutrients and healthy food is available to everyone regardless of geographic, economic, political, or social status. The term *sustainability* implies the ability to endure and do something forever. Sustainability reflects a society (local or global) that understands history, human behavior, and environmental change in contrast to a society that is too overwhelmed with the concerns of today to worry about tomorrow.

The USDA hosts the Sustainable Agriculture Research and Education (SARE) program, which defines sustainable agriculture as follows:

> Every day, farmers and ranchers around the world develop new, innovative strategies to produce and distribute food, fuel and fiber sustainably. While these strategies vary greatly they all embrace three broad goals, or what SARE calls the 3 Pillars of Sustainability:
>
> - **Profit** over the long term
> - **Stewardship** of our nation's land, air and water
> - **Quality of life** for farmers, ranchers and their communities
>
> There are almost as many ways to reach these goals as there are farms and ranches in America.
>
> ***www.sare.org***

The 3 Pillars of Sustainability can be interpreted as relating to economic, environmental, and human impact of agriculture, and these topics will be covered in this chapter. As noted by SARE, there are many ways to address, and many factors that influence, these pillars, and no single approach will ensure sustainability. SARE is focused on the agronomic aspect of sustainability, but the food system encompasses the humanitarian aspect as well. Missing from

Understanding Food Systems.
DOI: http://dx.doi.org/10.1016/B978-0-12-804445-2.00008-9

TABLE 8.1 Components of a Sustainable Food System

	Agronomic	Humanitarian
Economic	Diverse types of farming are profitable	Healthy food is affordable
Environment	Soil, air, and water quality; biodiversity; and natural resources are protected and enhanced for future generations; types and amounts of energy used are well managed	Nutrients are in balance for proper growth, development, and life-long health of the global population
Political	Regulations ensure farmers and the environment are in balance; optimize access to food globally	Access to food is equitable, fair, and compassionate; regulations are used to protect and advance human health
Social	Agriculture is valued and understood; food production and processing are safe and marketing is just	Food and nutrition are understood and culturally acceptable; healthy lifestyles are encouraged
Ethical	Rights of farmers and land owners are in balance with environmental and consumer rights	Safe, nutritionally appropriate, and healthy food to sustain well-being is ensured for all
Science	Technology and science are thoughtfully applied to enhance both agricultural production and protect the environment	Role of nutrition and food in human health are understood and form the basis of food policies

these pillars is the concept of producing enough food to feed all the people on Earth, producing the right types of food to ensure the health of all people, and equitable distribution and access to food. Human health and the viability of life are necessary components of the food system and must be addressed within the realm of sustainability. Hunger and starvation on the one hand, and overconsumption leading to chronic diseases on the other hand, suggest the amount and types of foods being produced and consumed are not sustaining human health. As discussed in Chapter 7, Nutrition and Food Access, obesity and related chronic diseases have multifaceted etiologies that include sedentary lifestyles, lack of knowledge about food and nutrition, personal choices, poverty, and limited access to healthy foods. Some argue that enough food is being produced to feed the world's population, and that the causes of hunger and starvation are due to social, political, and economic forces. Social justice is a core element of a sustainable food system, and solutions are needed to address fair and equitable access to food. With all these factors, defining a sustainable food system is complicated. Table 8.1 lists a few of the concepts that might be considered necessary elements for a sustainable food system.

Agriculture in the United States has evolved over the years of its existence as a nation. The great majority of citizens do not grow their own food or rely on their farming neighbors to provide food for them. Today less than 2% of the population produces foods for domestic and export markets. Technology and ingenuity have led to higher production and lower costs allowing Americans to spend less money on food (Fig. 8.1). Systems have been created to process, distribute, and market foods. Americans have come to expect grocery stores to be packed with thousands of food items that are consistently available year-round. The food industry has brought innovation, convenience, and enjoyment to consumers. More options are available about how and where food is produced. Farmers' markets are open in all parts of the country or consumers may choose to buy food directly from farmers.

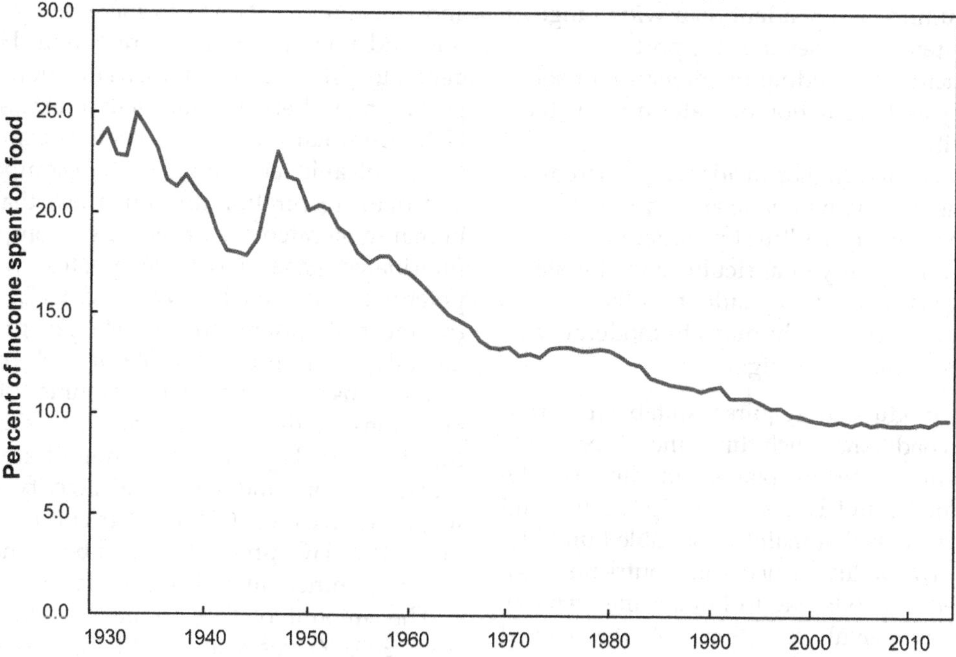

FIGURE 8.1 The percent of income Americans spend on food has decreased from about 25% in the 1920s to less than 10% today. The amount of money Americans spend on food is among the lowest in the world. *Source: USDA Economic Research Service, www.usda.ers.gov.*

Organic and locally grown food options are abundant and available, even from major retailers such as Walmart. Choices and options are plentiful in our food system, but there are consequences for these conveniences. The food system both depends on the environment and affects the environment and these aspects will be discussed in this chapter.

8.2 CLIMATE CHANGE AND THREATS TO THE FOOD SYSTEM

The EPA defines climate change as "any substantial change in measures of climate (such as temperature or precipitation) lasting for an extended period (decades or longer)" (EPA, 2016). There is clear evidence that significant changes in land and ocean temperatures, rainfall patterns, and extreme weather events are occurring. These changes have both long- and short-term impacts on food production. In 2012, the USDA Agricultural Research Service compiled a summary report on the impact of climate change on agriculture (Walthall et al., 2012) and provided these key messages:

- Increases in atmospheric carbon dioxide (CO_2), rising temperatures, and altered precipitation patterns will affect agriculture productivity
- Livestock production systems are vulnerable to temperature stresses
- Projections for crops and livestock production systems reveal that climate change effects over the next 25 years will be mixed
- Climate change will exacerbate current biotic stresses on agricultural plants and animals

- Agriculture is dependent on a wide range of ecosystem processes that support productivity including maintenance of soil quality and regulation of water quality and quantity
- The predicted higher incidence of extreme weather events will have an increasing influence on agricultural productivity
- The vulnerability of agriculture to climate change is strongly dependent on the responses taken by humans to moderate the effects of climate change

Food production requires suitable environmental conditions, including the appropriate temperature ranges for plants and animals; adequate, timely, and balanced precipitation; wind and air currents that maintain suitable humidity; and the availability of necessary nutrients from the water and soil. Due to human intervention, the environmental conditions of the United States today are not the same as they were 250 years ago. Intentional physical changes have been made to the land and water distribution systems across the United States to facilitate the current food production system. Each of these changes has consequences, both good and bad, for the environment. Many unintentional changes to the environment, including those leading to climate change, have occurred from human intervention and activity, and these have significant impact on the food system.

8.2.1 Greenhouse Gases and Temperature

Primary factors that affect climate change relate to emission of greenhouse gases (GHGs) that are produced from human activities and, to a much lesser extent, from natural sources. The main GHGs include carbon dioxide (CO_2), methane (CH_4), nitrous oxide (N_2O), and fluorinated gases (hydrofluorocarbons, perfluorocarbons, sulfur hexafluoride, and nitrogen trifluoride), which remain in the atmosphere and cause temperatures to increase by reemitting and retaining infrared radiation. Humans generate GHG from industrialization, transportation, and energy production. Emissions of GHG from natural sources such as from forest fires, volcanic eruptions, soil decomposition, and plant respiration are much less than from human-generated sources. By comparison, fluorinated gases have the greatest warming potential, followed by N_2O and CH_4. CO_2 is the most abundant greenhouse gas produced globally, accounting for about 75% of total GHG emissions. Relative amounts of GHG emissions in the United States are shown by the source in Fig. 8.2. Generation of electricity, transportation, and industrial activity are the major sources of GHGs. Agriculture contributes to GHG production at about the same rate as commercial and residential sources.

The amount of CO_2 in Earth's atmosphere from 800,000 years ago until the modern era fluctuated between about 200 and 300 ppm. From 1950 to 2015 the amount of CO_2 has climbed to over 400 ppm, which are levels that have never been recorded on Earth. Global warming is a consequence of the rise in GHGs, with direct effects on land and water temperatures. The EPA estimates that the Earth's average surface temperature has increased 0.15°F per decade since 1901. During that time the average temperature in the United States has increased at a higher rate of between 0.29 and 0.46°F per decade. More extremely hot temperatures during summer days have occurred, as well as hotter temperatures at night. This is a concern for farmers because air and soil temperatures directly influence plant growth and development, and the types of pests and diseases that affect crops. All plants have an ideal temperature range needed for their lifecycle. These ranges differ for specific plant species, and vary within the growth phases of the plants. When temperatures fall either below or above these defined ranges, plant growth will be inhibited. In general, higher temperatures

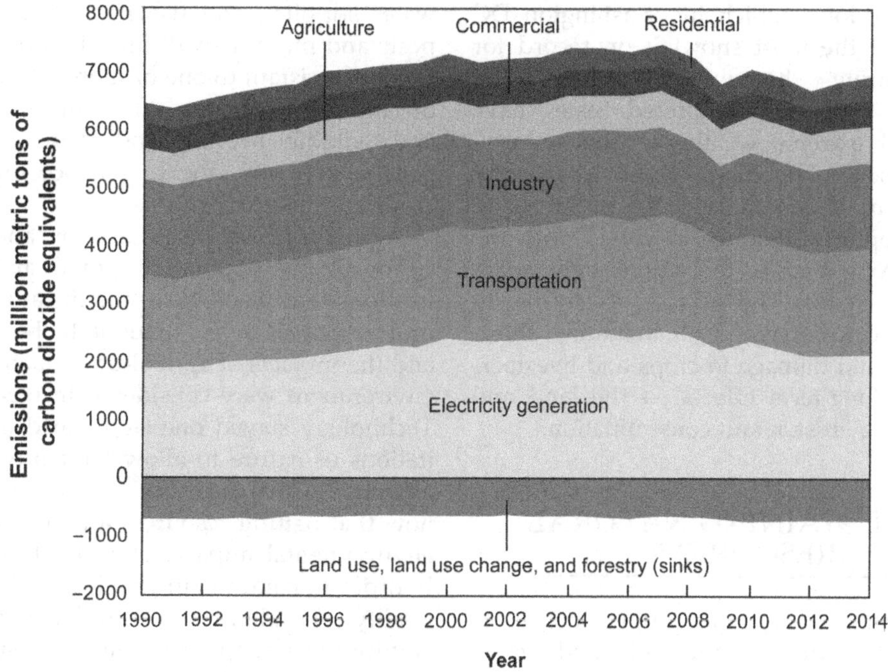

FIGURE 8.2 Greenhouse gas (GHG) emissions, expressed as CO_2 equivalents, come primarily from electricity generation, transportation, and industrial production. Agriculture, commercial, and residential activities produce lesser amounts of GHGs. *Source: EPA, www.epa.gov.*

tend to speed up the growth rate of plants, until the maximum is achieved and thereafter growth becomes inhibited. The effect of climate change on the global yield of crops over the time period of 1998–2002 was found to be positive for rice (+2.9%) and soybean (+1.3%) but negative for maize (−3.8%) and wheat (−2.5%), which demonstrates the complexity of predicting the impact of climate change on food production. Invasive weed species are affected by climate change, so different approaches to weed management will be needed. Reproduction and growth rates of livestock will be reduced when temperatures exceed optimum ranges, as will milk and egg production.

It has been proposed that, based on the increased amount of GHG, land temperatures should have increased even more, but the rising temperatures were buffered by the oceans that trap nearly 90% of the Earth's heat. Consequently, ocean temperatures have been steadily increasing since 1955, causing levels to rise (water expands in volume when heated), and providing energy for storms. Melting of the glaciers and polar ice caps is well documented within the past decade, adding to the higher ocean levels. The EPA estimates that the average sea level has increased at a rate of 0.11–0.14 in./year between 1880 and 2013. The combination of more powerful storms, warmer air and water temperatures, and higher ocean levels has resulted in unprecedented damage to the US coastal areas such as New Orleans (Hurricane Katrina in 2005), North and South Carolina (Hurricane Irene in 2011), and New York and New Jersey (Hurricane Sandy in 2012). Severe winter storms have also occurred, including the two blizzards referred to as

"Snowmageddon," which hit Washington DC in 2010, and the most snowfall on record for Boston during the winter of 2014–15. Noncoastal areas of the United States have experienced extreme weather events as well including severe flooding of the Mississippi and Missouri Rivers in 1993; an EF5 tornado that hit Joplin, Missouri in 2011; and the 2012–13 drought across the southern and western United States. The impact of storms on agriculture can be significant including short-term losses and damage to crops and livestock, as well as long-term effects on the land and soil including erosion and contamination.

8.3 SUSTAINING NATURAL RESOURCES

Agriculture is dependent upon access to suitable and sufficient land, soil, and water. Planting and harvesting crops and raising animals for food consumes these natural resources. Some types of agriculture may create damage to the environment through overuse or contamination. Producing food while protecting natural resources, biodiversity, and wildlife is the goal of sustainable agriculture.

Human activities, and specifically agriculture, have changed the natural environment. The cultivation of crops has reduced wilderness, shifted the balance of plant and animal species, and altered the hydrological cycle. A decline in soil productivity and a loss of topsoil due to wind and water erosion occurs when soil is not protected by plant cover. Overuse of surface and ground water for irrigation has caused water scarcity and salinization of soils in irrigated farming areas. Agricultural practices contribute to water pollutants such as sediments, salts, fertilizers (nitrates and phosphorus), pesticides, and manures. Pesticides are commonly found in groundwater beneath agricultural areas and reduced water quality impacts agricultural production, drinking water supplies, and fisheries. Over 400 insect pests and more than 70 fungal pathogens have become resistant to one or more pesticides. Use of land for agriculture has removed wetlands and wildlife habitats and there is reduced genetic diversity due to reliance on uniform crops and livestock breeds.

Agriculture was necessary for, and perhaps supportive of, population growth and the civilizations that resulted. In the distant past, natural resources were thought to be unlimited and the impacts of agricultural practices on the environment were considered inconsequential. Technology stayed one step ahead of the limitations of nature to allow food production to continually increase. But it is clearly evident now that natural resources are limited and the environmental impacts of agriculture are real. In order to continue to provide sufficient, high quality, abundant, and affordable food, agriculture must adapt. Returning to a subsistence approach to agriculture is not possible, so methods to advance food production sustainably are needed: "Today, any strategy to eliminate crop farming and domesticated livestock would have to include the means of eliminating most humans now on earth, for only a small remnant could live by hunting and gathering" (Conkin, 2008, p. 169).

The USDA Climate Change Program Office tracks the impact of climate change on agriculture, forests, grazing lands, and rural communities. A report that summarizes research on these was published in 2015 (U.S. Department of Agriculture, 2015a) and the main findings of the report are listed in Table 8.2.

8.3.1 Land Use in Agriculture

Only one-quarter of the total surface of Earth is dry land; the other three-quarters are covered with water. Only 1/32 is arable land, meaning that all of the world's food production occurs on this very small percent of

TABLE 8.2 Key Findings from the USDA Climate Change Report

Key finding	Impacts
Climate change is very likely to affect global, regional, and local food security by disrupting food availability, decreasing access to food, and making utilization more difficult.	Increased food prices, variability in food production, higher risk of food insecurity in low-resource populations.
The potential of climate change to affect global food security is important for food producers and consumers in the United States.	Changes in types and costs of imported foods to the United States, increased demand for food assistance to other countries, dissemination of US agricultural technology.
Climate change risks extend beyond agricultural production to other elements of global food systems that are critical for food security, including the processing, storage, transportation, and consumption of food.	Greater risk of food-borne pathogens and food spoilage, disruption in transportation routes, higher food costs.
Climate risks to food security increase as the magnitude and rate of climate change increase. Higher emissions and concentrations of greenhouse gases are much more likely to have damaging effects than lower emissions and concentrations.	Models that predict the effects of GHG on global food insecurity range from minimum change to significant increases that depend on how high GHG emissions rise.
Effective adaptation can reduce food system vulnerability to climate change and reduce detrimental climate change effects on food security, but socioeconomic conditions can impede the adoption of technically feasible adaptation options.	Technological solutions will address some of the impacts of climate change, but the capacity to implement these solutions will vary; continued research to develop solutions is needed.
The complexity of the food system within the context of climate change allows for the identification of multiple food security intervention points, which are relevant to decision makers at every level.	Applying current knowledge about climate change impacts is needed to provide food security; continued investment in research in advanced technologies for food storage and packaging is needed.
Accurately projecting climate change risks to food security requires consideration of other large-scale changes.	Ecosystems and land degradation, technological development, population growth, and economic growth must all be considered.

From U.S. Department of Agriculture. (2015). Climate change, global food security, and the U.S. food system. *Washington, DC: U.S. Department of Agriculture. <www.usda.gov/oce/climate_change/FoodSecurity.htm>.*

Earth's surface. The total land mass of the United States is nearly 2.3 billion acres and agricultural production occurs on about 51% of the land. The amount of land used for agriculture has declined nearly 13% since 1949. Gradual declines have occurred in crop, pasture and range land, while grazed forestland has decreased more rapidly. The distribution of land use in the United States in 2007 was:

- 671 million acres in forest
- 614 million acres in pasture and range
- 408 million acres in cropland
- 127 million acres in grazed forestland
- 60 million acres in urban use
- 12 million acres in farmsteads and farm roads

Cropland acreage increased by about 4 million acres between 2007 and 2012 after a steady decline over the previous 25 years. Most of this gain came from cultivating land that had been in the Conservation Reserve Program (USDA program that pays farmers to not use certain areas of their land). The higher prices for commodity crops during this time period was a motivating factor to farm more land because more money could be earned from the crops than from the government funding. The consequences of this loss of conservation land will likely include more soil erosion and reduced water quality and loss of wildlife habitat.

Nearly half of all US cropland is concentrated in the Midwest and the Northern Plains. Almost all cropland is privately owned, while 62% of grassland pasture and range is privately owned and 24% is owned and managed by federal grazing programs. With a value of $2.38 trillion, farm real estate (land and structures) accounted for 80% of the total value of US farm sector assets in 2014. Only 1.7% of privately owned land in farms or forest (22.8 million acres) was owned by foreigners in 2009. As the global demand for farmland and food production increases impacts are likely to affect the future dynamics of farm ownership, land prices, and agriculture in the United States.

Land use is influenced by history, quality and capacity, location, access to infrastructure, price, planning and zoning regulations, and market demands for crops or development. How land is used has important economic and environmental implications for commodity production and trade, access to open space, soil and water conservation, air quality, and atmospheric greenhouse gas concentrations. Population pressures are increasingly changing agricultural land access and use.

> The increasing mix of rural and urban land uses creates added social conflict and environmental impact. Farmers are faced with complaints about odor, dust, or noise, and perhaps experience more trespassing. Impermeable surfaces increase, directing rainwater to sewer drains rather than to the soil. New chemicals—from road salt to lawn care pesticides—are introduced into the environment. The total U.S. conversion of prime farmland to urban or builtup land between 1982 and 1992 translates into 45.7 acres every hour over those 10 years. An additional 266,000 acres (3 acres every hour) of unique farmland (soil and climatic conditions suitable for production of specific high-value food and fiber crops) was also lost to development. As prime farmland is being developed, less stable non-prime farmland in arid regions is being added to the base, leading to increased erosion rates and irrigation demands.
>
> *Heller and Keolian (2000)*

As the population of the United States grows, cities and towns expand creating a demand for more land to build houses, roads, malls, airports, hospitals, power and water treatment plants, and other infrastructures. A survey conducted by the USDA in the early 1980s found that millions of acres of farmland were being converted to urban use. To counter this, the Farmland Protection Policy Act was passed and implemented in 1994. This legislation mainly requires that the impact of irreversibly converting agricultural land to nonagricultural uses is considered and documented, but does not prohibit such conversion. Individual cities and counties determine how land will be used. The encroachment of urban and suburban areas into agricultural land prompted the passing of Right to Farm acts by all 50 states. These laws are meant to protect farmers and ranchers from lawsuits filed by people who move into agricultural areas and want to stop the existing farming operations. Concentrated animal feeding operations (CAFOs) are often the target of these lawsuits because of air quality and aesthetic issues, but all types of farming operations have been impacted. The statutes of these laws differ by state and many disputes end up being settled in court.

8.3.2 Soil Erosion

The USDA defines *soil* as "...a natural body comprised of solids (minerals and organic matter), liquid, and gases that occurs on the land surface" (USDA Natural Resources Conservation

Service). Soil composition that provides the right amounts of nutrients, retains the right amounts of water, and has sufficient depth to support plant root development is a key aspect of agriculture productivity. The breakdown of sedimentary rock and leaching of minerals into water defines the natural mineral composition of soils, which will vary depending on the geographic location. For example, the red soils that are characteristic of the southern United States are rich in iron oxides. Organic matter provides soil with nutrients and gases required for plant growth, and the ability to hold and filter water. Organic matter refers to living and decomposed leaves and roots, as well as animal and biotic material, including earthworms, insects, fungi, bacteria, protozoa, arthropods, and algae. Humus is the final stage of decomposed organic matter. Chemically, organic matter contains cellulose, hemicelluloses, starch, pectin, lignins, and protein from the plant residue. The amount of organic matter in soil fluctuates from season to season, depending on the continual process of decaying plant material and new growing plants or the application of manure.

Regions of the United States have different types of soil that directly influence the types of crops that can be grown, and these soils have evolved over millions of years. Human interaction with the land directly impacts soils. Soil loss through erosion and damage from overuse, as well as chemical or physical contamination, are significant concerns for sustaining agricultural production.

Soil erosion is the breakdown, transport, and redistribution of soil particles by forces of water, wind, or gravity. The Dust Bowl (Chapter 2: History of US Agriculture and Food Production) was a prime example of significant soil erosion caused by human and environmental factors, but less dramatic loss of soil occurs constantly. The average rate of US soil erosion on total cropland in 2012 was 4.60 tons per acre. Of that total, 1.94 tons per acre were eroded by wind and 2.66 tons per acre by rainfall and runoff. To put that into perspective, 4.6 tons would fill a dump truck.

Soil erosion is localized due to soil characteristics, topography, climate, landscape, and farming practices. As much as 50% of total water erosion of soil occurs in the Midwest and Northern Plains where there is high production of field crops. The majority, 90%, of wind erosion of soil occurs in the Northern Plains, Southern Plains, Mountain, and Great Lakes states. Water and wind erosion was lowest in the Northeast (2.7 tons per acre per year) and highest in the Southern Plains (8.8 tons per acre per year). The USDA encourages farmers to protect soils from erosion, which has resulted in a decrease in the total amount of soil erosion caused by water and wind from 3.06 billion tons in 1982 to 1.73 billion in 2007 (Fig. 8.3).

The consequences of soil erosion on agricultural production are significant. When soil is lost from croplands, plant nutrients including nitrogen, phosphorus, potassium, and calcium are reduced and have to be replaced. Adding back these nutrients costs farmers billions of dollars annually. Wind and water selectively remove the top layer of soil where the majority of organic matter is found. Loss of organic matter is correlated with lower crop yields. Irrigation may be needed in areas where there is insufficient organic matter to retain moisture in the soil. Fields with inadequate soil depth have fewer bacteria and biota that enhance soil quality. Earthworms, ants, and snails engage in soil cycling that aerates and mixes soil and improves water filtration. Beyond the farmer's land, soil erosion affects the environment by adding sediment and excessive nutrients (eutrophication) to streams and rivers, which negatively affect fish and wildlife. Algae blooms caused by nutrient runoff into waterways consume the oxygen (hypoxia) needed by fish leading to massive fish kills. High levels of soil-derived minerals, such as nitrogen, must be removed from public drinking water sources.

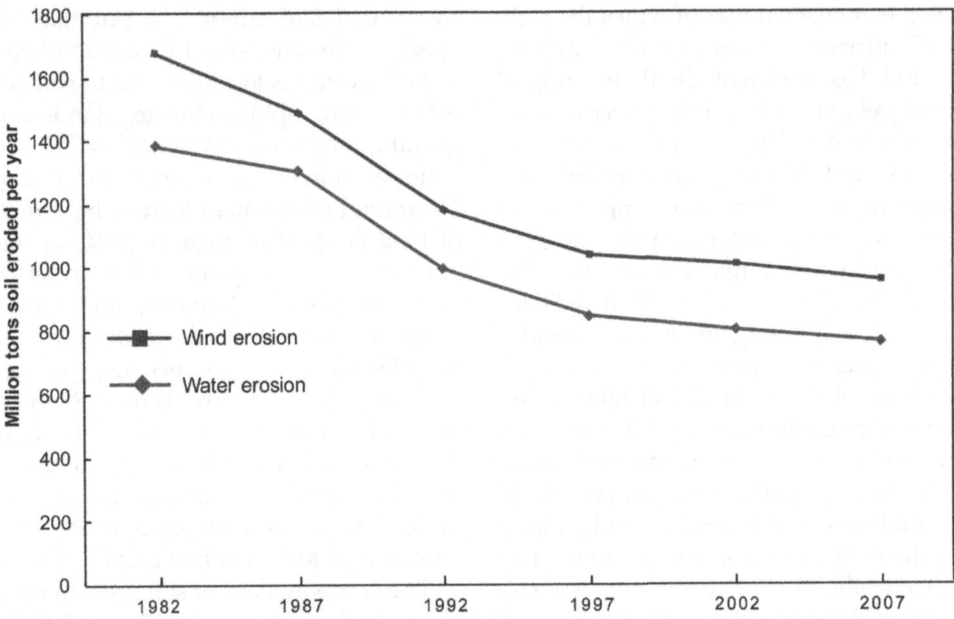

FIGURE 8.3 Soil erosion caused by both wind and water has decreased in the United States since 1982 when farmers were encouraged to implement conservation measures. Soil erosion continues to be a significant problem for agriculture. *Source: Natural Resources and Conservation Services, www.nrcs.gov.*

Once soil is lost from the land it cannot be easily replaced. Depending on the climate and organisms in the soil, organic matter can be regenerated within a few years. Replacement of sand, silt, and clay takes much longer—decades or more—because these are derived from decomposition of rock. Since the Soil Conservation Act was passed in 1935, the USDA Natural Resources Conservation Service (NRCS) has promoted ways to reduce soil erosion. Practices that were implemented included planting trees and shrubs as windbreaks, no-till, contour and terrace plowing, planting grass buffer strips between fields and streams, mulching, and planting of perennial crops. As a consequence of the farm crisis, the Food Security Act passed in 1985 created the Conservation Reserve Program (CRP), which paid farmers to convert some of their cropland to conservation areas. These areas provided natural filtration systems to protect streams and rivers, and increased biodiversity by providing habitat for birds and animals. Newer approaches to reduce soil erosion and enhance organic matter include planting cover crops, such as alfalfa and clover, that can be planted after the crop is harvested and then tilled into the soil before the next planting season. Dividing fields into alternating bands of row crops and hay or small grains, planting permanent strips of grass, or building embankments or structures across drainage areas are management tools that prevent soil erosion. Limiting agriculture, construction, and development on highly erodible land, fragile lands, riverbanks, and coastal lands also prevents loss of soil.

8.3.3 Water Use in Agriculture

Water plays an essential and complex role in agriculture. Rain supplies water for use directly by crops, but water obtained from lakes, rivers, oceans, and groundwater is required for crop and livestock production. Having the right amount of water during key points of crop development directly influences yield, and animals require substantial amounts of water throughout their lives. Waterways are used to transport agricultural products. Agriculture contributes to water contamination from chemical and nutrient runoff. This has negative effects on fish and plant life and drinking water supplies for cities and towns. Wetlands in the United States have been drained to make the land suitable for agriculture, leading to reduced environments for wildlife. Groundwater reserves are becoming depleted in some regions, creating problems for both agriculture and people living in those areas. In contrast, farmers have developed ways to mitigate water runoff from their lands using buffer zones, and returned parts of their land to the wild to restore natural areas. Technology has been applied to develop crops that need less water, planting systems that retain and conserve water, and irrigation systems that optimize water administration. Water conservation, management, and protection are core aspects of agricultural production.

In the United States about 355,000 million gallons of water are used per day. The top two uses for water are to generate thermoelectric power (45% of total water use), and irrigation (33%; Fig. 8.4). Over 98% of water used for power is recycled so losses are not of as great concern as is water used for irrigation. Irrigation, which includes agricultural as well as nonagricultural uses, consumes about 115 billion gallons of water per day in the United States. In 2010, about 62,400 acres were irrigated, and almost all of that occurred in western states. The main regions of irrigation are the Columbia and Snake River basins, the Central Valley of California, the Great Plains, and the Mississippi Delta. California, Nebraska, Texas, Arkansas, and Idaho which account for 52% of total

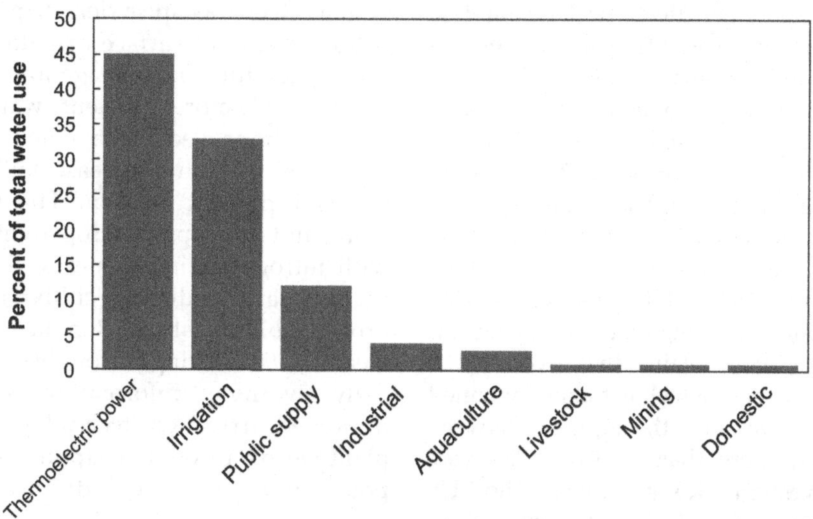

FIGURE 8.4 Thermoelectric power uses the most water of any sector, although this water is not consumed in the process. Surface and ground water used for irrigation of agricultural crops, golf courses, gardens, and lawns make up the largest consumption of water among all of the sectors. Water used in raising livestock is a relatively small use compared to public and industrial uses. *Source: From Maupin, M., Kenny, J. F., Hutson, S. S., Lovelace, J. K., Barber, N. L., & Linsey, K. S. (2014). Estimated use of water in the United States in 2010. In* U.S. Geologic Survey Circular 1405 *(p. 56). Available from <http://pubs.usgs.gov/circ/1405/pdf/circ1405> (Maupin et al., 2014).*

irrigated acreage. The states with the highest total amount of water used for irrigation (in thousand acre-feet per year) include:

- California—25,800 (surface water > groundwater)
- Idaho—15,700 (surface water > groundwater)
- Colorado—10,900 (surface water > ground water)
- Arkansas—9770 (groundwater > surface water)
- Montana—8030 (surface water > groundwater)
- Texas—7660 (groundwater > surface water)
- Nebraska—6340 (groundwater > surface water)

Surface water (rivers and lakes) contributes more than groundwater to irrigation in California, Idaho, Colorado, and Montana. These sources are dependent on rainfall and snow melt, and are usually regularly renewed. In the Central Plains, surface water is largely unavailable, and water deep underground is used for irrigation. The Ogallala aquifer, an underground reservoir that spans from western Texas to South Dakota, contains about 30% of the total groundwater used for irrigation and supports about 20% of the US wheat, cotton, corn, and cattle production. The aquifer was created millions of years ago and is a finite resource. The water in the aquifer cannot be quickly or easily replaced because of its geophysical structure. Growing crops or raising cattle on the dry, high plains that overlay the aquifer requires water, which is mainly obtained from wells that tap into the aquifer. Between 1937 and 1971, more than 65,000 wells were drilled in western Texas alone. The US Geological Survey found that the amount of water being removed from the aquifer was as much as 6 ft/year while only one-half inch was being replaced. Concerns over depletion of the aquifer were recognized in the past decade and the USDA established the Ogallala Aquifer Initiative to provide new approaches to reduce water use, improve conservation, and manage this valuable water resource. As the water in the Ogallala aquifer becomes less available, farmers who depend on that water source will need to reconsider the types of crops and methods of production they are using, perhaps moving away from water-intensive crops and livestock.

The intensity of water used in agriculture can be compared by estimating how much water is used to generate profit from the product. As shown in Fig. 8.5, grain farming utilizes more water per dollar of product generated than any other agricultural sector. Sugar production is also water-intensive, as is tree nut and fruit production. In this comparison, animal food production (poultry and egg production, cattle ranching, and milk production) is the least water intensive, when expressed relative to the economic return for these products.

Water scarcity may be the greatest threat to US agriculture in the future. Methods to reduce water use include improved irrigation systems (such as sprinkler, drip, or microirrigation instead of surface irrigation), conservation agriculture (less tillage and use of cover crops), and more efficient water allocation systems. Improved management is needed to retain soil moisture, increase soil organic matter, and prevent erosion. The use of cover crops and appropriate crop rotation methods with nitrogen-fixing legumes, limiting fallow periods, and reducing cultivation can help protect soil moisture. Remote sensing, local weather forecasting, drought-tolerant crops, early warning information systems, and improved irrigation technologies (based on plant needs through evapotranspiration) are potential techniques to reduce water use.

8.3.4 Precipitation and Climate Change

Measurable changes in the distribution of precipitation across the United States have

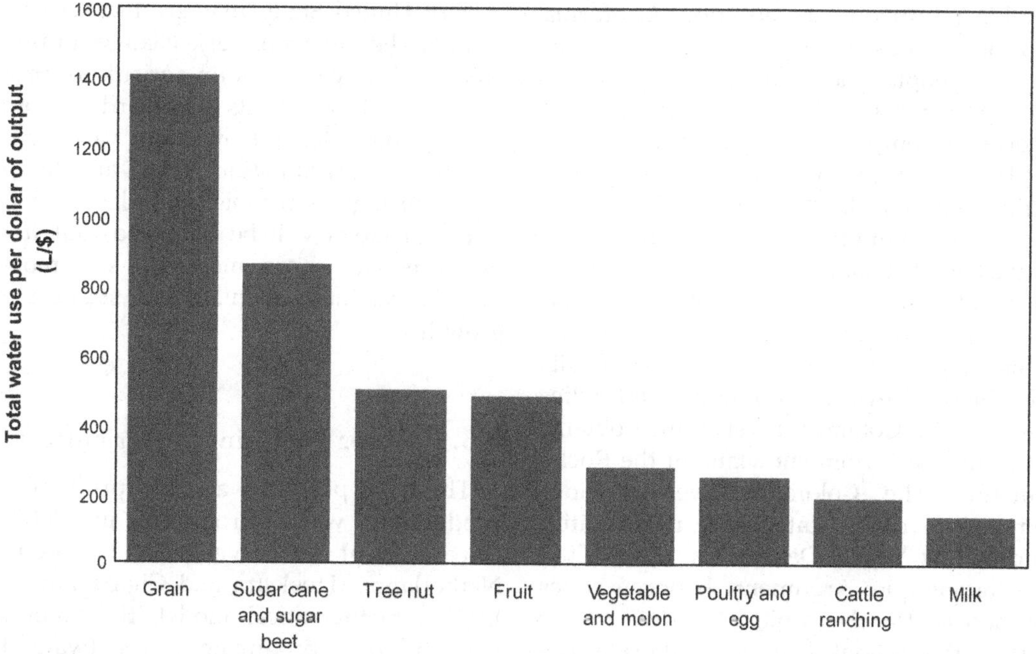

FIGURE 8.5 When water use is expressed relative to the amount of money generated from a product, grain production uses the most water per dollar of any agricultural system. Animal food production is at the low end of this scale because of the higher dollar value of these products relative to the amount of water used. *Source: From Maupin, M., Kenny, J. F., Hutson, S. S., Lovelace, J. K., Barber, N. L., & Linsey, K. S. (2014). Estimated use of water in the United States in 2010. In* U.S. Geologic Survey Circular 1405 *(p. 56). Available from <http://pubs.usgs.gov/circ/1405/pdf/circ1405>.*

occurred over the past 100 years. The northwest, central, and southern regions have received increased precipitation whereas the Eastern Seaboard, Rocky Mountains, and Southwest have had decreased precipitation. All regions have seen the intensity of precipitation events increase, associated with climate change. Larger amounts of water in a short period, as would be experienced in high-intensity events, are not conducive to agriculture. More erosion, runoff, and crop damage occur during such events, and the sediment and ground contamination that results can have long-term negative consequences on farmland. For example, flooding in the Midwest that occurred in the 1990s left significant amounts of silt and sand, as well as metal, rubber, wood, and other debris in the fields. Additionally, top soil was washed away. Removal of these contaminants required substantial cost and time, and the quality of the soil will take years to recover.

Drought is a major concern to agriculture as are periods of high temperature. Between 2012 and 2014 the United States experienced the worst drought since the Dust Bowl. The lack of precipitation was exacerbated by prolonged high temperatures causing 71% of the country to be classified as exceptionally dry in July 2012. California was significantly affected by this drought event, which lasted into 2016. California produces a large percentage of the fruits, vegetables, and tree nuts consumed in the United States and therefore has a high demand for water. Some 5.7 million acres of land are irrigated in the state, including nearly 100% of the orchard, berry, and vegetable production. In addition, California is one of the most populous states with over 30 million people who need water.

The history of water use and rights in California is complex and controversial. From the 1940s, the California State Water Project and the Central Valley Project have been managing dams, pumping stations, canals, and pipelines to distribute water. These systems direct water from a variety of rivers and streams to farmland and cities. California obtains about 30% of its water from snowfalls in the Sierra Nevada Mountains, and relies heavily on the Colorado River, which obtains most of its water from snowfalls in the Rocky Mountains. The Colorado River provides water to six other states and major cities including Las Vegas, Denver, Salt Lake City, and Albuquerque. Arguments between states over access to the Colorado River have occurred throughout history. California has tended to consume more than its share of the water and, during times of shortage, this has led to significant conflicts.

Lack of precipitation and low snowfalls over the past decade have exacerbated the drought in California with negative effects on agriculture. The state suffered a 22% decrease in agricultural revenue between 2014 and 2015, which represents a loss of about $9 billon. The precarious dependency of the US food system on California to produce the majority of fruits, vegetables, and nuts, which is in turn dependent on availability of water, raises concerns about sustainability. Climate change has increased the severity of weather events, and made weather forecasting more challenging, putting additional stress on farmers who must determine the types and quantities of food products they will produce.

The USDA ARS predicts that over the next 30–40 years, regions of the northwest and southern United States may see reductions of between 5% and 25% in summertime precipitation while regions in the north central and eastern United States may see increases of 5% to 15%. The impact of these changes in precipitation will have effects on agriculture production, but these effects are hard to predict because they depend on many variables. For example, if precipitation is low and temperatures are high, soil moisture is lost at a rapid rate and crops will be damaged, but if temperatures are more moderate soil moisture may be retained adequately to support plant growth.

8.3.5 Water Footprint in Agriculture

The concept of a water footprint for food production was introduced in 2007 by researchers at the University of Twente in the Netherlands (Hoekstra and Chapagain, 2007). In this mathematical model, the amount of water used in producing a crop (water footprint) is estimated by the total volume of crop produced and the water content of the crop. For animals, the model also includes the water content of feed and volume of water consumed during the animals' lifetime. From these calculations, a country's internal (water used for products generated by that country) and external (water used for products imported into that country) water consumption can be determined. The water footprint for US agricultural goods was estimated to be 1192 m^3/capita per year for internal production and 267 for external production, and the global averages were internal 907 and external 160 m^3/capita per year. Developed countries tend to have higher water footprints than developing countries. Some European countries, such as the United Kingdom and Italy, have high external water footprints because they rely heavily on imported foods.

To further define the ways water is used, the Dutch researchers classified water use as green, blue, or gray. Green water refers to rainwater use, blue water to ground or surface water use, and gray water estimates pollutants released into the water. In this model, animal products demand high water input, with pork and beef at the highest green water footprint of all animal products (Fig. 8.6). For plant foods, rice is the most water-intensive crop, consuming about 21% of the total volume of water used for all field crops. Wheat is the second highest crop for water use. In Fig. 8.7 a comparison of the global water footprint for some common foods is shown. Coffee, chocolate, and olive oil are among the most water-intensive crops, while fruits and vegetables have lower water footprints.

While a comparison of water footprint is helpful to determine water use, there are many factors that impact the total amount of water needed to produce foods. Air temperature, wind, and soil quality will dramatically alter the amount of water needed to raise a crop. Where food is grown becomes a very important consideration as water resources become more limited. Growing wheat in the Central Plains requires irrigation from the Ogallala aquifer, which is running low. The majority of the fruits and vegetables consumed by Americans are grown in drought-prone California. These regional food production systems have evolved over time in the United States based on economic and logistical factors. As climate change alters weather patterns and water resources become more vulnerable, it is likely food production patterns will also need to change. If these changes occur, the entire food system infrastructure will also need to adapt.

8.3.6 Water Contamination

In 1948 the Federal Water Pollution Control Act (FWPCA) authorized the Surgeon General of the Public Health Service to oversee reducing pollution of US waterways. Several amendments to the legislation were made through 1966, with authority for water quality under the Secretary of Health, Education, and Welfare. The Clean Water Restoration Act of 1966 authorized a comprehensive study of the effects of pollution on US waterways. The

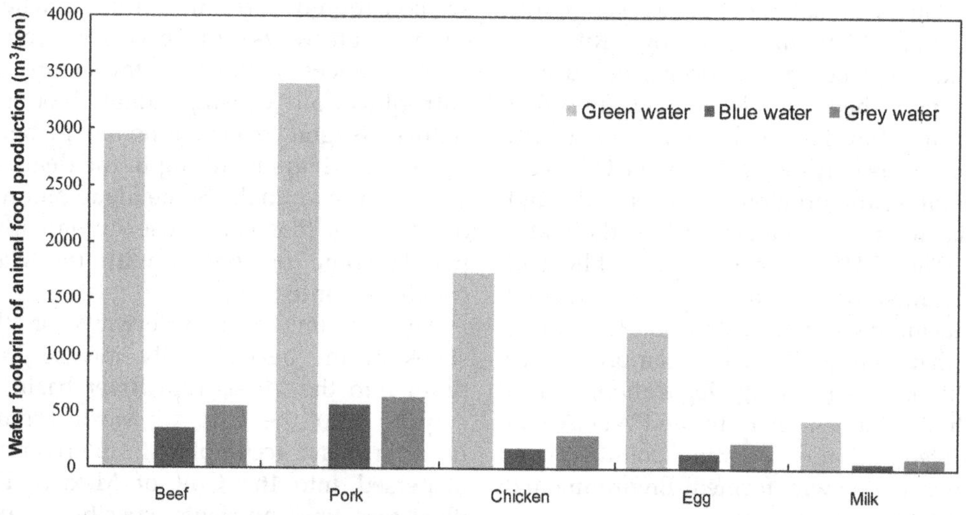

FIGURE 8.6 Beef and pork production have the highest green water footprint among animal food sources. Green water refers to rainwater used to produce the grains consumed by the animals, blue water is surface and ground water used in the animal's care, and gray water is the amount of water pollution as a result of animal production. *Source: From Mekonnen and Hoekstra (2011).*

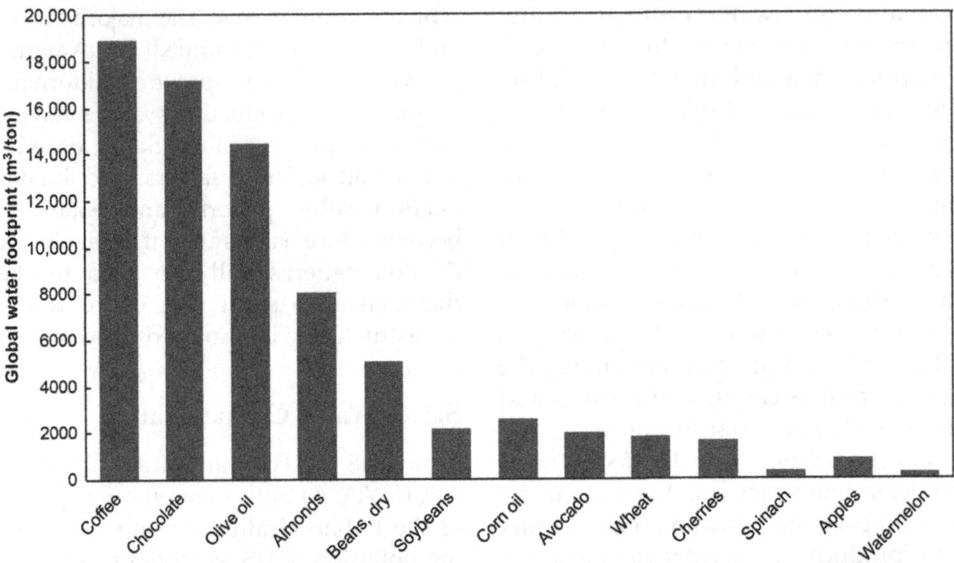

FIGURE 8.7 Coffee and chocolate have the highest water footprint among plant foods. These might be considered luxury foods that contribute little nutritional value, whereas dry beans, soybeans, and wheat, which have lower water footprints, are a valuable source of nutrients. *Source: From Mekonnen M.M. and Hoekstra A.Y., The green, blue and grey water footprint of crops and derived crop products,* Hydrologic Earth System Science *15, 2011, 1577–0600.*

report, published in 1969, found that half of the public drinking supply systems in the United States did not meet safety standards. Also in June 1969, the Cuyahoga River in Cleveland, Ohio became notorious because of a story published in *TIME* magazine that showed fires burning on the river. Fires had occurred on the Cuyahoga and other US rivers for several years previously, when oils and industrial wastes on the surfaces of the water ignited. The *TIME* story brought public and political attention to major environmental issues including water pollution. It became evident that there was a lack of comprehensive oversight of water quality legislation, which led to the 1972 amendment to the FWPCA that consolidated authority for water pollution control under the newly formed Environmental Protection Agency (EPA).

Agriculture operations can contribute to water contamination. The fertilizers and animal manure that are spread on farmland leach phosphorus, nitrate, fecal coliform bacteria, and suspended sediment to ground and surface water. Nitrogen and phosphorus are considered nutrient pollutants when they enter waterways in excessive amounts. Consequences of these contaminants include eutrophication causing algal blooms that reduce oxygen levels (hypoxia) in bodies of water and rivers resulting in the death of fish and aquatic animals. Some algae blooms produce toxins that can cause human illness if people come in contact with the water or consume tainted fish.

The streams and waterways of thirteen states in the agriculturally intense Midwest drain into the Mississippi River basin, which empties into the Gulf of Mexico. Nutrients from farmland accumulate in the river and are dispersed into the Gulf of Mexico. Fig. 8.8 illustrates how nutrients, specifically nitrates, accumulate in the Mississippi River from rivers and streams that drain the Midwest farmland. The EPA has identified 166 dead zones in US waterways, as a result of hypoxia, where

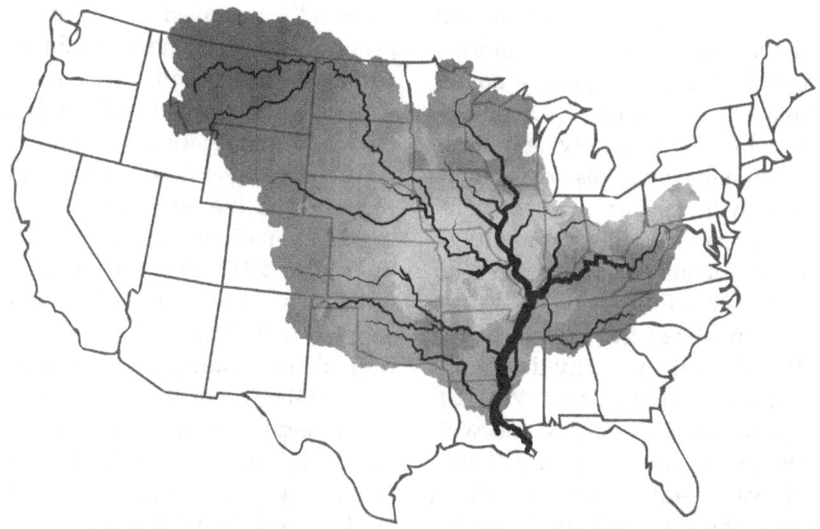

FIGURE 8.8 Rivers and streams throughout the Midwest drain into the Mississippi River and contribute nitrates from agricultural lands. These nitrates from the Mississippi River empty into the Gulf of Mexico, creating hypoxia and a large dead zone. *Source: Illustration by Robertson D.M., Saad D.A., and Schwarz G.E., Spatial variability in nutrient transport by HUC8, state and subbasin based on Mississippi/Atchafalaya river basin sparrow models,* Journal of the American Water Resources Association *50(4), 2014, 1–22.*

no fish or aquatic life survive. The Gulf of Mexico has the largest dead zone, at the mouth of the Mississippi River, which was estimated to be over 5800 square miles in 2013. About 40% of the US seafood supply ($82 million a year) comes from the Gulf of Mexico, and is threatened by these polluted waters.

The Chesapeake Bay on the East Coast has also experienced significant eutrophication and dead zones. The Chesapeake watershed covers six states (Delaware, Maryland, New York, Pennsylvania, Virginia, and West Virginia) with over 84,000 small farms. Because of the topography of the land in this region and the types of farming, runoff into the water systems is a significant problem. Eutrophication of these waterways has negative effects on the ecosystem and food production, and reduces the recreational quality of these areas. The USDA Natural Resources Conservation Service has been focused on restoring and protecting the Chesapeake Bay watershed for several years with positive results. Between 2006 and 2011 the amount of soil erosion has decreased in the region by 15.1 million tons per year; nitrogen levels have been reduced by 38% and phosphorus by 45%. The effort to clean up Chesapeake Bay is projected to cost nearly $19 billion.

Pesticides and herbicides that enter the water system are a concern for wildlife and humans. The EPA regulates all pesticides that are used in agriculture and has defined their proper use. But the Clean Water Act does not regulate agricultural pollution directly. Pesticide users must follow Federal Insecticide, Fungicide, and Rodenticide Act (FIFRA) regulations and secure the proper permits. Inspection and oversight of most pesticide use is conducted by regional or local NRCS staff. A wide range of agricultural chemicals, primarily herbicides and insecticides, are approved for use in the United States. Each of these chemicals has a defined mode of action, optimum effective concentration, toxicity threshold, and degradation rate. The USGS National Water Quality Assessment Program (www.usgs.gov) tracks the amounts of each of these chemicals in water and the environment and posts these data on their website. Monitoring the effects of these chemicals on aquatic animals, wildlife, and humans is an ongoing challenge.

Implementation of water quality standards for wetlands and drainage fields has occasionally stirred controversy with farmers who contend that the EPA's interpretation of clean water regulations will have detrimental economic effects on agriculture. In 2015 the EPA and US Army Corps of Engineers finalized the Waters of the United States (WOTUS) rule to clarify the definition of "waters of the United States." WOTUS makes it clear that streams, tributaries, and wetlands—essentially any water that channels into larger bodies of water—fall under the Clean Water Act. Critics of the rule argue that complying with Clean Water Act regulations for all of these waters would require planting extensive buffer strips along waterways, including rural drainage ditches, to reduce runoff from farm fields. Some farmers raised concerns that this could result in millions of acres of farmland potentially removed from cultivation, causing profit loss for farmers. Additionally, the rule would enforce Clean Water Act regulations on waters within private land, which would be a broader interpretation of the act than had ever been held. Several states filed lawsuits against the rule and the Sixth Circuit Court of Appeals granted a nationwide stay on the WOTUS rule in October 2015. The position of the EPA is that mitigation of water contamination from all sources, including field runoff, is expensive and difficult to achieve, and a better strategy is to prevent the contamination at the source. Because water seeps and runs through all land, it is necessary to address contamination wherever it occurs. Opponents of the rule, especially private land owners, argue that it gives the government authority to define how they manage their land and resources. Clearly, this is an ethical dilemma that pits individual rights against the utilitarian perspective of protecting the nation's water systems. A solution to this dilemma will require discussion and scientific thinking from both sides.

EXPANSION BOX 8.1

NITROGEN FERTILIZER

Nitrogen is essential for plant growth and reproduction. Nitrogen is used in photosynthesis so plants deficient in nitrogen are yellowish while those with adequate nitrogen are dark green. Plants take up simple inorganic nitrogen compounds from soil as either ammonium (NH_4) or nitrate (NO_3). There are various forms of nitrogen (nitrite NO_2, nitrate NO_3, nitrous oxides NO, NO_2 and HNO_3, and ammonia NH_3/ammonium NH_4) in the soil and these forms change depending on conditions of pH, moisture, temperature, and oxygen. Utilization of organic nitrogen by crops requires prior transformation to inorganic ammonium or nitrate by soil microbes.

Legumes, such as soybeans and alfalfa, have a type of bacteria in their root nodules that converts nitrogen gas (N_2) to usable forms (ammonia or NH_3) of nitrogen. These plants are referred to as "nitrogen-fixing" plants. Nonlegume crops such as wheat and corn do not fix nitrogen and must rely on biologically available nitrogen in the soil or nitrogen applied from commercial fertilizer or manure.

The most common form of nitrogen fertilizer is anhydrous (without water) ammonia (NH_3). During WWII, anhydrous ammonia was produced in large quantities to be used in munitions. When the war ended, it was recognized that it could be an effective and low-cost

EXPANSION BOX 8.1 (*cont'd*)

fertilizer. About one-third of the nitrogen fertilizer used in the United States is applied as anhydrous ammonia. Other forms of nitrogen fertilizer are ammonium nitrate, urea, and ammonium phosphates. Anhydrous ammonia is a hazardous chemical and must be handled with care during transport and application. Applied to the field as a liquid, anhydrous ammonia is converted to ammonium (NH_4) gas when it reacts with water, which is then readily absorbed by soil and plants.

Manure (animal feces and/or urine mixed with plant material or bedding) is organic matter that contains large amounts of nitrogen. Manure is composted for a period of time before being used as a fertilizer to allow any pathogenic microorganisms to decay. CAFOs collect manure into holding lagoons, which allows manure slurries to be applied to fields as liquid fertilizers. Dry manure is also collected and can be distributed onto fields with a manure spreader. Manure provides a valuable source of organic matter, which improves soil composition, increases the water-holding capacity of sandy soils, improves drainage in clay soils, provides a source of slow-release nutrients, reduces wind and water erosion, and promotes growth of beneficial soil organisms.

Nitrogen fertilizers such as anhydrous ammonia and manure are relatively inexpensive and easy to apply. These factors have allowed farmers to apply nitrogen fertilizers abundantly to their fields. In some parts of the US excessive use of nitrogen fertilizers has led to high levels of nitrates in public drinking water systems. High levels of nitrates in drinking water can have mild health effects on children and potentially severe effects on newborn infants. Blue baby syndrome is caused when infants, less than 6 months of age, consume high levels of nitrates. The nitrates are converted to nitrites by bacteria in the infant's stomach and are absorbed into the blood. Nitrites affect the oxygen-carrying capacity of the blood causing a condition called methemoglobinemia. The lack of oxygen causes the skin to look bluish. The condition is readily reversible by sequestering the nitrites, but if untreated can be fatal. Older infants and children are not susceptible to blue baby syndrome because they have developed sufficient stomach acidity to prevent the conversion of nitrates to nitrites. Nitrates in drinking water are not a significant direct health concern for adults. The Safe Drinking Water Act requirements set a maximum nitrate level of 10 parts per million in community drinking water supplies. Water systems that use rivers and streams that have come through agricultural land often have higher levels of nitrates. These must be removed using a special type of filter (ion exchange or reverse osmosis) that can be expensive to operate. Reducing nitrogen runoff from agricultural land is a high priority to protect water quality and avoid these problems. Soil management practices, such as using optimum fertilization rates or applying forms of nitrogen that are better retained in the soil, can be used to help optimize crop yields, nitrogen use efficiency, and water quality. Newer technologies that allow farmers to measure nitrogen levels in the soil in real time are being developed to make fertilizer application more precise.

Suggested reading: Powlson et al. (2008).

8.3.7 Carbon Footprint

There are over 10 million known carbon-containing compounds, and all living things are carbon-based. As the fourth most abundant element on Earth, carbon circulates throughout the biosphere of land, water, air, plants, and animals. Carbon exists in the atmosphere primarily as CO_2 and CH_4, the main greenhouse gases (GHGs). Carbon footprint is defined as "The quantity of GHGs expressed in terms of CO_2-e, emitted into the atmosphere by an individual, organization, process, product or event from within a specified boundary" (Pandey, Agrawal, & Pandey, 2011). CO_2-e refers to CO_2 equivalents, which are a measure of the impact of a GHG on climate change relative to the impact of CO_2 (e.g., methane has a high CO_2-e). To conduct a carbon footprint analysis, a lifecycle assessment (LCA) must be generated that includes measurements of CO_2-e at each step "from cradle to grave" for the product. For a food product, defining a carbon footprint takes into consideration CO_2-e during production, harvesting, processing, packaging, distribution, and consumption. LCA is a useful analytical method to evaluate resource consumption and the burdens on the environment associated with a product, a process, or an activity.

The large amount of CO_2 produced by human activities is a major contributor to climate change. High CO_2 levels cause the Earth's surface to retain heat, which affects weather patterns, so reducing these levels is a priority. The burning of fossil fuels produces 54% of the total CO_2 produced on the planet. When burned, one gallon of gasoline produces 20 pounds of CO_2 and 10 pounds of coal generates 29 pounds of CO_2. In contrast, plants consume CO_2 in the process of photosynthesis and sequester it in their biomass. It has been estimated that one tree can absorb 48 pounds of CO_2 in a year and, over a 40-year lifespan, retain 1 ton of CO_2. The US Forestry Service estimates that forests offset only about 15% of total GHG generated, and the amount of forested land has declined from about 1023 million acres in 1630 to 766 million acres in 2012 due to urbanization. Periods of high insect infestation, such as the pine beetle outbreak in the mid-2010s that killed millions of trees in the western United States, reduce the capacity of forests to remove CO_2. Further loss of trees from widespread forest fires was exacerbated by the drought conditions in the western states and the high number of dead trees from insect damage. Forest fires are doubly damaging because they take away the carbon sequestration activity of trees, and release massive amounts of CO_2. Loss of forest land is a significant global concern, especially the dramatic loss of rainforests that are being cut down to grow crops.

Agricultural crops consume CO_2 from the atmosphere as they grow. A Michigan State University study found that 1 acre of corn absorbs about 36,000 pounds of CO_2 during the growing season. Most of this will be returned to the atmosphere through animal or human consumption of the grain, conversion to and burning of ethanol, or degradation of the plant biomass, so in this way crops are carbon-neutral. The processes involved in growing agricultural crops, however, are not carbon-neutral. Soil that contains organic matter is a sink for CO_2 but CO_2 is released when soils are tilled and the organic matter is exposed to air and water. Tilling releases methane from the soils, which is a significant GHG. Tractors, powered by engines burning fossil fuels that generate CO_2, are used to till, plant, spread fertilizers, and harvest crops. And fertilizers and chemicals used on the fields generate CO_2 and other GHGs. Overall, raising crops produces more GHG than are consumed, contributing about 9% to the total GHG emissions in the United States.

Livestock production generates GHGs, mainly CO_2, CH_4 and nitrous oxide (N_2O), that are directly produced by the animals, and

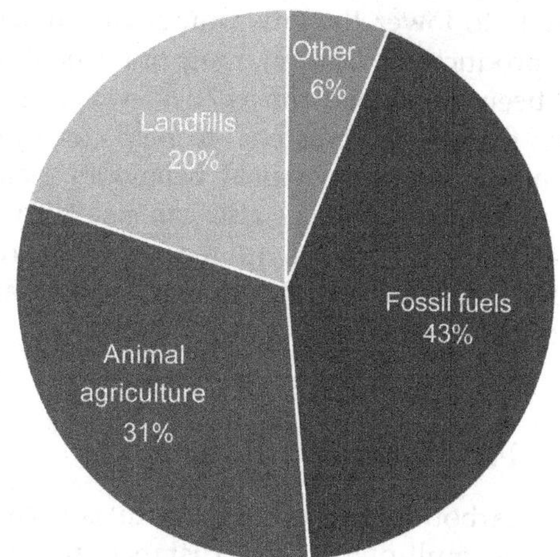

FIGURE 8.9 Fossil fuel use (coal and petroleum) and animal agriculture generate the majority of methane that contributes to greenhouse gases. Landfills are also a source of methane generated from food wastes. *Source: Environmental Protection Agency, www.epa.gov.*

indirectly from the growing of the feed they consume, the manure they produce, and the energy used in the animal management operations. About one-third of the methane production in the United States comes from animals (enteric fermentation plus manure). Fossil fuels (natural gas, petroleum, and coal) account for 42% of methane production and landfills add 20% (Fig. 8.9). Beef cattle produce about 65% of the total livestock GHG emissions, dairy cattle 20% and swine 8%. How animal waste is handled determines the amount of GHG produced. CH_4 generation is high when manure is stored in liquid slurries, and low if stored in dry lots. In contrast, N_2O generation is high when manure is stored in dry lots, and low if stored in liquid slurries.

Production of biogas from waste products including animal manure, human sewage from waste water treatment plants, and food wastes is an approach to lower GHG emissions and generate a renewable source of fuel. From a process of anaerobic digestion, the organic matter in these waste-streams can be converted to biogas containing a high amount of methane. Methane can be used as an energy source to replace petroleum in vehicles, or the biogas can be further processed to generate pure hydrogen. Hydrogen is used in fuel cells that power electric cars and engines. It has been estimated that the amount of biogas that could be generated from landfills, waste water treatment plants, and animal agriculture could be substantial. While the technology to generate biogas is fairly simple and inexpensive on a small scale, it is complex and costly to implement on a commercial scale. California has taken the lead nationally to implement statewide clean air standards, which makes biogas more economically viable as a source of energy. By converting waste materials to biogas, the California Energy Commission generated 3.44% of the state's electricity in 2015. The dairy industry has also been proactive in developing biogas digesters that can be installed on the farm to generate energy for the operation. The digesters use manure and other biomass waste to generate biogas. By capturing the GHG from manure to generate fuel, less fossil fuel is needed and the impact on the environment is reduced.

It is difficult to calculate the carbon footprint of individual food products because of the many variables involved from production to consumer. Efforts to standardize this assessment have been proposed. Table 8.3 lists estimates of GHG emission from production of 100 kcal of food by category. Foods derived from animals top the list for the highest GHG production. It is helpful to further define the sources of GHG within a production system. Using milk production as an example, the amount of CO_2 contributed at each stage can been estimated (Fig. 8.10). The highest sources of GHG from milk production are the enteric methane production and manure management, followed by feed and transportation. Using these types of analyses, farmers and researchers can identify where to implement conservation

TABLE 8.3 Greenhouse Gas (GHG) Estimates for 100 kcal Portions of Foods

Food category	GHG (kgCO_2e)
Meat from ruminants	857
Fish	517
Mixed dishes	312
Pork, poultry, eggs	308
Fruits and vegetables	290
Dairy foods	216
Snacks and sweets	91
Starches (cereals, grains, bread)	61
Food fats	55

From Vieux, F., Soler, L.-G., Touazi, D., & Darmon, N. (2013). High nutritional quality is not associated with lowest greenhouse gas emissions in self-selected diets of French adults. American Journal of Clinical Nutrition, 97(3), 569–583 (Vieux, Soler, Touazi, & Darmon, 2013).

methods to lower the environmental impact of food production. Carbon footprint calculators have been developed (https://www3.epa.gov/carbon-footprint-calculator/) to help consumers evaluate their own personal behaviors regarding energy use. Farmers also can use software modeling to find ways to reduce the carbon footprint of their operations (www.cometfarm.nrel.colostate.edu).

8.3.8 Food Miles

The carbon footprint for food miles is calculated by multiplying the distance the food (or each food ingredient) has traveled by the carbon emission of the transportation type, airplane, truck, barge, or railroad. The distribution patterns of food in the United States have clearly changed since the days when 40%

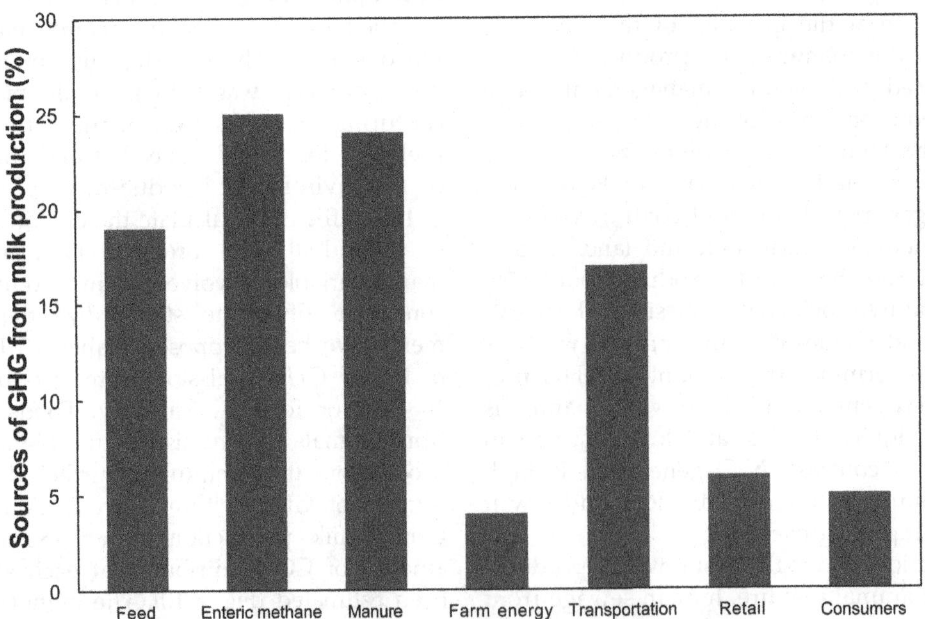

FIGURE 8.10 The main sources of greenhouse gases generated during milk production are from feed production, enteric methane release, and decaying of manure. Transportation of products by truck also generates significant greenhouse gases. *Source: From Thoma, G., Popp, J., Nutter, D., Shonnard, D., Ulrich, R., Matlock, M., … Adom, F. (2013). Greenhouse gas emissions from milk production and consumption in the United States: A cradle-to-grave life cycle assessment circa 2008.* International Dairy Journal, 31, S3–S14 *(Thoma et al., 2013).*

of Americans lived on farms and produced the majority of their own food. As discussed in Chapter 3, Innovations in US Agriculture, food production occurs in concentrated regions, with fewer but larger operations. More food is also imported from other countries. Distribution of foods from their site of production to consumers involves a complex network of handlers and modes of transport. Air transportation is by far the most energy-intensive method of food distribution and barge transport the lowest:

- Airplane—10.0 MJ/km
- Truck—2.7 MJ/km
- Railroad—0.3 MJ/km
- Barge—0.2 MJ/km

There are efficiencies of scale associated with larger food production systems that make them more economical, such as more mechanization to reduce labor costs, and higher volume generating more net income. There is a tradeoff in these efficiencies by adding the costs, both economic and environmental, to distribute the product. The mode and volume of transportation of food defines the net environmental impact. Food transported by road contributes 60% of the world's food transportation emissions because vans and trucks move smaller amounts of food, compared to railroads or barges. A smaller carbon footprint and less energy may be needed to deliver milk by bulk rail transport several hundred miles from a dairy to a distribution center, than for one consumer to drive 25 miles for 3 gallons of milk.

The purpose of "Buy Fresh Buy Local" (BFBL) initiatives is to reduce transportation miles and to support rural economies for the benefit of farmers and small town businesses. The sociological aspects of our current food system are very different than 50 years ago when more people were engaged in agriculture. Many consumers have lost connectedness with food production. The desire to regain this connection has fostered the buy local movement. Defining how buying local affects the environmental impact of farming is complicated. With an increased interest in local foods, there will be smaller deliveries made by more farmers. A consumer that drives to one large grocery store, once a week to purchase food will use less fuel than a consumer who drives to a farmer's market for produce, a dairy farm for milk, and a butcher for chicken. In contrast, the net amount of energy needed to deliver all the food to the large grocery store is high, whereas the farmer has essentially no transportation costs for the food s/he produces.

Limiting foods to those that are grown or produced locally would have an impact on those living in temperate climates, which is most of the United States. There would be limited access and availability of tropical fruits such as oranges, bananas, kiwi, pineapple, and avocado. Consumption of fresh fruits and vegetables would be limited to the summer months or to preserved or greenhouse-grown products during the rest of the year. The cost of all foods would likely increase. Consumers have come to expect all fresh fruits and vegetables to be available during the entire year and seasonality is a concept of the past. Based on today's economy and lifestyles, reliance on only foods produced locally cannot be achieved in most parts of the country. Quantifying these economic, social, and environmental impacts of food production and distribution will be needed to optimize the food system. It is not sufficient to simply assume a local food production model will correct the environmental impact without considering all aspects of the food system.

8.3.9 Invasive Weeds, Insects, and Diseases

The types and quantities of pests that affect agricultural production are influenced by the temperature, amount of sunlight, water, and humidity conditions. Climate change is predicted to have significant effects on the environmental conditions that define the survival and proliferation of agricultural pests. In 1999, President Bill Clinton signed Executive Order 13112, called the Invasive Species Act, which created an Invasive Species Council to oversee planning and implementation of approaches to mitigate invasive species. Invasive species refers to any seeds, eggs, spores, or other biological material capable of propagating a species that is not native to the ecosystem. Palmer amaranth is an invasive weed that negatively impacts crop production across the lower half of the United States. This plant is an example of a highly adaptable and invasive species that evolves quickly to survive in varying environments, and develops resistance to herbicides. The plant produces tiny seeds that are transported by humans, equipment, or in animal feed. Controlling the spread of the seeds is difficult. It is thought that the spread of Palmer amaranth infestation into Indiana, Michigan, and Wisconsin may have started when local beef and dairy cattle were fed grain from amaranth-infested fields in the southern United States. The seeds were disseminated onto croplands when the manure from the animals was spread for fertilizer. Several other invasive weeds have expanded their territory as they have adapted to new climate conditions, requiring farmers to learn quickly how to manage these pests.

Plant pathologists, who study crop diseases, have found that wind and precipitation patterns, as well as seasonal temperature changes, allow certain plant diseases to spread into new areas. The soybean rust fungus is such an example. Soybean rust was not present in the western hemisphere until 2001 when it first appeared in South America. By 2004 soybean rust was reported in the United States, first in Louisiana and thereafter in other southeastern states. The locations of soybean rust followed the path of Hurricane Ivan, which had impacted that region in September 2004, suggesting that the fungus may have been transported from South America in the wind and water of that storm. Soybean rust causes severe economic losses to soybean farmers and the presence of the disease in the United States raised great concern. In response, the USDA implemented a coordinated soybean rust monitoring and tracking program soon after the disease was discovered that involves farmers, extension educators, agribusinesses, and USDA-APHIS. Thus far, the spread of soybean rust has been contained to the Southeast and not spread into the Midwest soybean farming areas, mainly because the fungus is not able to survive the freezing temperatures of the Midwest. Plant pathologists are concerned that as climate change causes winters in the Midwest to become warmer, soybean rust may spread into these areas.

There is evidence from agricultural research data in the United States that some insects have increased their geographical ranges and others are surviving through the winter as temperatures have increased. Corn flea beetle causes damage to young corn plants and transmits a bacteria to the plant that causes Stewart's wilt. The beetle and the bacteria combine to cause significant damage to corn production. The corn flea beetle's survival is dependent on the temperatures during December, January, and February. When temperatures are warm during those months, as has occurred in recent years, the pest survives and can cause damage to the spring crop. Because of the large variety of agricultural pests, and their close interactions with the environment and crops, tracking each species will be necessary to fully understand how they are responding to climate changes.

The spread of invasive weeds, insects, and plant diseases will be continually influenced by weather and agronomic practices. Climate change increases the chaos in predicting where and when these pests will spread. Regions of the country that were protected from some pests due to climate or soil conditions will find them occurring more frequently, while other areas may see a reduction in some pests. The effectiveness of pesticides and herbicides is likely to become increasingly reduced as pests develop resistance.

8.3.10 Biodiversity

The number of living species, not including bacteria, on Earth has been estimated to be between 2 and 8 million, but only about 1.3 million of these have been catalogued. The precise number of species is in constant flux as new species arise and others become extinct. Determining how many species come (diversification) and go (extinction) over time is difficult to measure. There is evidence that the number of species extinctions is higher in the modern era compared to prehuman era, by perhaps as much as 1000 times. The Convention on Biological Diversity (1992) recognized global diversity as a public resource. The loss of species diversity as a form of harm was noted in the US Endangered Species Act (1973).

Genetic diversity, which defines biodiversity, is necessary for adaptation to various environmental stresses. Agriculture, through selection and breeding, tends to reduce genetic diversity within plant and animal species. The top producing species become dominant and crowd out other species. A monoculture results when a single species of crop, plant, or livestock type is grown at one time. Reduced biodiversity has occurred with agricultural domestication. Over 80% of the world's calories are provided by only 10–20 crops. In the United States, 42% of the soybeans, 43% of the corn, and 38% of the wheat were dominated by the top six varieties. The hundreds of corn hybrids grown in the United States are based on about 12 inbred lines that originated from a few open-pollinated varieties of a single race, although there are some 200 known races of corn. Uniformity in planting, cultivation, and harvesting means reduced competition for nutrients, sunlight, and space and allows efficient management of the crop, resulting in high yields. Specialization in one crop reduces the amount of labor and equipment needed for farm operations. As noted in Chapter 4, Animals in the Food System, modern farm operations typically specialize in 1–3 products, unlike farms in the past that produced many different crops and animals. Animals bred for selective traits of high growth, high feed efficiency, and optimum milk, meat, or eggs are more cost effective for the producer.

The risks of this reduced genetic diversity are a loss of disease and pest resistance, or growth and reproductive problems. In field crops, planting the same species repeatedly in the same fields drains the fields of selected nutrients and allows pests to become adapted to the crop and the pest-control methods. Broilers that have been bred for high breast meat have skeletal and metabolic disorders. For these reasons, variety in crop rotations and genetic diversity in animal species needs to be maintained. Alternatives to monocultures are systems of polycropping and agroecology. Polycropping is the planting of more than one crop in the same space at the same time. It also includes companion planting and intercropping to increase the diversity of plants, primarily to decrease the incidence of disease and insect damage. Agroecology or agroforestry is polycropping with trees and shrubs in permanent strips or rows. Retaining varieties of genetic species of livestock is important as well.

Further limiting crop diversity is the small number of agribusinesses that produce seeds and animal stocks. The seed corn market is controlled by four companies and, globally, 10 companies control 32% of the commercial seed trade. Lack of competition in seed markets keeps prices high and reduces development of new products and adoption of new technologies. It reduces farmers' access to diverse genetic resources and often leads to the elimination of specialized varieties for local conditions, which will become more important to respond to climate changes. A criticism of genetic engineering is the control of the process by a few companies and economic barriers to entering the market adding to reduced genetic diversity in agriculture.

Agriculture affects biodiversity by reducing wildlife habitat. Fragmentation of preserved areas, diversion of water for irrigation, and destruction of prairie and woodlands have negative effects on native plants, animals, and insects. Drainage of wetlands for conversion to cropland reduces wildlife population and diversity. Runoff and chemicals used in agriculture negatively affect water quality, which hinders aquatic wildlife. Mitigation of the impacts of agriculture on wildlife while increasing the amount and quality of food produced is a wicked problem to solve. In a global view, it may be most efficient, economical, and ethical to concentrate the growing of crops or raising of animals in certain areas so that other lands may be preserved for parks, forests, and wildlife reserves. Deciding where to produce food to optimize wildlife and natural areas, as well as support the expanding human population, is and will continue to be challenging.

8.4 CREATING A SUSTAINABLE FOOD SYSTEM

The economic impact of US agriculture is significant. Agriculture and agriculture-related industries (forestry, fishing, food, beverage, tobacco, textiles, apparel, leather, and foodservice) contributed $789 billion to the US gross domestic product (GDP) in 2013. The output of America's farms generated $167 billion of this sum, or about 1% of the GDP. Agriculture and agriculture-related industry provide 9.2% of US employment or about 16.9 million full-time and part-time jobs. Americans consume more than 37 million tons of meat annually, worth $100 billion, and US fisheries harvest 5 million metric tons of fish and shellfish, which contributes more than $1.4 billion to the economy annually. Sustainability of this production system is essential for the economic viability of the United States.

Traditional and diversified farms of the 1930s and 1940s may be viewed as being ideally sustainable (if soil erosion was controlled). But these farms were inefficient, had demanding labor inputs, low production volume, and generated low income for farmers with high food costs for everyone. Modern, conventional farming, which is often referred to as "industrial farming," generates a negative image of destroying the environment to produce cheap, unhealthy foods. Finding a sustainable food production system lies somewhere in the middle of these extremes. Sustainable farming means an integrated or mixed farm pattern with a concern for the environment and minimal external inputs. Consideration of where and how food is produced, with size and scale of operations interspersed with market demands is needed. Any food production system "...must be efficient, which in most cases will require that farming remain mechanized, scientifically informed, and chemically supported. Only such an agriculture will be able to feed 9 billion people by 2050" (Conkin, 2008).

Being able to clearly predict the impact of climate change on weather patterns, the environment, and ultimately agricultural production is unlikely given the wide range of factors that are involved. Vulnerability of agriculture to these changes will be dependent upon not only the natural systems that arise, but on the ways humans respond to these changes. The key drivers as defined by the USDA (Technical

Bulletin Number 1935 (Walthall et al., 2012)) that will impact agriculture include:

- Increased variability in growing conditions (changes in seasonal temperature and precipitation patterns)
- Increased soil degradation (increased erosion reduces soil quality)
- Increased pest pressure, novel pests
- Increased number, length, and/or intensity of drought events
- Increased number and/or intensity of flood events
- Shift in optimum zones for current production systems
- Government climate change policy
- Economic (e.g., carbon markets)
- Consumer behavior (e.g., diet change)
- Perception of climate risk

Identifying ways to increase resilience and find adaptive solutions will be essential. The USDA has defined a national adaptation strategy to address the impacts of climate change on agriculture (Table 8.4). It will be necessary to monitor local climate changes closely and to adapt agricultural practices early and efficiently. Early and efficient adaptation is possible in large, midlatitude countries such as the United States where crops can shift within national borders. Adaptation will be difficult in developing countries with high population densities. Poor countries are the least flexible and most likely to experience famine.

TABLE 8.4 Proposed Adaption Strategies to Mitigate Climate Change Impacts on US Agriculture

Integrating adaptation into federal government planning and activities	Agencies are taking steps to manage climate impacts to federal agency missions, programs, and operations to ensure that resources are invested wisely and federal services remain effective for the American people. Agencies are developing climate adaptation plans to identify their vulnerabilities and prioritize activities that reduce climate risk.
Building resilience to climate change in communities	Recognizing that most adaptation occurs at the local level, Federal agencies are working with diverse stakeholders in communities to prepare for a range of extreme weather and climate impacts (e.g., flooding, drought, and wildfire) that put people, property, local economies, and ecosystems at risk.
Improving accessibility and coordination of science for decision making	To advance understanding and management of climate risks, the federal government is working to develop strong partnerships, enhance regional coordination of climate science and services, and provide accessible information and tools to help decision makers develop strategies to reduce extreme weather impacts and climate risks.
Developing strategies to safeguard natural resources in a changing climate	Recognizing that American communities depend on natural resources and the valuable ecosystem services they provide, agencies are working with key partners to create a coordinated set of national strategies to help safeguard the nation's valuable freshwater, ocean, fish, wildlife, and plant resources in a changing climate.
Developing strategies to safeguard natural resources in a changing climate	Recognizing that American communities depend on natural resources and the valuable ecosystem services they provide, agencies are working with key partners to create a coordinated set of national strategies to help safeguard the nation's valuable freshwater, ocean, fish, wildlife, and plant resources in a changing climate.

From Federal Actions for a Climate Resilient Nation. <https://www.whitehouse.gov/sites/default/files/microsites/ceq/2011_adaptation_progress_report.pdf>.

8.4.1 Organic Farming

The Organic Foods Production Act of 1990 authorized the USDA to define specific criteria for organic food production and to certify producers and food products as "organic." Foods produced following these standards bear the organic seal. To obtain the organic certification growers must follow a 3-year transition during which only approved crop and livestock practices are used. Table 8.5 describes the crop, livestock, and food processing standards that must be met for the organic certification.

Organic food production practices include the use of animal manure, fish meal, peat, compost, and cover crops as fertilizer; the use of mulch, cover crops, cultivation, oils, vinegar, copper sulfate, sodium hypochlorite, and soaps for weed control and the use of boric acid, lime sulfur, copper sulfate, oils, and approved compounds such as minerals, botanicals (pyrethrins), and bacteriological compounds (*Bacillus thuringiensis*) for pest control. Irradiation treatment, use of genetically modified seeds and use of human sewage sludge for fertilizer are not approved. A primary goal of organic production methods is to improve the health of soil and reduce soil erosion. The addition of organic matter from manure and cover crops is required in organic farming practice, which enhances the water-holding capacity of the soil and enhances beneficial microorganisms and biota. The planting of cover crops, limited tillage, retention of crop residue, and specific irrigation methods are additional approaches used in organic farming to protect and promote soil health. Organic farming practice requires farmers to employ strategies to mitigate pests, including using mulches to suppress weeds, rotating crops to reduce insects and diseases, and using natural rather than synthetic chemicals when needed

TABLE 8.5 Defined Organic Standards by the USDA

Organic crop production	Organic livestock production	Organic processing practices
Soil fertility: no sewage, use cover crops, soil conservation practices	Living conditions: natural behavior; access to outdoors; protection from extreme temperatures	Organic ingredients: organic processors must use 95% organic ingredients; no ingredients produced by GMO, from sewage, sludge, or ionizing radiation
Seeds and planting stock: organic seeds; not GMO; not treated with prohibited chemicals	Grazing: ruminants access to pasture; not continuously confined	Commingling and contact: no mixing organic with nonorganic ingredients; clean and sanitize equipment between sources
Crop rotation: different crop each season; wait years before planting same crop	Animal health: vaccination and prevention strategies; no antibiotics or growth hormones	Managing pests: may use synthetic pest management if needed, but cannot be in contact with products
Pest management: prevention, avoidance, monitoring, and suppression; approved pesticides	Organic feed: feed, pastures, bedding must be certified organic; nonorganic vitamins and minerals allowed	
Identity and integrity of crops: separate organic production from conventional to prevent contamination; wait 36 months after prohibited chemicals used	Animal origin: raised organically from last third of gestation; birds for poultry or egg production raised organically from second day of life	

From USDA Organic Standards. <www.ams.usda.gov/grades-standards/organic-standards>.

to treat plants. In this manner, organic farming promotes healthy soils and reduced environmental contamination.

Organic foods have been well accepted by consumers, which has driven increased production. Between 1992 and 2011, the number of organic farmers and acres of crop land converted to organic increased (Fig. 8.11), but the total amount of organic production remains a small fraction of total agricultural output. By 2011 organic crops accounted for only 0.63% of total US crop land and 0.49% of pasture and rangeland. During this time period, organic food sales grew from almost nothing to over $25 billion. Sales of organic foods show no sign of decreasing; in 2014 organic foods represented 4% of total food sales, estimated at over $35 billion a year. The most popular food items are fruits and vegetables, followed by dairy, packaged/prepared foods, breads and grains, snack foods, then meat, fish, and poultry. Although organic agriculture began as small farm enterprises, consumer demand has led major retailers to enter the market. Nearly 93% of organic foods is sold in grocery stores or supermarkets, and the majority of organic milk, lettuce, and spinach is marketed by large corporations such as Walmart. Consequently, organic producers have had to expand their production capacity to meet the demands of supplying these major retailers. In 2013, US producers exported $537 million in organic produce, primarily apples, lettuce, and grapes. Imported organic bananas, coffee, olive oil, mangoes, wine, soybeans, and other foods were valued at $1.4 billion. Organic is the

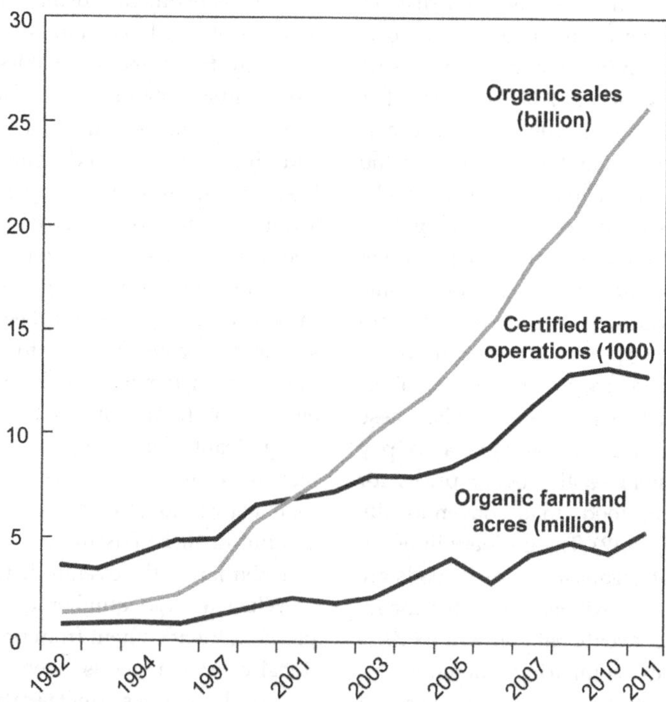

FIGURE 8.11 The amount of land under organic operation, and the number of organic farm operators, have both increased since 1992. Sales of organic foods have increased more steeply than other food categories, creating high price demands for organic products. *Source: USDA Economic Research Service, www.usda.ers.gov.*

dominant sustainable food label and is the largest component of the global sustainable food industry partly because of the high investment by governments and private sector businesses.

8.4.2 Sustainable US Food Production

Food production in the United States will be impacted by climate changes but with regional effects. Some areas near climate thresholds

EXPANSION BOX 8.2

ORGANICALLY GROWN FOODS AND HEALTH

For the past several years, most grocery stores and markets prominently feature organic products. Fruits and vegetables are displayed in separate locations in the produce aisle labeled with banners indicating that they are organically grown. Packaged food products that contain 95% organically produced ingredients display the USDA Organic Seal, and those that contain 70% organic ingredients are labeled as "made with organic ingredients." Such separate and distinct labeling and marketing tends to generate a "better for you" perception in consumers. From surveys of consumers, most will respond that organic foods are healthier, safer, and taste better; help small farmers; and are better for the environment than commercially grown foods. Foods that are grown using conventional agricultural practices do not receive comparable market attention. The rapid increase in sales of organic foods indicates consumers see a perceived benefit. Compared to conventionally grown foods, organically produced foods are more labor-intensive, and tend to have lower yields. These factors, plus the consumer willingness to pay more, has led to significantly higher prices for organically produced foods than conventionally produced foods (Fig. 8.12). There is very little scientific evidence that organically grown foods are safer, or have special nutritional value for consumers compared to conventionally grown foods.

Research studies comparing methods for growing plant foods are difficult to conduct because of the many variables that must be controlled. The nutrient content of a fruit or vegetable is determined by the cultivar or genotype, and the environment in which it is grown, including soil and climate. The degree to which the plant is stressed from temperatures, too much or too little water, or pests affects how much and which nutrients are retained. The length of the growing period, harvest time, storage conditions, and length of storage determines which nutrients are available when the food is consumed. Further complexity arises because of the large number of nutrients provided by fruits and vegetables, some of which are essential (vitamins and minerals) and others that are nonessential (bioactive compounds) and have no defined human requirement. Research studies have reported a higher or lower content of one or more nutrients or bioactive compounds in organically grown versus conventionally grown fruits and vegetables but these differences have not been consistent from season to season or year to year. The overall impact on a person's health from small differences in nutrient content of one food may be insignificant. For example, some grass-fed beef may have higher levels of omega-3 fatty acids than conventionally raised beef, but the total amount of these fats in beef adds only a minor contribution to the overall diet.

When a large number of studies that compared nutrient content from organic and conventionally grown foods were integrated using statistical analyses (metaanalysis), vitamin C, phenolic compounds, magnesium, calcium, potassium, zinc, and copper content were

EXPANSION BOX 8.2 (*cont'd*)

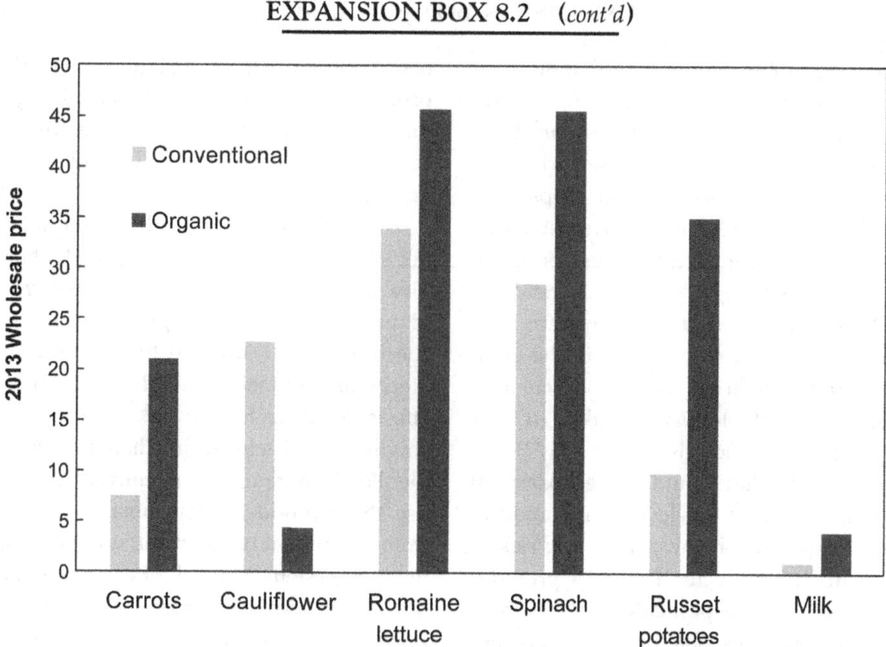

FIGURE 8.12 The price of organically produced foods is generally much higher than conventionally grown foods. Some consumers are willing to pay higher prices for these foods even though there is no scientific evidence that they are safer or more nutritious than conventionally grown foods. *Source: USDA Economic Research Service, www.usda.ers.gov.*

equally available from both types of food production. Bioactive compounds (such as phenolic or flavonoid compounds that act as antioxidants) in organic compared to conventionally grown strawberries, plums, peaches, pears, and cabbage were present in similar amounts. Higher levels of some bioactive compounds such as the flavonoids have been reported for organic compared to conventionally grown foods. Flavonoids include a large number of different chemicals that function to protect the plant from pests and sun-derived radiation damage, and provide the color pigments responsible for attracting pollinators and seed dispersers. When plants are stressed the concentration of flavonoids tends to increase. Year-to-year variations in various flavonoids in apples, blueberries, tomatoes, and peppers have been documented regardless of production method. Because there are a wide variety of flavonoid compounds in foods, and their specific function in the body is not well-defined, there is no clear amount of flavonoid intake that correlates with a health outcome. Most nutritionists agree that consuming more fruits and vegetables, regardless of how they are grown, provides beneficial nutrients to the diet.

A majority of consumers (70%) base their decision to purchase organic foods on the belief that organic foods contain fewer pesticides and chemicals than conventionally produced foods. The fear of chemical exposure is a major consideration that drives the purchasing of organic foods. The USDA prohibits synthetic pesticides from use in organic food production but does allow over 195 chemical compounds in the production, processing, and handling of organic foods. Conventional farming also follows

EXPANSION BOX 8.2 (*cont'd*)

regulations regarding the types and amounts of pesticides that are allowed. The EPA reviews the scientific data on all pesticide products before they can be registered for use and establishes a tolerance for use on food crops. The FDA enforces these tolerances on most foods while the USDA enforces tolerances for meat, poultry, and certain egg products. The FDA and USDA set enforcement guidelines for residues of pesticides that may remain in the environment and in the food. Based on scientific evidence, an Acceptable Daily Intake (ADI) is defined for each pesticide that is set at 1/100 of the exposure level that would cause toxicity in lab animals. This tolerance level is intended to be several magnitudes below where any risk to human health would occur. The FDA provides an annual report of the pesticide residues measured in food samples obtained from grocery stores (Table 8.6). In recent surveys, the great majority of foods contained no measurable residues, and those that did contain residues were below the safety threshold. Only a very small percent of foods had levels of chemical residues that violated the standards. Imported foods, particularly grains, vegetables, and fruits had a higher rate of violations than domestic products. Most health professionals agree that the risk to health from pesticide residues on foods, both conventionally and organically grown, is insignificant.

By regulation, organically raised animals are not exposed to antibiotics or growth hormones. This is frequently used as a marketing strategy to promote organically produced foods. Most consumers are concerned that exposure to antibiotics or hormones in foods will have negative effects on their health and the health of their children. As discussed in Chapter 4, Animals in the Food System, when conventional farmers use these products they must be fully cleared from the animal before meat, milk, or eggs enter the food system. The risk of exposure to antibiotics or artificial hormones from conventionally raised animals is extremely low and essentially not different from organically raised products.

The decision to purchase organically produced foods is sometimes based on the consideration of taste and flavor. The assumption is that organically produced foods have better flavor than conventionally grown foods. Flavor compounds in fruits and vegetables and animal foods

TABLE 8.6 Pesticide Residues in Foods Analyzed by the FDA

	Domestically produced			**Imported**		
	No residue	**Safe range**	**Violation**	**No residue**	**Safe range**	**Violation**
			Percent of total			
Grains	64.7	33.8	1.5	62.3	13.4	24.3
Dairy/eggs	95.5	4.5	0.0	85.7	14.3	0.0
Fish	68.8	27.5	3.8	90.5	9.1	0.5
Fruits	19.0	80.3	0.7	54.3	37.0	8.7
Vegetables	55.0	39.5	5.5	55.1	34.8	10.1

From Pesticide Monitoring Program Fiscal Year 2013 Pesticide Report, U.S. Food and Drug Administration. <http://www.fda.gov/downloads/Food/FoodborneIllnessContaminants/Pesticides/UCM508084.pdf>.

EXPANSION BOX 8.2 (*cont'd*)

are influenced many factors, and perception of desired flavor is defined by personal experience and preference. Freshness is another factor that affects flavor. A tomato picked from the garden and consumed immediately will have a different flavor profile than the same tomato eaten after being held for a few days on the grocer's display table. Organic farmers tend to grow different varieties of fruits and vegetables than commercial growers. Some of these may be heirloom, or varieties that have unique taste and texture profiles. There will be inherent differences in taste and texture in these products, not mainly derived from the organic production practices but more so because of the cultivar. When the same cultivars, grown under organic or conventional methods, are compared no consistent differences in flavor preference have been found. Organically raised animals consume more plant material and less grain than conventionally raised animals, which affects the flavor profile of the meat, milk, and eggs. Some people will prefer the taste profiles of organic foods while others may not.

Organically and commercially grown fruits and vegetables and animal foods provide similar nutrient value, and neither pose a health risk from pesticide, antibiotic, or hormone residues. Organic methods do support practices for improved soil quality and integrated methods of crop and animal production. Organically grown products receive higher prices in the marketplace, which may benefit some small farmers. Consumer demand for organic foods is strong and continues to increase, which will demand a new paradigm for organic food production. To meet demand, organic farmers will need to increase the size of their operations, or more farmers will need to enter the organic production market. Another strategy will be for conventional farming to adopt some of the organic approaches to better meet consumer expectations. The latter approach has been proposed by Secretary of Agriculture Tom Vilsack in the *Enhancing Coexistence* report. Coexistence is defined as "... the concurrent cultivation of conventional, organic, IP [identity preserved], and genetically engineered (GE) crops consistent with underlying consumer preference and farmer choices" (Coexistence Report, 2012). Bringing together the best practices from organic and conventional farming, rather than forming an "us versus them" mentality, will be better for consumers, farmers, and the environment.

Suggested reading: Economic Research Service. (2015). Organic agriculture. *Available from <http://www.ers.usda.gov/topics/natural-resources-environment/organic-agriculture.aspx> and U.S. Department of Agriculture Advisory Committee on Biotechnology and 21st Century Agriculture (2012).*

such as wine-growing regions of California and the grain-producing Great Plains are likely to experience decreases in yield and quality while other areas with limited growing seasons such as the Great Lakes region may benefit by being able to grow more products. Decreased yields of wheat, rice, corn, soybean, barley, and sorghum occur when the average high temperatures during the growing season increase by as little as 2 degrees. Increased stress from weeds and plant pests may mitigate increased crop yields as these also will change in response to higher CO_2 and temperatures. Crop yield directly impacts the economic balance of farming, with lower yields tending to generate lower profits. But there are many other variables that affect that equation. Overall crop availability and demand define crop prices, so a low crop yield with high prices may net the same income as a high yield with low prices. Predicting how climate change will drive yields and commodity costs requires development of models that consider all of these variables.

More extreme weather events, such as floods, droughts, and hurricanes are projected to occur leading to economic losses for farmers and higher government subsidies for crop losses and damage. Some predictions indicate a 60%–90% decrease in snowfall in the Sierra Nevada Mountains over the next decade, which could result in a lack of irrigation water in California's Central Valley and desertification in some locations. The Central Valley produces more than 25% of US fruits, vegetables, and nuts. Apricots, almonds, artichokes, figs, kiwis, olives, and walnuts require some chilling temperature threshold for dormancy and to set fruit. With increasing temperatures, a 30%–40% decrease in production of wine and table grapes, almonds, oranges, walnuts, and avocados could occur. Some predictions suggest Colorado will be drier resulting in less grass for cattle grazing while Florida will see an increase in storms and hurricanes. There will be less rainfall throughout the Midwest. Building models to predict climate change influences on agriculture economics is complicated by the many factors that influence production, prices, and markets.

8.4.3 Sustainable Global Production and Trade

Climate change and population growth are the most serious concerns for future food production. The world's population increased to 4.5 billion people in the last 100 years and is expected to reach 9.5 billion by 2050. By 2050, the world's food needs will be 14.3 trillion pounds, which is a 44% increase from 2011. In the past, expanded populations were fed because of the inputs of fossil fuels for fuel, fertilizer, and pesticides, and water and crop improvement technologies from the Green Revolution. From 1930 to 2000, global agricultural output of most commodities including corn, soybeans, oats, rice, and wheat increased. Looking toward the future, climate variability is expected to decrease the global production of these commodities, with the possible exception of wheat due to enhance yield with higher CO_2 (Fig. 8.13) As discussed in Section 8.3.9 climate change will impact pests, plant, and animal diseases and soil quality, which threaten global food production. Globally, the cost of food is a higher percent of personal income than it is in the United States. Reduced food production will increase prices and drive more people into food insecurity. As was observed during the Arab Spring of 2011, food scarcity and high prices lead to civil unrest and war. Most models predict that agriculture production will decline in developing countries under climate change scenarios, likely increasing the risk of political and social turmoil.

The US food system is closely intertwined with the global food system. Agriculture trade and food security will be affected as climate change alters areas of production. Growth in US agricultural exports has exceeded the growth in imports, leading to a positive trade balance. Rising global demand, primarily in developing countries, along with the dollar's competitive exchange rate, helped US exports grow faster than imports during the past decade. As a result, the US agricultural trade surplus widened to $38.8 billion in 2014. Nearly 40% of the food and feed crops produced in the United States were exported in 1995. While the United States is the second largest grain producer in the world and holds the world's largest grain surplus, grain exports from the United States

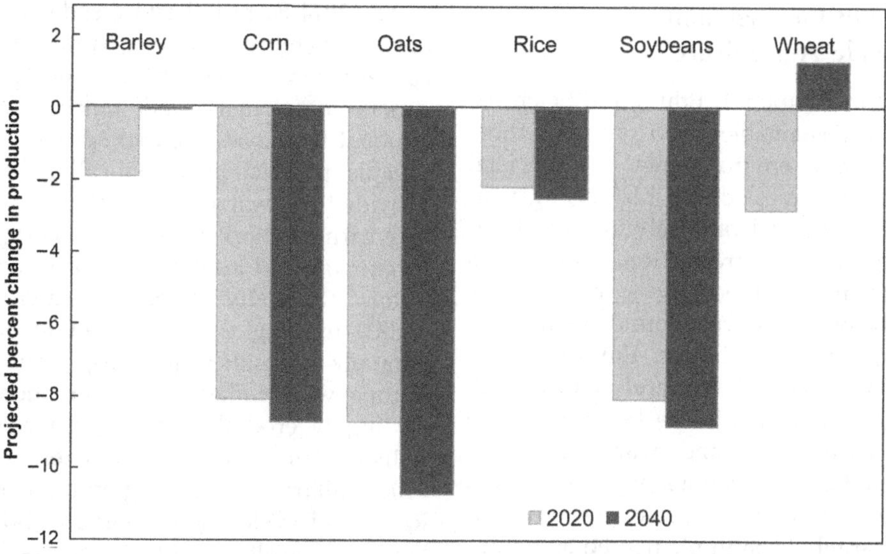

FIGURE 8.13 The effects of climate change on agriculture production are difficult to predict. Some projection models suggest that staple crop production in 2020 and 2040 will be decreased as much as 10% as a result of climate change, whereas some crops such as barley and wheat may be less affected. *Source: Environmental Protection Agency, www.epa.gov.*

amount to only about 5% of global production and are primarily fed to animals in Europe and Japan. It has been predicted that as the economic status of global populations increase the demand for animal-derived foods will also increase. This will put higher demand on corn, soybean, and alfalfa crops to provide feed for these animals. Debate around this scenario has arisen with concerns about the net environmental effects of raising crops in the United States that will be shipped to other countries to be fed to animals to provide food for the people of those countries. Consideration of the ethical, environmental, and economic balance of these choices will clearly be needed.

An example of how global food sustainability can be approached from the economic and human dimensions are the Fair Trade efforts. Fair Trade is a partnership between producers and consumers that creates greater equity in the global marketplace. Fair Trade targets the social, and sometimes the environmental, dimension of sustainability by guaranteeing a "fair" price for local producers. The price takes production costs and profit into account, rather than simply market demand forces. Standards for wages are set for small farmers (cooperatives) and workers on plantations and in processing factories (organized workers) and incorporated into the market price. Fair Trade labeling and certification is usually conducted by a network of independent, nonprofit organizations and the Fair Trade symbol is internationally recognized and monitored by the Fair Trade Labeling Organization International (FLO). Other organizations focused on sustainable production include the Marine Stewardship Council, The Rainforest Alliance, and Food Alliance, which work in the local environment to prevent environmental damage and worker abuse, and connect consumers with products that meet sustainable production standards.

8.4.4 Role of Government in Sustainable Agriculture

As noted in Chapter 1, Ethics and Scientific Thinking, a sustainable food system must include economic, environmental, and societal acceptability. For most of US history, agriculture has been focused on applying new technologies to increase production. Consumers benefited from low food prices, consistently high quality products, year-round availability and diversity in food choices. Environmental impacts of agriculture were rarely considered. This changed somewhat in the 1970s when the EPA was established, and water, air, and ground pollution regulations were enforced. Farming regulations that protect the environment have largely been implemented as voluntary or best practices, perhaps because of the strong farm lobby, which resists regulatory policies. Since the 1990s, following the introduction of the Organic Foods Production Act, demand for changes in the way agriculture is done began to enter the public arena, and government responded to these concerns.

The USDA established the Low Impact Sustainable Agriculture (LISA) program in 1985 to address research needs for alternative agricultural methods. LISA was replaced by the SARE program, which supports $19 million for agricultural systems research for long-term, interdisciplinary study of components that influence food and farming outcomes.

Sustainable agriculture was defined by Congress in the 1990 Farm Bill as:

> an integrated system of plant and animal production practices having a site-specific application that will, over the long term: satisfy human food and fiber needs; enhance environmental quality and the natural resource base upon which the agricultural economy depends; make the most efficient use of nonrenewable resources and on-farm resources and integrate, where appropriate, natural biological cycles and controls; sustain the economic viability of farm operations; and enhance the quality of life for farmers and society as a whole.

The 2014 Farm Bill addressed several aspects of sustainability including rural development, conservation, renewable energy, next-generation farmers and ranchers, local and regional food systems, and specialty crops and organic production. Specifically, funding was provided for water and wastewater infrastructure, farmers markets and local food production promotion, and local and regional food systems grants. Specialty Crop Research Initiative (SCRI) funding was increased to $80 million annually, subsidies for crop insurance premiums were linked to conservation practices for highly erodible lands and wetlands, $880 million was earmarked for energy programs, $100 million for the Beginning Farmers and Ranchers Development Program, and $72.5 million for specialty crop block grants. These programs will encourage new approaches to agriculture and provide funding for crops that have not been well supported previously.

The USDA, EPA, and related government agencies are well aware of the urgency to address climate change and environmental impacts of agriculture. Several comprehensive documents have been developed by scientists in these agencies that outline the projected problems and proposed actions. Policy recommendations include reforestation, conservation of energy, decreased fossil fuel combustion, development of renewable fuels, increased energy efficiency of farm implements, increased water use efficiency, and decreased soil loss. Finding the right balance of agricultural policy that addresses environmental protection without incurring high economic costs will be a challenge. Organic food production has promise to reduce chemical use and enhance soil and water quality, but these approaches may not be fully adaptable to all types of agriculture on the scale necessary to generate food at reasonable costs. Conventional food production is moving more toward precision systems that accurately detect and deliver chemicals to the fields that will lessen the environmental impact. It is essential that both types of agricultural systems are valued so that the optimum food system can be developed.

The food system is multidimensional with complex social interactions, which requires food systems research to be cross-disciplinary. Areas that need to be studied include the quantity and quality of food for the world's population (by input/output and mass balance research); evaluation of the environmental impacts of food systems (by LCA, carbon footprint, and water footprint studies); methods to distribute value in the food chain (by costing, value chain, and GDP analyses); mechanisms to create employment and corporate social responsibility; and methods to ensure adaptability to crises (by cost–benefit, impact, and risk assessments).

There are currently many economic and social barriers to creating a fully sustainable food system. As has been discussed in previous chapters of this text, there is widening disparity among farmer incomes and escalating concentration of agribusiness involved with manufacture, processing, and distribution of farm products into only a few companies. Market competition is limited and farmers have little control over farm prices. Farmers receive only a small portion, about 10%, of the consumer dollar spent for food. The reduction in the number of farms and the number of young farmers entering agriculture has contributed to the disintegration of rural communities and local marketing systems. Also, productive farm land has been lost to urban and suburban sprawl. Less than 2% of Americans produce food for all US citizens yet consumers, most with no agricultural knowledge, are increasingly involved in driving decisions about farming practices. It is critical that discussions about changes in the US food system occur with respect for the farmers who produce food and with engagement of nonfarmers with creative ideas for improvement in agricultural practices to protect the environment.

US food production currently is involved in the production of food, fiber, biofuels, and renewable energy but a sustainable future will require agriculture to expand its responsibility to provide clean water, reduce soil erosion, foster carbon sequestration, and enhance wildlife habitat. While some farmers have already adopted these environmental responsibilities, it is likely to require government regulation and oversight to enforce a national, and perhaps global, change in agronomic practices. It must also be noted that environmental protection will come with added costs. Consumers will need to be prepared to pay higher taxes and more for their food at the grocery store. Such changes will impact rates of food security, as more people are likely to fall below the poverty level.

8.4.5 Role of Farmers in Sustainable Agriculture

Farmers will apply sustainable practices and innovations to their operations when they provide an economic or competitive advantage. It is not economical to overuse pesticides and fertilizers, to till the soil more than necessary or to damage the land or water that provides the farmer with an income. Just as with other behaviors related to future benefits (such as recycling and limiting water usage), adoption of sustainable farming practices occurs along a bell-shaped curve with a few farmers using many sustainable practices, the majority of farmers using some and a few farmers using none. Enforcing compliance generally requires political action and government regulation.

The dairy industry provides an example of how modern agriculture has become more environmentally sustainable. Today the dairy industry produces 1 billion kilograms of milk using 21% fewer animals, 23% less feed, 35% less water, and 10% less land than in 1944. The net environmental impact of these improvements include 24% less manure, 43% less CH_4, and 56% less NO_2 generated per billion kilogram milk. The carbon footprint of milk produced in 2007 was 37% less than an equivalent milk production in 1944. These advances were achieved from better dairy cattle genetics and nutrition leading to cows that are more efficient in converting feed to milk, improved manure management strategies, and efficiencies of scale.

Field crops have also become more productive. According to USDA surveys, energy use per bushel of corn fell 43% between 1980 and 2011 due to farming methods that reduce chemical, water, and diesel fuel use. Some of these reductions were associated with the use of GMO crops. Efficiencies have occurred in overall crop production, including 30% less land, 67% less soil erosion, 53% less irrigation and 36% less GHG emissions. To be sustainable from an economic perspective, farmers will continue to improve production efficiency while reducing resource use and mitigating adverse environmental impacts.

Use of precision farming techniques, practiced since the 1980s, may be the solution to some of the expected variations in precipitation, day/night temperature differences, and summer and winter temperature averages caused by climate change. As discussed in Chapter 3, Innovations in US Agriculture, precision farming uses global positioning systems (GPS) to make applications of nutrients precisely to locations throughout a field so little or no excess fertilizer will be leached to groundwater or washed to surface waters. Inputs can be directly linked to soil conditions and past production. The use of drones and computer mapping systems allow real-time assessment of field conditions. Practices such as no-till farming use less fuel and reduce soil erosion and water evaporation while sequestering carbon. Only about 10% of US cropland is in no-till production and 25% is tilled selectively leaving room for improvement in tillage methods. Agriculture can contribute more to carbon sequestration by applying no-till methods, planting drought- and weed-resistance crops, reducing the use of chemical fertilizers and pesticides, conserving water, and improving the management of grasslands and livestock. GHG emissions can be reduced with better manure management practices and biogas production systems. Integrated pest management (IPM) also uses current and comprehensive monitoring of the life cycles of pests to limit damage to crops by appropriate application of pesticides and other physical control mechanisms. The expected occurrence of insects and disease in a very localized area can be weighed against the likely production losses and cost of treatment. Use of computerized data, development of low-carbon fertilizers, genetic modification of crops for reduced water and fertilizer needs, drip irrigation, use of buffer strips, wetland restoration, and other developments in agricultural science will continue to help farmers decrease inputs while increasing yields. Research will be needed to define a whole systems analysis approach that can be used to improve management decisions, weigh costs and benefits, and evaluate productivity and environmental impacts.

8.4.6 Role of Consumers in Sustainable Agriculture

The chapters of this book describe the history and progression of the US food system over time and the ways food is produced, processed, and consumed. In the United States, the production of food shifted from self-sufficiency to commercial in less than four generations. US farmers have been highly successful in providing nutritious, inexpensive, convenient, and plentiful food for consumers. The agricultural section has expanded the US economy with many people working in agribusinesses, although only a few are involved in direct food production. In contrast, the evolution of our food production system has produced negative impacts on the environment, human workers, and rural communities. These impacts include pollution and depletion of groundwater, loss of soil and reduced soil quality, and increased global warming due to GHG emissions and use of fossil fuels. The food system has also been blamed for creating health problems including obesity and related chronic diseases. Hot-button issues such as high fructose corn syrup, lean finely textured beef, GMO crops, chemicals (pesticides, hormones, antibiotics, and food additives) migrant farm workers, and animal housing systems have brought attention from consumers to the food system. Conflict between farmers and consumers around these issues is plentiful, and fueled by social media and activists.

An ethical approach to sustainability suggests that society has an obligation to refrain from wasteful uses of resources among the affluent, and a special obligation to foster economic development for the poor, while maintaining environmental resources. Sustainable production must be defined by what is to be sustained, for whom, and for how long. Sustainability is not an absolute condition, but always partial. Sustainability, like justice, occurs along a continuum. A main concern of the food production industry is economics. Food producers attempt to manage the market and government regulations to best suit their economic gain. US farm policy has supported increased productivity by providing long-term financial security and technological advances to farmers. The result has been great efficiency and increased production, but at potentially high environmental costs. Consumers also have an economic drive for low-cost foods, but want to feel good about how food is produced (not hurting animals, not taking advantage of people, not polluting the environment).

Alternative food production practices such as organic; marketing networks such as Fair Trade, local, and community sponsored agriculture; and movements such as slow food, eco-gastronomy, and agrotourism grew from this collective desire to feel good about how food is produced. Personal choices such as a vegetarian diet and buying local and organic foods are sometimes made from an ethical perspective or to foster a perceived healthier lifestyle. These social initiatives become economic issues when consumer demand drives production, and can lead to political change. A society that is interested in sustaining a healthful and sustainable food system must consider the entire breadth of the food system, not just the local food environment and personal food behaviors.

Innovations in food production are continually arising. New systems for growing food include high-tunnels and greenhouses, hydroponics, and vertical, roof, and urban gardening. These methods can move agriculture out of the fields and closer to where people live. In some cities, unused industrial warehouses are being converted into food production facilities that grow food indoors. Fish and shrimp can be raised in tanks housed in abandoned big-box stores. Simultaneously growing fish with hydroponic vegetables is a way to recycle fertilizer (from the fish excrement) for the plants, with water purification (by the plants) for the fish. Scientific approaches to generate new foods is also underway. Plant proteins and fats have been modified to produce an egg substitute (www.thevegg.com) and meat has been cultured from starter cells (www.modernmeadow.com). While currently too expensive for widespread application, these and other approaches to expand the ways and locations where food is produced will add to the sustainability of the food system.

8.4.7 Management of Food Waste

Food that is produced but not consumed is a significant waste of energy, natural resources, and money. Globally, one-third of food produced, about 2.9 trillion pounds per year, is wasted. In developing countries, the majority of food waste occurs during production and postharvest. Pest damage of crops, inadequate storage and refrigeration, and lack of transportation are some of the reasons for food waste. In developed countries, where food production and storage are very efficient, retail and consumers account for the majority of food waste. In the United States, 30%–40% of food produced (133 billion pounds a year) is discarded. Almost all of this uneaten food ends up in landfills. About 25% of total municipal solid waste (MSW) is organic material that generates a significant amount of GHG as it decays and ferments. Food is the single largest component of MSW (21%) going to landfills. Food scraps, which can be composted, represent 11.9% of MSW by weight.

The USDA defines food loss as the edible amount of food, postharvest, that is available for human consumption but is not consumed for any reason. This includes cooking loss and shrinkage from moisture loss; damage by mold, pests, and inadequate storage; food discarded by retailers; and plate waste by consumers. The main sources of food loss include consumer food waste at restaurants and at home (21% of the food supply, or 90 billion pounds) and convenience stores and other retail outlets (10% of the food supply or 43 billion pounds). Dairy products, vegetables, and grain products are the top three food groups in terms of food loss, while meat, poultry, fish, vegetables, and dairy products are the largest categories of loss in terms of value.

Reducing food waste would have a direct immediate positive impact on the food system by reducing the amount of food that needs to be produced. Environmental impacts of reduced food waste include less GHG generated in producing food and from food waste decaying in landfills. There are many ways food waste can be reduced at the level of food processing, retail and restaurants, and by consumers (Table 8.7). As discussed in Chapter 7, Nutrition and Food Access, efforts to provide food to those in need can be combined with reducing food waste. Gleaning efforts and redistribution of edible foods from restaurants, schools and colleges, and retail outlets help connect food-insecure people with excess foods.

Many people associate fresh foods with being more healthful, and perhaps more sustainable because they have been less processed. As consumers' demand for fresh foods increases, food waste and negative impacts on the environment increase. Fruit and vegetable waste accounts for 38.4% of total food waste (Fig. 8.14). Fruits and vegetables have a limited shelf-life, so they must be transported quickly and with refrigeration. In order to maintain a year-round supply, fruits and vegetables that are not in season in the United States are imported from other

TABLE 8.7 Ways to Reduce Food Waste

	Reduce excess food	Reduce food spoilage	Reduce food loss
Food processing	Grow and produce foods for defined markets	Harvest and process foods close to the farm	Store and transport foods efficiently
Food retailers	Match food stocks with customer demand	Limit size of fresh food displays	Donate unused food to those in need
Restaurants	Balance food production with customer demand	Manage food handling and storage	Reduce portion sizes, offer limited menus
Consumers	Buy only what you need	Store and use food efficiently, read and understand "best by" dates	Plan meals to use food efficiently

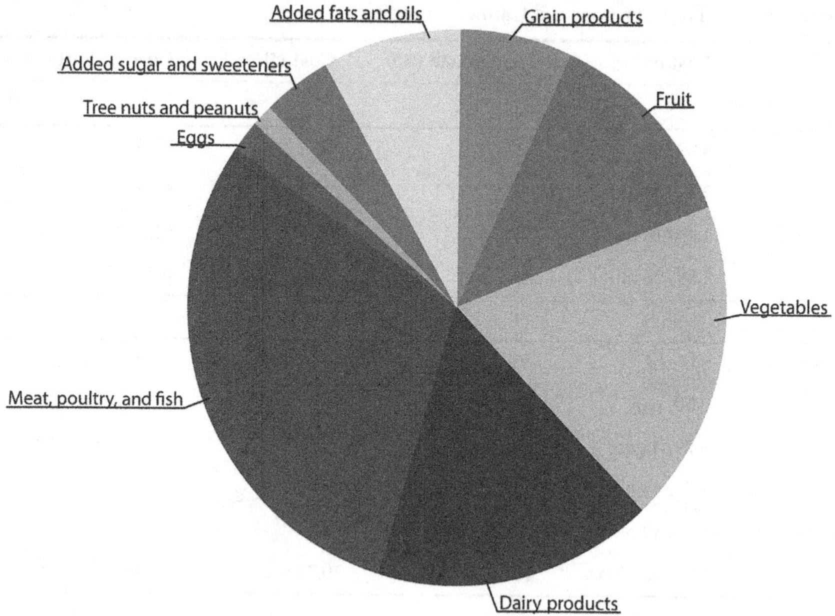

FIGURE 8.14 Food waste in the United States occurs mainly at the consumer level. Fresh products including, meat, poultry and fish, vegetables, and dairy products are the major food groups most commonly discarded. *Source: USDA Economic Research Service, www.usda.ers.gov.*

countries. Consuming more fresh fruits and vegetables imposes higher fuel use and contributes to GHG production. Retail stores and restaurants throw out a great majority of fresh fruits and vegetables each day because they spoil quickly, adding to landfill GHG. Fresh fruits and vegetables are more expensive than canned or frozen versions (Table 8.8).

The nutrient content of frozen and canned fruits and vegetables are similar to fresh and there is no unique health benefit to consuming fruits and vegetables in the fresh state. While canned foods are often regarded as less nutritious than fresh or frozen products, this is generally not true. Canning does reduce the content of water-soluble and thermally labile nutrients but the processing methods are set to ensure these losses are minimal. The nutritional content of fresh fruits and vegetables prepared at home may be reduced during storage and cooking and these losses are not controlled. Carotenoids, vitamin E, minerals, and fiber are generally similar in comparable fresh and processed products. The canning and freezing processes do consume energy and require water, which both have environmental impacts. Consumers would need to wash, cook and store fresh fruits and vegetables as well, and commercial processing is done more efficiently than individual consumer processing. Based on the high nutritional value of processed fruits and vegetables, and the lower food waste incurred by processing them at the peak of their freshness, consuming fresh fruits and vegetables is less sustainable and worse for the environment than consuming processed fruits and vegetables.

8.4.8 Sustainability of Animal Foods

The nutritional benefits of meat, milk, and eggs to human health, as well as the potential health risks of these foods, were presented in

TABLE 8.8 Comparison of Fresh, Frozen, and Canned Vegetables

Vegetable	Cost ($)	Food waste (%)	Cost ($)/oz.	Cost ($)/Serving (cooked ½–2/3 c)
PEAS				
Fresh in pods	3.99/lb	62	0.24	2.43
Frozen	2.97/28 oz	0	0.11	0.33
Frozen microwave pkg	1.25/10 oz	0	0.12	0.42
Canned	1.59/15 oz	10	0.10	0.43
CORN				
Fresh, on cob, in season	$6/doz 0.50/ear	46	—	1.09/ear (1/2 c)
Frozen, on cob	3.79/4 ears	46	—	0.95/ear (1/2 c)
Frozen	1.25/14.4 oz	0	0.09	0.28
Frozen, microwave pkg	1.99/12 oz	0	0.17	0.50
Canned	1.59/15.25 oz	10	0.11	0.45
GREEN BEANS				
Fresh	3.19/lb	12		0.59
Frozen	1.25/16 oz	0	0.08	0.21
Frozen, microwave pkg	1.25/12 oz	0	0.10	0.36
Canned	0.59/15 oz	10	0.04	0.02
BROCCOLI				
Fresh	2.69/lb	19	0.17	0.80
Frozen	2.85/28 oz	0	0.10	0.31
Frozen, microwave pkg	1.25/12 oz	0	0.10	0.34

Chapter 7, Nutrition and Food Access. The production of animal foods has been discussed in Chapter 4, Animals in the Food System, and the environmental issues associated with animal production were presented in Section 8.3.7. Defining the sustainability of animal food production requires engagement of all of these aspects of these food sources. From a sustainability perspective, ruminant animals can generate high-quality foods for humans (meat and milk) from grasses that are not food sources for humans. They also generate manure that is an effective fertilizer for growing crops. In that manner, grazing cattle are highly sustainable. When cattle are fed grains, the equation is modified because grains require significant water, energy, and chemical inputs. The amount of grain required to produce 1 kg of animal product has been estimated to be highest for beef cattle and lowest for dairy cattle (milk):

- Beef cattle = 13 kg
- Eggs = 11 kg
- Swine = 5.9 kg
- Turkeys = 3.8 kg

- Broilers = 2.3 kg
- Dairy cattle (milk) = 0.7 kg

The inputs of grain required for animal food production, and the consumption of water, production of manure, and required energy are factors that must be considered in defining sustainability. It is estimated that the amount of water required to produce 1 kg of animal protein is 100 times greater than to produce 1 kg of plant protein. Producing meat and milk from cattle fed grain would be considered less sustainable than from grazing animals, but the nutritional quality of meat and milk for human health is high. Determining what foods should be produced and how foods should be produced are not simple equations.

The United Nations Food and Agriculture Organization defines a sustainable diet as "diets protective and respectful of biodiversity and ecosystems, culturally acceptable, accessible, economically fair and affordable; nutritionally adequate, safe and healthy; while optimizing natural and human resources." Meeting all aspects of this definition is clearly challenging, especially for a complex food system such as in the United States or other developed countries.

The term *nutrition ecology* was first introduced by German nutrition scientists in 1986, as an interdisciplinary scientific discipline that considers the entire food chain from production to consumption, including waste disposal, as well as its interactions with health, the environment, society, and the economy. A tenet of nutritional ecology is that

> vegetarian diets are well suited to protect the environment, to reduce pollution, and to minimize global climate changes. To maximize the ecologic and health benefits of vegetarian diets, food should be regionally produced, seasonally consumed, and organically grown. Vegetarian diets built on these conditions are scientifically based, socially acceptable, economically feasible, culturally desired, sufficiently practicable, and quite sustainable.
>
> ***Leitzmann (2003)***

The current production methods to generate animal foods are energy, land, and water intensive. In a comparison of the use of land and energy resources needed to produce a meat-based diet or a lacto-ovo-vegetarian diet, the meat-based food system required more energy, land, and water resources than the lacto-ovo vegetarian diet (Pimental & Pimental, 2003). In this limited sense, the lacto-ovo vegetarian diet is more sustainable than the average American meat-based diet. The amount of fossil fuels used to generate a 3600-kcal diet that included 1000 kcal from animal products required 35,000 kcal of fossil energy, a vegetarian diet required 18,000 kcal energy, and a lacto-ovo vegetarian diet required 25,000 kcal energy. Measures of the amounts of GHG emissions produced by selected diets consumed by French citizens have also been completed as another measure of sustainability (Vieux et al., 2013). Diets with higher nutritional quality with more animal-based products tended to generate more GHG than plant-based diets. By comparing GHG production on a per-calorie basis, fruits and vegetables were similar to animal foods (exclusive of beef) in the amount of GHG emissions. The production of some fruits and vegetables can be very water and energy intensive, and some such as rice generate methane at high levels. Consuming a vegetarian diet did not necessarily guarantee a low GHG emission rate, and in fact diets that contained modest amounts of animal products were among the lowest in GHG emissions because of the higher nutrient quality. A similar finding was made in a modeling study in which ideal diets to promote health and minimize GHG emissions were tested (Macdiamond et al., 2012). The study found that it was not necessary to eliminate meat or dairy products from the diet to create a low-GHG diet. Using such computer-based tools it is possible to design and test diet composition to meet nutritional and environmental ideals.

The production of animal foods is unlikely to end in the United States or other countries, and in fact the demand for these foods is increasing especially in countries with rapidly growing populations with higher economic status. The positive impact of animal foods to provide important nutrients must be considered in the overall assessment of a sustainable food system. Using environmentally sound approaches to animal husbandry, and optimizing waste management, will be necessary to ensure sustainability of these food sources. Balancing animal food intake with nutrient needs, especially by affluent societies may be recommended to limit current environmental effects.

8.4.9 Sustainability in the Food Processing Industry

As discussed in Chapter 5, Human Resources in the Food System, processed foods are an essential component of a sustainable food system. Food companies must be responsive to customer requests for information and concerns about sustainability because their business depends on customers who will purchase their products. Most US businesses have statements of responsibility that address sustainability and provide examples of sustainability practices. Food companies address environmental sustainability by sourcing ingredients from responsible producers, reducing carbon emissions and energy usage in processing and transportation, reducing landfill waste, using less packaging material, and conserving water. Similarly, restaurants target sustainability goals by reducing energy and water use, using energy-efficient equipment, developing strategic transportation, establishing standards for products in order to reduce waste, and implementing standards for animal welfare. In contrast, some approaches, such as refusing to accept GMO technology, or the use of approved food additives, may reduce sustainability potential of the food system by limiting productivity and increasing food waste.

Walmart and Ben & Jerry's are two food processors known as leaders in corporate sustainability initiatives. Walmart is the world's largest grocer and is able to establish policies and practices for other segments of the food system. Walmart works with farmers to optimize fertilizer and tilling practices for soy and corn, sourcing goods from small- and medium-sized farmers and sustainably sourcing palm oil, beef, and seafood. The Walmart statement about sustainability reads:

> Walmart is committed to a sustainable food supply chain, which means offering customers choices and transparency into how their food is grown and raised, helping to further the humane treatment of animals, and always working to lessen the environmental impact of our agricultural practices. We believe that it is our responsibility to identify the challenges that impact our supply chain and our customers, and be a part of the solution. We are working with our suppliers, government agencies, academics, NGOs, animal health companies and veterinary experts. As part of this effort, Walmart U. S. and Sam's Club U.S. announced new positions around animal welfare and the responsible use of antibiotics in farm animals in the U.S.
>
> *Walmart (www.corporate.walmart.com/global-responsibility/environment-sustainability/sustainable-agriculture)*

Ben & Jerry's issued a Social and Environmental Assessment Report and has initiatives to promote GMO labeling, a carbon reduction program, a Fair Trade–certified supply chain and a "Caring Dairy" program. Learning more about a food company's position on sustainability is helpful when comparing foods to purchase or companies to support.

8.5 PERSONAL DECISIONS AND CHOICES

Responsible food behaviors for individuals include learning about nutrition and food, recognizing the relationship between food and health, thoughtful purchasing of food, and preparing healthful meals. Reduction of waste, resisting advertising and overconsumption, eating foods in season, and maintaining a healthy weight are practices that contribute to a healthy food environment and support sustainability goals. Consumers who understand the importance of their food behaviors will make efforts to change buying and eating habits. Table 8.9 lists several ways consumer behaviors affect economic, societal, and environmental issues. Reading food labels and understanding the role of food additives in maintaining freshness and safety are also important tools, as is using scientific thinking when making food choices.

Small changes, made by many people, will have a great impact on the sustainability of the food system. It is not necessary to buy only organic foods from local vendors, or eliminate animal foods from the diet to contribute to sustainability. Reducing food waste is one of the most important aspects of food sustainability that can be readily achieved by everyone. Remembering that canned and frozen foods are healthful and can be stored longer than fresh foods can save money and reduce waste. Buying only the quantity of food that can be consumed and storing it properly will reduce waste.

TABLE 8.9 Consumer Behaviors and Their Impacts on a Sustainable Food System

Consumer behavior	Economic effect	Societal effect	Environmental effect
Buy organic foods	High-income consumers can afford; higher prices for some producers	Groups can influence policy and availability in market	Lower environmental impact
Eat less meat	Consumer saves money on food purchased, meat producers may lose money, producers of other products may earn more money	Some benefits for personal health	May reduce production of grains, use of water, need for manure disposal
Buy Fair Trade foods	High-income consumers can afford; higher prices for some producers	Strengthen connections between US consumers and international market	Production of crops is more sustainable
Buy from farmers' markets or CSAs	High-income consumers can afford; higher prices for some producers	Strengthen relationship between consumer and producer	May increase GHG emissions from transportation to markets
Eat fresh (raw) foods	High-income consumers can afford; higher prices for some producers	Perceived as a healthier diet pattern	May generate more GHG emissions than some processed foods, more waste
Reduce wasted food	Consumers, business, taxpayers save money	Increased respect for food costs	Reduced need for waste disposal, less GHG from landfills
Buy directly from local producers	High-income consumers can afford, keeps small producers viable	Strengthen relationship between consumer and producer	May increase GHG emissions from transportation to markets

8.5.1 Learning About Agriculture, Food, and Nutrition

There are many ways to learn about food and the food system. Agricultural production information is available on the USDA, EPA, FDA, and land-grant university websites. Fun events such as county fairs and festivals celebrate all kinds of food such as wild rice, watermelon, pork, bacon, apples, grapes, turkey, BBQ, chili, ribs, chocolate, wine, cheese, pumpkins, and more. There are technical demonstrations of current farming practices at fairs and agricultural shows offered in most states. Historical agriculture festivals such as Steam Thresher's Reunions (in Illinois, Ohio, Minnesota, Iowa); ethnic events that showcase Polish, Indian, German, Asian, Latin, Greek, and other cultures with their foods; agricultural museums; and living history farms (see the Association for Living History Farms and Museums) provide education and entertainment around US foods.

There are many ways for individuals to become involved in food production through gardening themselves or in community gardening projects. Community gardens are collaborative projects on shared open spaces where participants contribute to the maintenance of the garden, and receive a portion of the products. School gardens were encouraged by Michelle Obama's Let's Move campaign for healthier lives through exercise and balanced eating. Gardening offers many benefits in addition to the consumption of fruits and vegetables. Gardeners are engaged in physical activity, learning new skills, and creating green space. Gardeners also learn first-hand the trials and tribulations of crop production and appreciate the labor required to produce good food.

Of course, learning about agriculture, food science, and human nutrition through coursework is a formal way to understand more about the food system. For more in-depth study of agriculture, land-grant colleges offer majors in agronomy, horticulture, animal science, food science, human nutrition, plant pathology, and other areas of study related to food. There are also colleges that offer work experience on farms such as School of the Ozarks and many universities have student-run organic farms such as those at Michigan State University, Clemson University, University of California–Santa Cruz, California Polytechnic State University, Berea College, Iowa State University, University of California–Davis, and many others. There are international volunteer farm work programs. The Worldwide Opportunities on Organic Farms (WWOOF) links volunteers with farmers around the world and in the United States. Some offer room and board in exchange for work on the farm; some offer apprenticeship programs as well as volunteer opportunities.

There are agrotourism and food experiences offered by travel agencies. Some food processing plants offer tours and sampling of their foods to the public. Celestial Seasonings in Boulder, Colorado; Ben & Jerry's Ice Cream in Waterbury, Vermont; Cabot's Cheese in Cabot, Vermont; Eli's Cheesecake in Chicago, Illinois; Hershey's in Hershey, Pennsylvania; and Coca-Cola in Atlanta, Georgia are popular tourist destinations. Visits to wineries, orchards, cheese factories, candy stores, and even the local supermarket can provide insights about food.

Reliable nutrition information and instructions for cooking healthful meals can be found on many websites, including WebMD Living Healthy (http://www.webmd.com/living-healthy), Food and Nutrition Information Center (http://fnic.nal.usda.gov/), Academy for Nutrition and Dietetics (http://www.eatright.org/), land-grant university extension programs such as http://www.extension.umn.edu/family/health-and-nutrition/, and Best Food Facts (http://www.bestfoodfacts.org).

Consumer information about food is everywhere, including cooking shows on television, websites with food information, nutrition labeling websites, extension service programs, community cooking classes, and church and charity groups. Use scientific thinking and verify the reliability of the sources to ensure the information is based on scientific evidence.

Some recommended books with interesting and accurate information about food science are *On Food and Cooking: The Science and Lore of the Kitchen* (by Harold McGee, Scribner, New York, 2004) and *Why Do Donuts Have Holes: Fascinating Facts About What We Eat and Drink* (by Don Voorhees, Kensington Publishing Corp, New York, 2004).

High quality cookbooks include *Cookwise: The Hows & Whys of Successful Cooking with Over 230 Great-Tasting Recipes* (by Shirley O. Corriher, William Morrow and Company, Inc., New York, 1997), *The Joy of Cooking* (by Irma Rombauer, Marion Bombauer Becker, & Ethan Becker, Simon & Schuster Inc, New York,

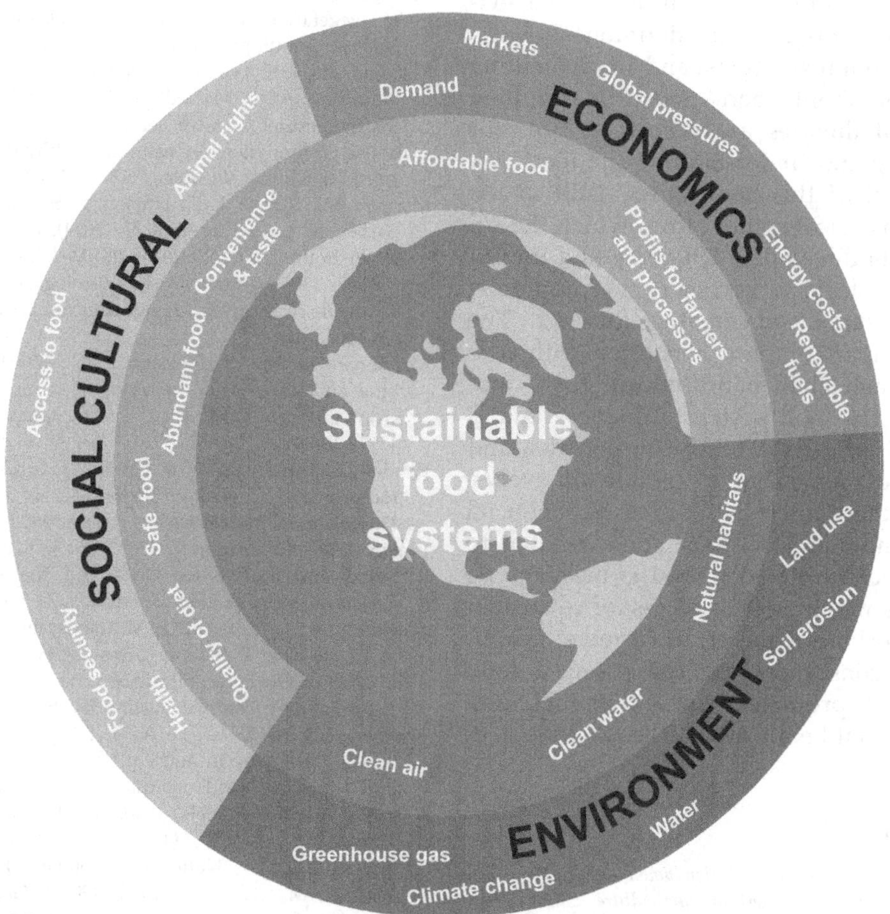

FIGURE 8.15 Components of a sustainable food system must include economic, environmental, and social–cultural considerations. The goal to produce healthy, abundant, and affordable food that promotes health and well-being while protecting the environment and natural resources will require the unified efforts of consumers, producers, and the government. Sustainability will be achieved by working together to address the challenges and problems of future food production. *Source: Illustration by Reannon Overbey.*

first published 1931, latest edition 2006), and any version of the *Betty Crocker Cookbook*, which is particularly helpful for beginners.

Development of a healthy and sustainable food system is a priority for the United States. Some of the key components of a sustainable food system presented in this book are shown in Fig. 8.15. By learning about food, and how it is produced, processed, and affects the body, consumers will be better able to make decisions about the food system. Relying on accurate, scientifically based evidence is essential when defining policies, making recommendations, and implementing regulations about food. Scientific thinking and ethical theories are necessary tools for interpreting and understanding the food system. A goal of this book is to instill in the reader a historical context from which the US food system developed. Knowing and understanding history is necessary to provide a solid foundation for moving forward. The food system is complex, and filled with controversial and challenging issues. Each component of the food system influences and is influenced by all the other components, and changes to one create a cascade of events. Realizing these interrelationships is beneficial when even small decisions, such as what to eat for lunch, are made. Food is meant to be enjoyed as much as it is necessary to sustain life. By working together, as farmers, producers, and consumers, we can create a food system that promotes human, animal, and environmental health and sustainability.

References

Conkin, P. K. (2008). *A revolution down on the farm: The transformation of American agriculture since 1929.* Lexington, KY: The University Press of Kentucky, 240 p.

Convention on Biological Diversity (1992). *Sustaining life on earth.* Montreal, QU: United Nations Environmental Programme. Available from <www.unep.ch/conventions>.

Environmental Protection Agency. (2016). *Agriculture and food supply.* Available from <http://www3.epa.gov/climatechange/impacts/agriculture.html>.

Heller, M. C., & Heolian, G. A. (2000). *Life cycle based sustainability indicators for assessment of the U.S. food system.* Ann Arbor, MI: University of Michigan Center for Sustainable Systems, School for Natural Resources and Environment. 59 p. Available from <http://css.snre.umich.edu/css_doc/CSS00-04.pdf>.

Hoekstra, A. Y., & Chapagain, A. K. (2007). Water footprints of nations: Water use by people as a function of their consumption pattern. *Water Research Management, 21*(1), 35–48.

Leitzmann, C. (2003). Nutrition ecology: The contribution of vegetarian diets. *American Journal of Clinical Nutrition, 78* (Supplement), 657S–659S.

Macdiamond, J. I., Kyle, J., Horgan, G. W., Loe, J., Fyfe, C., Johnstone, A., & McNeill, G. (2012). Sustainable diets for the future: Can we contribute to reducing greenhouse gas emissions by eating a healthy diet? *American Journal of Clinical Nutrition, 96*(3), 632–639.

Maupin, M., Kenny, J. F., Hutson, S. S., Lovelace, J. K., Barber, N. L., & Linsey, K. S. (2014). Estimated use of water in the United States in 2010. *U.S. Geologic Survey Circular 1405.* Reston, VA: U.S. Geological Survey. 56 p. Available from <http://pubs.usgs.gov/circ/1405/pdf/circ1405.pdf>.

Mekonnen, M. M., & Hoekstra, A. Y. (2011). The green, blue and grey water footprint of crops and derived crop products. *Hydrologic Earth System Science, 15,* 1577–1600.

Pandey, D., Agrawal, M., & Pandey, J. S. (2011). Carbon footprints: Current methods of estimation. *Environmental Monitoring and Assessment, 178,* 135–160.

Pimental, D., & Pimental, M. (2003). Sustainability of meat-based and plant-based diets and the environment. *American Journal of Clinical Nutrition, 78,* 660S–663S.

Powlson, D. S., Addiscott, T. M., Benjamin, N., Cassman, K. G., de Kok, T. M., van Grinsven, H., ... van Kessel, C. (2008). When does nitrate become a risk for humans? *Journal of Environmental Quality, 37,* 291–295.

Robertson, D. M., Saad, D. A., & Schwarz, G. E. (2014). Spatial variability in nutrient transport by HUC8, state and subbasin based on Mississippi/Atchafalaya river basin sparrow models. *Journal of the American Water Resources Association, 50*(4), 1–22.

Thoma, G., Popp, J., Nutter, D., Shonnard, D., Ulrich, R., Matlock, M., ... Adom, F. (2013). Greenhouse gas emissions from milk production and consumption in the United States: A cradle-to-grave life cycle assessment circa 2008. *International Dairy Journal, 31,* S3–S14.

U.S. Department of Agriculture (2015). *Climate change, global food security, and the U.S. food system.* Washington, DC:

U.S. Department of Agriculture. Available from <http://www.usda.gov/oce/climate_change/FoodSecurity2015Assessment/FullAssessment.pdf>.

U.S. Department of Agriculture Advisory Committee on Biotechnology and 21st Century Agriculture. (2012). *Enhancing coexistence: A report of the AC21 to the Secretary of Agriculture*. Washington, DC. Available from <http://www.usda.gov/documents/ac21_report-enhancing-coexistence.pdf>.

Vieux, F., Soler, L.-G., Touazi, D., & Darmon, N. (2013). High nutritional quality is not associated with lowest greenhouse gas emissions in self-selected diets of French adults. *American Journal of Clinical Nutrition, 97*(3), 569–583.

Walthall, C. L., Hatfield, J., Backlund, J., Lengnick, L., Marshall, E., Walsh, M., ... Ziska, L. H. (2012). *Climate change and agriculture in the United States: Effects and adaptation. USDA Technical Bulletin 1935*. Washington, DC: United States Department of Agriculture, Agricultural Research Services, 186 p.

Further Reading

ADA (2007). Position of the American Dietetic Association: Food and nutrition professionals can implement practices to conserve natural resources and support ecological sustainability. *Journal of the American Dietetic Association, 107*(6), 1033–1043.

Agricultural Marketing Service (2016). *Local food research & development*. Washington, DC: U.S. Department of Agriculture. Available from <https://www.ams.usda.gov/services/local-regional>.

Anwar, M. R., Liu, D. L., Macadam, I., & Kelly, G. (2013). Adapting agriculture to climate change: A review. *Theoretical Applications in Climatology, 113*, 225–245.

Backlund, P., Janetos, A., & Schimel, D. (2009). *Effects of climate change on agriculture, land resources, water resources, and biodiversity in the United States*. New York, NY: Nova Science Publishers.

Baldwin, C. J. (Ed.), (2009). *Sustainability in the food industry* Ames, IA: Wiley-Blackwell and IFT Press, 257 p.

Beck, J. (June 28, 2015). Don't worry so much about whether you food is "processed". *The Atlantic*. Available from <http://www.theatlantic.com/health/archive/2015/06/is-fresh-food-better-than-processed-or-frozen/397079/>.

BFBL (2015). *Buy fresh buy local program*. Millheim, PA: FoodRoutes Network LLC. Available from <http://foodroutes.org/>.

Blackhurst, M., Hendrickson, C., & Vidal, J. S. (2010). Direct and indirect water withdrawals for U.S. industrial sectors. *Environmental Science and Technology, 44*, 2126–2130.

Bloom, J. (2010). *American wasteland: How America throws away nearly half its food (and what we can do about it)*. Boston, MA: De Capo Press, 360 p.

Boote, K. J., Allen, L. H., Prasad, P. V. V., Baker, J. T., Gesch, R. W., Snyder, A. M., ... Thomas, J. M. G. (2005). Elevated temperature and CO_2 impacts on pollination, reproductive growth, and yield of several globally important crops. *Journal of Agricultural Meteorology, 60*(5), 469–474.

C2ES (2016). *Science and impacts*. Arlington, VA: Center for Climate and Energy Solutions. Available from <www.pewclimate.org>.

Capper, J. L., Cady, R. A., & Bauman, D. E. (2009). The environmental impact of dairy production: 1944 compared with 2007. *Journal of Animal Science, 87*, 2160–2167.

Center for Sustainable Systems (2016). *U.S. food system factsheet*. Ann Arbor, MI: University of Michigan. Pub. No. CSS01-06. Available from <http://css.snre.umich.edu/factsheets/us-food-system-factsheet>.

Chassy, A. W., Bui, L., Renaud, E. N., Van Horn, M., & Mitchell, A. E. (2006). Three-year comparison of the content of antioxidant microconstituents and several quality characteristics in organic and conventionally managed tomatoes and bell peppers. *Journal of Agricultural and Food Chemistry, 54*, 8244–8252.

Crinnion, W. J. (2010). Organic foods contain higher levels of certain nutrients, lower levels of pesticides, and may provide health benefits for the consumer. *Alternative Medicine Review, 15*(1), 4–12.

Dangour, A. D., Lock, K., Hayter, A., Aikenhead, A., Allen, E., & Uauy, R. (2010). Nutrition-related health effects of organic foods: A systematic review. *American Journal of Clinical Nutrition, 92*, 203–210.

Del Grosso, S. J., Ogle, S., Wirth, J., & Skiles, S. (2008). *U.S. agriculture and forestry greenhouse gas inventory: 1990-2005. United States Department of Agriculture Technical Bulletin 1921*. Washington, DC: U.S. Department of Agriculture. Available from <https://www.ars.usda.gov/research/publications/publication/?seqNo115 = 243953>.

DeVos, J. M., Joppa, L. N., Gittleman, J. L., Stephens, P. R., & Pimm, S. L. (2014). Estimating the normal background rate of species extinction. *Conservation Biology, 29*(2), 452–462.

Economic Research Service (2015a). *Irrigation and water use*. Washington, DC: U.S. Department of Agriculture. Available from <http://www.ers.usda.gov/topics/farm-practices-management/irrigation-water-use/background.aspx>.

Economic Research Service (2015b). *Water use*. Washington, DC: U.S. Department of Agriculture. Available from <http://www.ers.usda.gov/publications/tb-technical-bulletin/tb1909.aspx>.

Environmental Protection Agency. (2012). *Sustainability.* Available from <http://www.epa.gov/agriculture/tsus.html>.

Environmental Protection Agency. (2015). *DDT: A brief history and status.* Available from <https://www.epa.gov/ingredients-used-pesticide-products/ddt-brief-history-and-status>.

Environmental Protection Agency (2016a). Climate change indicators in the United States, 2016 (4th ed., EPA 430-R-16-004. Available from <http://www.epa.gov/climate-indicators>). Washington, DC: United States Environmental Protection Agency.

Environmental Protection Agency. (2016b). *Overview of greenhouse gas emissions.* Available from <https://www.epa.gov/ghgemissions/overview-greenhouse-gases>.

Environmental Protection Agency. (2016c). *Reducing wasted food at home.* Available from <https://www.epa.gov/recycle/reducing-wasted-food-home>.

Esnour, C., Russel, M., & Bricas, N. (2013). *Food system sustainability.* Cambridge, UK: Cambridge University Press, 303 p.

Etherton, T. (June 27, 2011). *How much food will the world need in 2050?* Terry Etherton Blog on Biotechnology. Available from <http://sites.psu.edu/tetherton/?s = future + food + needs>.

Foley, J. A., Ramankutty, N., Brauman, K. A., Cassidy, E. S., Gerber, J. S., Johnston, M., … … Zaks, D. P. M. (2011). Solutions for a cultivated planet. *Nature, 478,* 337–342.

Food and Agriculture Organization of the United Nations. (2010). *Dietary guidelines and sustainability.* Available from <http://www.fao.org/nutrition/education/food-dietary-guidelines/background/sustainable-dietary-guidelines/en/>.

GlobalChange.gov. (2016). U.S. Global Change Research Program, Washington, DC: U.S. Available from <www.globalchange.gov>.

Goodman, M. K., & Sage, C. (Eds.), (2014). *Food transgressions: Making sense of contemporary food politics* Surrey, UK: Ashgate Publishing Ltd, 250 p.

Henson, R. (2014). *The thinking person's guide to climate change.* Boston, MA: American Meteorological Society, 497 p.

Intergovernmental Panel on Climate Change. (2016). *IPCC.* World Meteorological Organization (WMO) and United Nations Environment Programme (UNEP). Available from <www.ipcc.ch>.

Iowa Department of Natural Resources (2016). *Climate change.* Des Moines, IA: Iowa Department of Natural Resources. Available from <http://www.iowadnr.gov/Conservation/Climate-Change>.

Iowa State University Extension (2016). *Nitrogen use in Iowa corn production.* Ames, IA: Iowa State University, Extension Publication CROP 3073.

Juroszek, P., Lumpkin, H. M., Yang, R.-Y., Ledesma, D. R., & Ma, C.-H. (2009). Fruit quality and bioactive compounds with antioxidant activity of tomatoes grown on-farm: Comparison of organic and conventional management systems. *Journal of Agricultural and Food Chemistry, 57,* 188–1194.

Kunkel, K. E., Easterling, D. R., Hubbard, K., & Redmond, K. (2004). Temporal variations in frost-free season in the United States: 1895-2000. *Geophysical Research Letters, 31,* 1–4.

Lobell, D. B., Schlenker, W., & Costa-Roberts, J. (2011). Climate trends and global crop production since 1980. *Science, 333,* 616–620.

Marlow, H. J., Hayes, W. K., Soret, S., Carter, R. L., Schwab, E. R., & Sabaté, J. (2009). Diet and the environment: Does what you eat matter? *American Journal of Clinical Nutrition, 89*(1), 96S–703S.

Marshall, E., Aillery, M., Malcolm, S., & Williams, R. (2015). Climate change, water scarcity, and adaptation in the U.S. fieldcrop sector. *Economic Research Report 201.* Washington, DC: U.S. Department of Agriculture. Available from <http://www.ers.usda.gov/media/1951525/err-201.pdf>.

Michigan State University (2015). *Economic analysis of sustainable ag. & food systems.* East Lansing, MI: Department of Agricultural, Food, and Resource Economics. Available from <http://www.afre.msu.edu/centers_services/economic_analysis_of_sustainable_ag._food_systems>.

Miller, D. A. (2011). In M. Mann (Ed.), *Farming and the food supply.* Farmington Hills, MI: Gale Cengage Learning, Greenhaven Press, 135 p.

Mitchell, A. E., Hong, Y.-J., Koh, E., Barrett, D. M., Bryant, D. E., Denison, R. F., & Kaffka, S. (2007). Ten-year comparison of the influence of organic and conventional crop management practices on the content of flavonoids in tomatoes. *Journal of Agricultural and Food Chemistry, 55,* 6154–6159.

National Aeronautics and Space Administration. (2015). *Climate change.* Available from <www.climate.nasa.gov>.

National Geographic. (March 2016). *Waste not want not* (pp. 30–55). Available from <www.theplate.nationalgeographic.com>.

National Oceanic and Atmospheric Administration (2016a). *2015 state of the climate: Carbon dioxide.* Washington, DC: U.S. Department of Commerce. Available from <https://www.climate.gov/news-features/featured-images/2015-state-climate-carbon-dioxide>.

National Oceanic and Atmospheric Administration (2016b). Climate information. *National Centers for Environmental Information.* Washington, DC: U.S. Department of Commerce. Available from <http://www.ncdc.noaa.gov>.

National Resources Conservation Service. (2007). *Soil erosion on cropland 2007*. Available from <www.nrcs.usda.gov/wps/portal/nrcs/main/national/technical/nra/nri/>.

National Resources Inventory. (2015). *2012 NRI summary report*. Available from <http://www.nrcs.usda.gov/wps/portal/nrcs/main/national/technical/nra/nri/>.

Nickerson, C., Ebel, R., Borchers, A., & Carriazo, F. (2011). Major uses of land in the United States, 2007. *Economic Research Service*. Washington, DC: U.S. Department of Agriculture.

Nye, B. (2015). *Unstoppable: Harnessing science to change the world*. New York, NY: St. Martin's Press, 341 p.

Paarlberg, R. (2013). *Food politics: What everyone needs to know* (2nd ed., 260 p). Oxford, UK: Oxford University Press, .

Pesek, J., Standord, G., & Case, N. L. (1971). Nitrogen production and use. In R. A. Olson, T. J. Army, J. J. Hanway, V. J. Kilmer, & R. C. Dinauer (Eds.), *Fertilizer technology & use* (2nd ed.). Madison, WI: Soil Science Society of America, Inc. 611 p.

Piccolo, A. (Ed.), (2012). *Carbon sequestration in agricultural soils: A multidisciplinary approach to innovative methods* Berlin Heidelberg: Springer-Verlag.

Pimental, D., & Burgess, M. (2013). Soil erosion threatens food production. *Agriculture, 3*, 443–463.

Pollan, M. (2008). *In defense of food: An eater's manifesto* (p. 256). New York, NY: Penguin Press.

Reynolds, M. P. (Ed.), (2010). *Climate change & crop production* Oxfordshire, UK and Cambridge, MA: CAB International, 292 pp.

Richardson, J. (2009). *Recipe for America: Why our food system is broken and what we can do to fix it*. New York, NY: Ig Publishing, 208 p.

Rickman, J. C., Barrett, D. M., & Bruhn, C. M. (2007a). Nutritional comparison of fresh, frozen and canned fruits and vegetables. Part I. Vitamins C and B and phenolic compounds. *Journal of the Science of Food and Agriculture, 87*(7), 930–944.

Rickman, J. C., Barrett, D. M., & Bruhn, C. M. (2007b). Nutritional comparison of fresh, frozen and canned fruits and vegetables. Part II. Vitamin A and carotenoids, vitamin E, minerals and fiber. *Journal of the Science of Food and Agriculture, 87*(7), 1185–1196.

Riddle, J., & Markhart, B. (2010). *What is organic food and why should I care?* Lamberton, MN: Organic Ecology Research and Outreach Program, University of Minnesota Extension. Available from <http://www.ctnofa.org/documents/Organic%20Food%20handout%20for%20web.pdf>.

Sabaté, J., & Soret, S. (2014). Sustainability of plant-based diets: Back to the future. *American Journal of Clinical Nutrition, 100*(1), 476S–482S.

Schutske, J. M. (2005). *Using anhydrous ammonia safely on the farm*. St. Paul, MN: University of Minnesota Extension Publication. Available from <http://www.extension.umn.edu/agriculture/nutrient-management/nitrogen/using-anhydrous-ammonia-safely-on-the-farm/>.

Sikora, E. J., Allen, T. W., Wise, K. A., Bergstrom, G., Bradley, C. A., Bond, J., ... Zidek, J. (2014). A coordinated effort to manage soybean rust in North America: A success story in soybean disease monitoring. *Plant Diseases, 98*(7), 864–875.

Strack, B. A., Ruger, C. E., Weibel, F. P., Bub, A., & Watzl, B. (2009). Three-year comparison of the polyphenol contents and antioxidant capacities in organically and conventionally produced apples (*Malus domestica* Bork. cultivar "Golden Delicious"). *Journal of Agricultural and Food Chemistry, 57*, 4598–4605.

Sustainable Agriculture Research and Education. (2012). *What is sustainable agriculture?* Available from <http://www.sare.org/Learning-Center/SARE-Program-Materials/National-Program-Materials/What-is-Sustainable-Agriculture>.

Thompson, P. B. (2015). *From field to fork: Food ethics for everyone*. Oxford, UK and New York, NY: Oxford University Press, 329 p.

U.S. Department of Agriculture. (2011). *Assessment of the effects of conservation practices on cultivated cropland in the Chesapeake Bay region*. Washington, DC. Available from <http://www.nrcs.usda.gov/Internet/FSE_DOCUMENTS/nrcsdev11_023934.pdf>.

U.S. Department of Agriculture. (March 2014). *2014 Farm bill highlights*. Washington, DC. Available from <http://www.usda.gov/documents/usda-2014-farm-bill-highlights.pdf>.

U.S. Department of Agriculture. (2015). *Sustainable agriculture: Definition and terms*. Washington, DC. Available from <http://afsic.nal.usda.gov/sustainable-agriculture-definitions-and-terms-1#toc2>.

U.S. Department of Agriculture. (2016). *Sustainability in agriculture*. Washington, DC. Available from <http://afsic.nal.usda.gov/sustainability-agriculture-0>.

U.S. Environmental Protection Agency. (2015). *Summary of the U.S. endangered species act*. Washington, DC. Available from <https://www.epa.gov/laws-regulations/summary-endangered-species-act>.

U.S. Environmental Protection Agency. (2016). *Cutting food waste and maintaining food safety*. Washington, DC. Available from <https://www.foodsafety.gov/blog/2016/09/cutting-food-waste>.

U.S. Geological Survey (2016). *Irrigation water use*. Washington, DC: U.S. Department of the Interior. Available from <http://water.usgs.gov/edu/wuir.html>.

Verge, X. P. C., Dyer, J. A., Worth, D. E., Smith, W. N., Desjardins, R. L., & McConkey, B. G. (2012). A greenhouse gas and soil carbon model for estimating the carbon footprint of livestock production in Canada. *Animals, 2*, 437–454.

Vermier, I., & Verbeke, W. (2006). Sustainable food consumption: Exploring the consumer "attitude-behavioral intention" gap. *Journal of Agricultural and Environmental Ethics, 19*, 169–194.

Wang, S. Y., Chen, C.-T., Sciarappa, W., Wang, C. Y., & Camp, M. J. (2008). Fruit quality, antioxidant capacity, and flavonoid content of organically and conventionally grown blueberries. *Journal of Agricultural and Food Chemistry, 56*, 5788–5794.

Wogan, G. N., Generoso, W., Koller, L. D., Smith, R. P., & Tannenbaum, S. R. (1995). *Nitrate and nitrite in drinking water*. Washington, DC: National Academy Press.

WorldWatch Institute. (2016). *Food and agriculture*. Available from <www.worldwatch.org>.

Zhao, X., Chambers, E., Matta, Z., Loughin, T. M., & Carey, E. E. (2007). Consumer sensory analysis of organically and conventionally grown vegetables. *Journal of Food Science, 72*, S87–S91.

Zolin, C. A., De, R., & Rodrigues, A. R. (Eds.), (2015). *Impact of climate change on water resources in agriculture* Boca Raton, FL: CRC Press, 224 p.

Appendix A: Important Dates in US History and Their Impact on the Food System

Date	Event	Impact
1607	Settlers arrived in US Colonies	Europeans settled along Atlantic coast
1775–83	Revolutionary War	Freedom from British rule opened new markets for agricultural products
1789	US adopts Constitution	Current political system initiated, George Washington, a farmer, elected president
1803	Louisiana Purchase	Added 828,000 square miles to US territory
1804–06	Lewis and Clark expedition	Mapped the northwest passage from St. Louis, MO to the Pacific Ocean
1812–14	War of 1812	Freedom from British restrictions on trade
1846	Oregon territory acquired	Northwestern border defined at 49th parallel
1848	Guadalupe Hidalgo Treaty	Southwestern territory acquired
1849	Gold rush to California	Increased population of California
1860	Abraham Lincoln elected president	Opposed slavery, supported agriculture education
1861–65	Civil War	North versus South, ended slavery and reunited the US
1862	USDA created Homestead Act Morrill Act	Provided cabinet office for agriculture, granted public land to settlers, granted public land for agricultural colleges (land-grant colleges)
1866–69	Completion of railroad from Omaha, NE to Ogden, UT	Opened trade routes from West to Midwest
1876	Telephone created	Enhanced communication for business transactions and reduced isolation of rural communities
1879	Light bulb created	Allowed work to occur after sunset, which lengthened the workday
1883	Harvey Wiley appointed chief chemist at USDA	Campaigned for Pure Food and Drug Act
1884	Bureau of Animal Inspections created	Prevented diseased animals from being used as food
1887	Hatch Act	Funded Agricultural Experiment Stations at land-grant colleges Bureau of Agriculture became cabinet level
1890	Second Morrill Act	Funded 1890 colleges for African American students

(Continued)

Date	Event	Impact
1892	General Electric company founded	Replacement of steam power and manual labor with mechanized equipment
1893	Office of Road Inquiry in USDA	Built roads for transport of agricultural products
1894	USDA released first diet recommendations	Provided information to consumers to improve health through food choices
1902	National Reclamation Act	Funded major water projects to irrigate western states
1905	*The Jungle* by Upton Sinclair published	Depicted poor sanitation in Chicago meat packing industry
1906	Pure Food and Drug Act	Prevented sale of adulterated food and drugs
	Federal Meat Inspection Act	Ensured sanitary conditions for meat
1908	Model T introduced	Opened possibility for rapid transportation
1912	Concept of "vitamins" proposed	Initiated discovery of micronutrients in foods
1913	Haber–Bosch process to fix nitrogen	Created ammonia fertilizer, a less costly form of plant nutrient
1914	Smith–Lever Act	Funding created extension programs at land-grant colleges
1916	Federal Farm Loan Act	Federal land banks established to provide long-term credit to farmers
1917–19	US in World War I	High demand for agriculture products led to increased farm production
1920	Radio introduced	Enhanced communication
1921	Hybrid corn introduced	Improved crop yields, introduced scientific method to agriculture
1927	USDA Bureau of Chemistry renamed Food, Drug, and Insecticide Administration (FDIA)	Increased regulation and oversight of food safety
1929–42	Great Depression	Banks failed and US economy crumbled
1930	FDIA became Food and Drug Administration	Continuation of federal oversight of food safety and consumer protection
1933	Farm Credit Act	Farm credit system established to support all typed of agriculture operations
1935	Rural Electrification Act	Brought electricity to rural communities
1934–37	Severe drought and poor farming practices created the Dust Bowl	Brought attention to the need for oversight of farming practices and environmental stewardship
1936	Soil Conservation Act	Implemented regulations to reduce soil erosion
1938	Federal Food, Drug, and Cosmetic Act (FFDCA)	FDA given authority for food safety standards
	Agriculture Adjustment Act	Subsidies paid to farmers to reduce crop acreage
1939	Food Stamp Program part of New Deal	USDA bought commodities from farmers and gave to needy Americans
1940	FDA moved from under USDA to Department of Health, Education, and Welfare	Regulation of food safety became separate from agency that oversaw food production

Date	Event	Impact
1941–45	US in World War II	Great Depression ended, industrialization expanded into agriculture, number of farmers began to decrease
1941	Food and Nutrition Board released Recommended Dietary Allowances	Defined nutrient requirements for healthy people
1944	Federal-Aid Highway Act	National system of interstate highways created allowing food transportation by truck
1947	Federal Insecticide, Fungicide, and Rodenticide Act (FIFRA)	Required registration of pesticides
1950–60	Green Revolution	Improved crop traits and agronomic practices provide food for millions
1950–53	US in Korean War	Reduced commodity stockpiles, created higher demand for food
1950–70	85 new pesticides introduced	Chemical management of agriculture increased crop yields
1953	BAI and Bureau of Dairy became Agricultural Research Service (ARS)	More structured agricultural research programs
1953	Structure of DNA discovered	Field of biotechnology began, and applied to agriculture
1958	Food Additive Amendment to FFDCA	Safety of ingredients in processed foods
	Humane Slaughter Act passes	Standards for animal slaughter
1961–73	US in Vietnam War	Increased discontent with government and military, financial drain on US economy
1961–64	Pilot Food Stamp Program began	Provided food assistance to needy Americans
1962	*Silent Spring* by Rachel Carson published	Raised awareness of environmental damage from chemical pesticides; DDT banned in 1972
1964	Food Stamp Act	Made Food Stamp Program permanent
1969	Neil Armstrong walked on moon	Science and technology advances by NASA applied to consumer products
1969	White House Conference on Food, Nutrition, and Health	Raised awareness of food insecurity and role of nutrition in disease
1970	Environmental Protection Agency created	Oversight of pesticides and land use
	Egg Products Inspection Act	ARS authorized to inspect eggs
1972	Animal and Plant Health Inspect Service (APHIS) created within USDA	Consolidated activities to oversee plant and animal health within one agency
1973	Gasoline rationing and high prices due to OAPEC oil embargo	Raised awareness of energy production issues and began alternative fuel research
1974	Glyphosate introduced	Roundup herbicide used for corn and soybean production
1976	Apple computer company started	Information technology for business and personal use rapidly implemented
1977	Dietary goals for the United States released	Provided dietary recommendations to reduce the risk of chronic diseases; DG are updated every 5 years

(*Continued*)

Date	Event	Impact
1978	Humane Slaughter Act	Amendment included regulating practices for slaughter of poultry
1980–90	Low commodity prices and high farmer debt created a farm crisis	Many farmers went bankrupt and rural populations decreased, consolidation of farm operations
1982	Recombinant insulin approved by FDA	First drug produced using GM biotechnology and accepted by consumers
1986	Coordinated Framework for the Regulation of Biotechnology	Authorized USDA, EPA, and FDA to oversee genetically modified organisms
1988	Recombinant chymosin approved	GMO enzyme for cheese production replaced chymosin obtained from young calves
1990	Organic Foods Production Act	Created standards for organic food production and marketing
1992	USDA released the Food Guide Pyramid	Graphic illustration of food intake recommendations
1994	Equity in Education Land-Grant Status Act	Granted land-grant status to Native American colleges and created an endowment for continued funding
1994	Flavr Savr tomato entered market	First GMO food marketed to consumers
1995	GM corn, cotton, canola, and soybeans approved	Introduced herbicide-tolerant traits to agriculture, farmers rapidly adopted the technology
1996	Hazards Analysis Critical Control Points (HACCP) legislation	Enhanced food safety by defining process to monitor food production systems
	Food Quality and Protection Act	Required EPA to ensure safety of pesticides
2000	Golden rice developed	GMO rice with enhanced beta carotene content to improve vitamin A nutrition
2001–11	US in Iraq War	Enhanced food production technology in military meals
2005	Energy Policy Act Renewable Fuel Standard	Promoted development and use of biofuels
2010	Healthy, Hunger-Free Kids Act	Revision to school lunch nutrition standards
2010	Affordable Care Act	Required nutrition information on restaurant and vending machine foods
2011	Food Safety Modernization Act passed	Major reform to food safety laws with focus on preventing food contamination
2011	USDA released MyPlate	Replaced MyPyramid as a visual diet recommendation for consumers
2014	2014 Agriculture Act (Farm Bill)	Reduced direct payments to farmers, increased funding for specialty and organic crops and local foods, reduced nutrition support funding
2015	GMO salmon approved	First animal with bioengineered trait approved for human food

Index

Note: Page numbers followed by "*f*" and "*t*" refer to figures and tables, respectively.

S

T

Printed in the United States
By Bookmasters